HANDBOOK OF CONTAMINATION CONTROL IN MICROELECTRONICS

MATERIALS SCIENCE AND PROCESS TECHNOLOGY SERIES

Editors

Rointan F. Bunshah, University of California, Los Angeles *(Materials Science and Technology)*

Gary E. McGuire, Microelectronics Center of North Carolina *(Electronic Materials and Processing)*

DEPOSITION TECHNOLOGIES FOR FILMS AND COATINGS; Developments and Applications: by *Rointan F. Bunshah et al*

CHEMICAL VAPOR DEPOSITION FOR MICROELECTRONICS; Principles, Technology, and Applications: by *Arthur Sherman*

SEMICONDUCTOR MATERIALS AND PROCESS TECHNOLOGY HANDBOOK; For Very Large Scale Integration (VLSI) and Ultra Large Scale Integration (ULSI): edited by *Gary E. McGuire*

SOL-GEL TECHNOLOGY FOR THIN FILMS, FIBERS, PREFORMS, ELECTRONICS, AND SPECIALTY SHAPES: edited by *Lisa C. Klein*

HYBRID MICROCIRCUIT TECHNOLOGY HANDBOOK; Materials, Processes, Design, Testing and Production: by *James J. Licari* and *Leonard R. Enlow*

HANDBOOK OF THIN FILM DEPOSITION PROCESSES AND TECHNIQUES; Principles, Methods, Equipment and Applications: edited by *Klaus K. Schuegraf*

Related Titles

ADHESIVES TECHNOLOGY HANDBOOK: by *Arthur H. Landrock*

HANDBOOK OF THERMOSET PLASTICS: edited by *Sidney H. Goodman*

HANDBOOK OF CONTAMINATION CONTROL IN MICROELECTRONICS; Principles, Applications and Technology: edited by *Donald L. Tolliver*

HANDBOOK OF CONTAMINATION CONTROL IN MICROELECTRONICS

Principles, Applications and Technology

Edited by

Donald L. Tolliver

Motorola Semiconductor Sector
Motorola, Inc.
Phoenix, Arizona

np | **NOYES PUBLICATIONS**
Park Ridge, New Jersey, U.S.A.

Library of Congress Catalog Card Number: 87-31533
ISBN: 0-8155-1151-5
Printed in the United States

Published in the United States of America by
Noyes Publications
Mill Road, Park Ridge, New Jersey 07656

10 9 8 7 6 5 4 3 2

Library of Congress Cataloging-in-Publication Data

Handbook of contamination control in microelectronics.

Bibliography: p.
Includes index.
1. Integrated circuits--Design and construction.
2. Contamination (Technology) I. Tolliver, Donald L.
TK7836.H34 1988 621.381'73 87-31533
ISBN 0-8155-1151-5

Preface

The semiconductor industry has witnessed the birth of contamination control in the microelectronics industry within the last ten years. Prior to 1980 the importance and future impact of this independent technology was limited to a relatively few individuals scattered across a limited number of companies in the United States, Japan and Europe. As the industry entered the higher complexities of 64K DRAM manufacturing in the early 1980's the importance of defect control and particle reduction in the manufacturing process began to receive more specific focus from the supplier side and the semiconductor manufacturer.

Today a whole industry exists and continues to expand in order to serve the growing needs of contamination control for the microelectronics industry. Many semiconductor and microelectronic firms to this day do not recognize contamination control technology as an independent ingredient of the world wide semiconductor business. It nonetheless exists in one form or another and is called by a variety of different terms, but regardless of the name, it still is contamination control.

Originally, most of the contamination control suppliers and users saw the technology of contamination control as the exclusive territory of the clean room and only the clean room. This led to some ambiguity and confusion as to what specific categories constituted contamination control and which belonged to other better known processing categories. A technology such as liquid filtration, which was by definition exclusive of the concepts of clean rooms, at that time was focused toward one of the other complex technology segments which went into the manufacturing of semiconductors. Often contamination control, but not necessarily referred to by that name, would be designated as a sub-set of the diffusion/oxidation technology, the lithography and etch operations, the deposition technology, ion implantation, the device engineering and yield enhancement engineering groups, and very often the quality control organization that supported a particular manufacturing organization. The contamination control wheel described below summarizes most of the major categories which actually fall under the umbrella of contamination control.

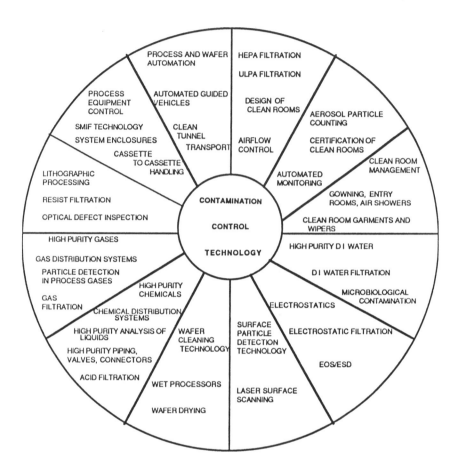

Where is contamination control technology today? The answer is still not clear cut, but there is no denying the fact that it has emerged as a separate and distinct technology within the mainframe of semiconductor manufacturing. It still experiences some confusion in definition and may very well be referred to as particle reduction, yield enhancement, defect prevention, and perhaps even quality control. But certainly contamination control is now recognized on a more independent basis and is receiving independent budgeting and independent resources many times greater than we could have imagined in 1975 and 1976. Liquid filtration is certainly a well accepted part of contamination control technology along with many other technologies such as laser particle counting, process gas purification, and high purity chemical distribution.

Members of this highly ubiquitous group of multifaceted technologies are continuing to format the boundaries and the specific applications which may fall into the arena of contamination control. It is certainly clear that clean room technology is only one subset of the entire family of requirements in contamination control. Very important today and in the future of our industry are the nonparticle contamination requirements which clearly fit into contamination control, but are not always found there. Sub ppb metallic ion contamination in DI water, process chemicals and process gases are now heavily in focus for contamination control requirements in the next five years.

The purpose of this handbook is to attempt to distill into one volume a synopsis of many of the major categories which go into the subject of contamination control. I have purposely focused on the microelectronics applications rather than those of other industries. I am best prepared to deal with the issues of the semiconductor industry rather than those topics which more precisely fall into industries such as pharmaceutical, aerospace, biomedical and lasers. In many cases we share the same concerns and the same solutions but for different reasons. Certainly it is not my goal to exclude excellent and very worthwhile subjects and contributions in these sister industries. However, the need is clear to provide a reference for the microelectronics industry to help define the needs, solutions and requirements to build upon for the future.

Included in this handbook is a broad look at most of the major subjects which impact and interact with contamination control technology in the microelectronics industry. This is only a beginning and there will be future editions with new subjects and updated technological requirements.

We have included a fairly basic treatment of the technology of aerosol filtration and aerosol particle measurements methods. Within the operations of clean rooms the very essence of clean room management and clean room garments is given an indepth discussion.

Electrostatics in contamination control is a difficult arena to target for contamination control. More often the reader will find this subject tied to the area of EOS/ESD and not in contamination control. The editor belives that the treatment of this subject is worthwhile as so many of the inherent particle mechanisms which must be dealt with on a daily basis are inherent to electrostatic technology.

The subject chapters under the materials categories such as DI water, gases, chemicals, and lithographic processing are now recognized as major contributors to device yield loss and are well entrenched inside contamination control requirements. Associated closely with our materials categories are the chapters on surface particle detection and liquid particle monitoring technology. These subjects would not have even surfaced had this book been published prior to 1980. The problems were not defined and their actual contribution to the success of semiconductor processing would probably have gone unnoticed.

Near and dear to the hearts of active contamination control engineers in the semiconductor industry, are the subjects relative to contamination control in processing equipment. We as members of this everchanging industry totally missed that one also, prior to about 1978. Today it is clear that in spite of all other efforts that must go on in contamination control, we must learn how to analyze and control the contamination from the processing equipment itself or our other efforts are futile. Two chapters are dedicated to the requirements of equipment analysis via wafer particle detection and automation in processing equipment as it relates to wafer and cassette automation.

A glossary, formally titled *A Glossary of Terms and Definitions Related to Contamination Control* has been included in this printing as an additional reference for the reader. This glossary has been produced and is under the jurisdiction of RP-11 Working Group of the Standards and Practices Committee of the Contamination Control Division, Institute of Environmental Sciences. The editor and publisher gratefully acknowledge the Institute of Environmental Sciences for permission to reprint this glossary. Comments and suggestions about this glossary should be made directly to the Institute of Environmental Sciences, Mt. Prospect, Illinois.

It is certainly the belief of the editor at this writing that the role and impact of contamination control in the microelectronics industry will be a dominant one and possibly the key factor to success in meeting the manufacturing challenges of the microelectronics industry of the future.

Phoenix, Arizona   Donald L. Tolliver
January, 1988

Contributors

Mauro A. Accomazzo
Millipore Corporation
Bedford, MA

Ann Marie Dixon
Cleanroom Management
 Associates Inc.
Tempe, AZ

Robert Donovan
Research Triangle Institute
Research Triangle Park, NC

David Ensor
Research Triangle Institute
Research Triangle Park, NC

Terry L. Faylor
T.L. Faylor & Associates
Santa Cruz, CA

Gary Ganzi
Millipore Corporation
Bedford, MA

Peter Gise
Tencor Instruments
Mountain View, CA

Bennie W. Goodwin
Angelica Uniform Group
St. Louis, MO

Jeffrey J. Gorski
Oxnard, CA

Robert Kaiser
Consultant to Millipore
Argos Associates, Inc.
Winchester, MA

Gerhard Kasper
Liquid Air Corporation
Countryside, IL

Robert G. Knollenberg
Particle Measuring Systems Inc.
Boulder, CO

Benjamin H.Y. Liu
University of Minnesota
Minneapolis, MN

Mary L. Long
University of Arizona
Tucson, AZ

Rollin McCraty
Static Control Services
Palm Springs, CA

Mike Naggar
K.T.I. Chemicals Inc.
Sunnyvale, CA

Mihir Parikh
Asyst Technologies, Inc.
Milipitas, CA

David Y.H. Pui
University of Minnesota
Minneapolis, MN

Donald L. Tolliver
Motorola Semiconductor Sector
Motorola Inc.
Phoenix, AZ

Barclay J. Tullis
Hewlett-Packard Laboratories
Palo Alto, CA

H.Y. Wen
Liquid Air Corporation
Countryside, IL

NOTICE

Contents

1

Aerosol Filtration Technology

David Ensor and Robert Donovan

1. INTRODUCTION

Filtration processes permeate every phase of modern technology[1] and especially impurity-sensitive manufacturing processes such as microelectronics. Filters remove particles from the air and process gases and from process chemicals and water. Manufacturing success in microelectronics, as measured by, for instance, chip or magnetic head assembly yield, depends critically upon high quality filtration. This chapter considers only aerosol filtration in microelectronics, although liquid filtration (for example, of acids, solvents, water and photoresist) is clearly every bit as important to successful microelectronic manufacturing.

1.1 Importance of Aerosol Filtration in Microelectronics

Perhaps no major industry places a higher premium on aerosol filtration than microelectronics. The total ambient aerosol particle concentration is probably lower in a microelectronic manufacturing area than in any other similarly sized manufacturing area on earth. As measured by a condensation nuclei counter, the total concentration of particles of size 0.02 μm and above in a contemporary microelectronics manufacturing area at rest (no process actvitity, no people in the room) is often in the 10- to 100-ft^3 (3.5 x 10^{-4} to 3.5 x 10^{-3} cm^{-3}) range.[2]

The need for such high quality air stems from the small dimensions of the devices built in today's microelectronic chips and the expectation that these dimensions will continue to shrink. Table 1 lists dimensions typical of the 0.5 μm technology forecast to be used in production of 4 megabit silicon chip memories in the 1990s. (These dimensions, however, are only 40 to 50 percent smaller than those of contemporary state-of-the-art chips.)

1

Table 1: Dimensional Requirements for a Half-Micrometer Technology[3]

Lateral dimensions:
    Pattern size   0.5 μm
    Pattern tolerance   0.15 μm
    Level-level registration   0.15 μm

Vertical dimensions:
    Gate oxide thickness   10 nm
    Field oxide thickness   200 nm
    Film thicknesses   0.25-0.5 μm
    Junction depth   0.05-0.15 μm

Particles appearing on the chip surface during manufacturing can cause chip failure by disrupting surface geometry, by creating defects in thin layers or films, and by introducing impurities. What constitutes a killer particle—a particle of sufficient size to cause chip failure when located in a critical region of the chip—depends on device dimensions. Arbitrarily, the size of a killer particle can be taken to be one quarter to one half the line width of a chip lateral dimensions and one half the thickness of the various vertical dimensions.[3] By this criterion, particles as small as 5 nm are potentially killer particles of the silicon chips of the 1990s. (The 1985 state-of-the-art in chip manufacturing uses dimensions only two to two-and-a-half times those listed in Table 1 so that 10- to 15-nm particles deposited in certain critical areas can be killer particles to today's production.) Thus, particulate contamination in microelectronic device production represents and warrants unprecedented control methods, of which high quality filtration is a necessary (but not of itself a sufficient) part.

The high quality air now achieved routinely in microelectronic manufacturing areas results from continual recirculation of room ambient air through banks of HEPA (High Efficiency Particulate Air) filters which sometimes make up one entire wall of the manufacturing area or, as is more common now, all or most of the ceiling. In the latter configuration, vents for the return air are at the floor level along at least two sides of the room or, in some designs, in the floor itself which then consists of a grating placed above the return plenum—the "raised floor" design. This design promotes vertical laminar flow of air through the room whereby clean air from the ceiling HEPA filters flows unidirectionally downward to the return grates in the floor, sweeping any particles emitted by room activities out of the room and into the return air being fed back to the HEPA filters. The linear velocity of this air wash through the room is typically 90 ft/min. Obstructions, such as apparatus or furniture or movement in the room, create eddies that upset the laminar flow so turbulent pockets often exist. However, the major flow pattern remains that of vertical laminar flow.

Controlling aerosol particles in the ambient air of the manufacturing area is just one aspect of controlling aerosol particles in microelectronics manufacturing. Equally important is control of aerosol particles in the many process gases used during chip fabrication. These gases create desired ambients surrounding the wafers during high temperature processing; high pressure gases are often used to introduce desired impurities for wafer doping; inert gases are used to clean off wafers or "blow dry" wafers following various wet process steps. Clearly, the quality of these gases to which process wafers are deliberately exposed is as important as room ambient.

Typically, these process gases are delivered through point-of-use filters consisting of high quality membrane filters rather than the fibrous HEPA filters used to clean the ambient room air. A membrane filter consists of a porous sheet rather than a packed bed of independent capture sites such as a fibrous filter. One can think of a membrane filter as a solid sheet in which many penetrating holes have been punched and, indeed, a certain subclass of membrane filters (irradiated polycarbonate sheets) are constructed in a process that matches that description closely. However, the more common membrane filter is a solvent cast layer which, when dried, leaves a porous but continuous one-piece substrate behind. Variables in this manufacturing process include drying rate and temperature in addition to substrate and volatile solvent properties.

Membrane filters are generally classified as surface filters, while fibrous filters are depth filters. This designation reflects the mechanisms thought to dominate particle capture by these differently labeled filters—a surface filter removes particles from the air stream primarily by interception (the particles are simply too big to fit through the filter pores); a depth filter depends upon diffusion and inertial impaction, as well as interception, to remove particles from the air streams as they wind their tortuous paths through the filter.

The size of the particle being captured and the air stream velocity determine which mechanism dominates under which circumstances. For the sub-micron aerosol particle regime to be considered in this chapter and the filter construction employed to control them, the distinction between surface filtration and depth filtration becomes fuzzy. Indeed, one of the major recent insights in filtration research is recognition and demonstration of the similarity of fibrous and membrane filter performance in the filtration of sub-micron aerosol particles.[4] In this particle size regime, the same equations that describe fibrous filter performance also predict porous membrane filter performance better than the capillary tube models traditionally used to model membrane filters.

Other roles of filtration in microelectronics include: (1) the collection of aerosol particle samples from the ambient air in order to subsequently analyze the composition and properties of those aerosol particles, and (2) the protection of workers from hazardous or toxic aerosols. In this latter role, the filter element is typically part of a respirator canister that also contains sorbent to capture vapors. Sampling filters are most frequently membrane filters made from irradiated polycarbonate sheets, while respirator filters are always fibrous filters because of their low aerodynamic resistance.

Neither of these two functions is unique to microelectronics, and microelectronics requirements are not the dominant force in the marketplace for such filters. Therefore, these roles will not be amplified further. The major application of filters in microelectronics are the fibrous HEPA filters used to create the clean rooms in which manufacturing is carried out and the membrane filters used to filter many process gases, some reactive, at their point of use. In both these applications, the needs of the microelectronics industry do represent a major market and, thus, significantly influence product development.

1.2 Filter Media Description

The two generic types of aerosol filters used in microelectronics are fibrous filters and membrane filters. Fibrous filters are a collection of randomly oriented fibers of varying diameters and lengths which are packed, compressed, or otherwise held together as mats (sometimes with the aid of fiber binders) in a fixed volume through which the air to be filtered passes. Fibrous filters vary from low efficiency furnace-type filters consisting of a loose fiber web held between two support grids to high efficiency "papers" made by the Fourdrinier process. The furnish used in the latter includes both fiberglass fibers of submicron diameter for high efficiency submicron particle capture and larger diameter fibers (up to 5 μm) for strength. The void volumes of fibrous filters are in the range of 85 to 99 percent (Figure 1).

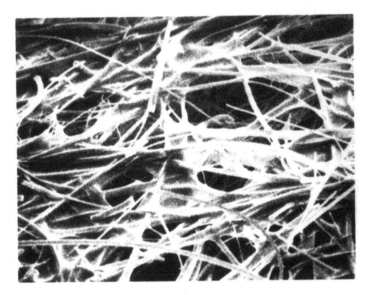

Figure 1: Type GC-90 glass fiber filter—with binder. 1000X magnification. (Courtesy, Micro Filtration Systems)

Membrane filters, on the other hand, are homogeneous, solid sheets into which penetrating holes or pores have been introduced by various methods including irradiation and etching, biaxial stretching, and solvent volatilization. The latter process is the most common. Solutions of appropriate plastics containing volatile solvents are cast into sheets for subsequent curing during which time the various components evaporate at different rates to leave behind a porous but continuous sheet. Typical void volumes of solvent cast membrane filters are 70 to 85 percent (Figure 2).

Irradiated and etched membrane filters made from polycarbonate sheets look like the classical membrane filter model (Figure 3)—neatly defined circular holes in an otherwise homogeneous flat sheet. While the void volume of this type

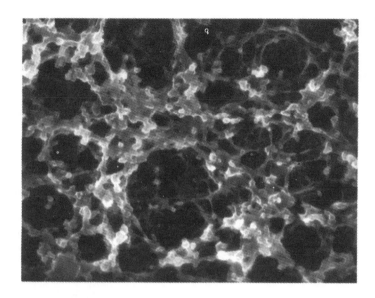

Figure 2: Cellulose nitrate membrane filter—pore size, 0.45 μm. 4800X magnification. (Courtesy, Micro Filtration Systems)

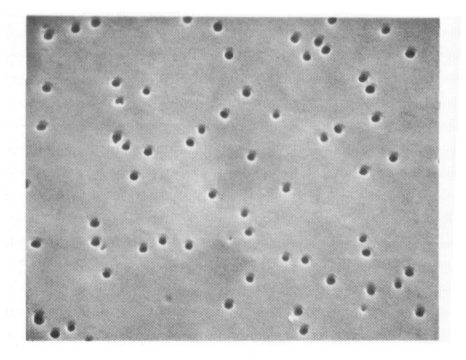

Figure 3: Standard Nuclepore filter. (Courtesy, Nuclepore Corporation)

of filter is only 10 to 20 percent, its almost purely surface collection feature makes this type of filter nearly ideal for examining collected particles by microscopy; and it is in this role that it is primarily used. It is also a good physical model of a capillary pore filter.

The holes in the polycarbonate sheet are defined by exposing the sheet to fission fragments from radioactive isotopes. The damaged region left in the wake of the irradiation particles reacts readily with an appropriate etch solution. Hole size is controlled somewhat by etch time.

The final membrane filter to be shown is that made from biaxial stretching of polytetrafluoroethylene (PTFE). While its chemical inertness is a major advantage of this type of filter, its structure is also noteworthy (Figure 4) because it illustrates better than any other membrane filter the similarity between membrane and fibrous filters. Under biaxial stretching, the PTFE membrane remains a single solid entity but tears into a delicate lattice of submicron fibrils that more closely resemble the fibrous filter structure shown in Figure 1 than the membrane filters of Figures 2 and 3. This similarity in appearance makes plausible the claim that similar equations can predict the performance of both types of filters.

That membrane filters and fibrous filters are generally manufactured by different organizations means competition between the two filter types.

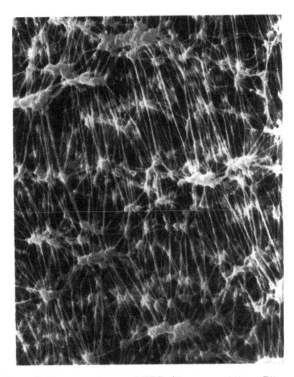

Figure 4: 0.5 μm Teflon membrane—PTFE. (Courtesy, Micro Filtration Systems)

At present, fibrous filters of the HEPA or ULPA type are used exclusively in the filtration of room air. Their combination of high efficiency, low pressure drop, and good loading characteristics make them well suited for this application.

For the filtration of high pressure process gases, membrane filters are more commonly used. They also possess high efficiency, and this application can tolerate a high pressure drop. An advantage widely touted by membrane filter manufacturers is their freedom from shedding in which parts of the filter break loose and become reentrained in the filtered gas stream. Fibrous filters are more vulnerable to this shortcoming so, while acknowledging the superior loading properties of fibrous filters, membrane filter manufacturers recommend their use only as prefilters in high pressure gas filtration, a view not totally shared by fibrous filter vendors.

1.3 Historical Background

Historically, filtration can be traced to efforts over the last 2000 years to protect workers from toxic dusts and smoke. For a detailed summary of the history of aerosol filtration, see Reference 5. The 20th century has seen a burst of activity in the development of filters. The two World Wars stimulated significant advances in filtration in preparation for anticipated chemical attacks. In particular, the development of respirators has stimulated a development of good understanding of the physics of filtration and the fabrication of media.

Ventilation systems using high efficiency filters were developed during World War II. Before 1950, small quantities of HEPA filters were built by the United States Army Chemical Corps for the containment of radioactive aerosols. These filters were called super-impingement or super-interception filters. Later they were called absolute filters and in the 1960s became known as HEPA filters. These first filters were made from esparto grass, and later asbestos fibers were added. In the 1950s, glass fiber paper was developed and used exclusively as the media. The asbestos was removed in the late 1970s because of the potential health risk.

In the late 1950s, the first horizontal laminar flow clean bench was developed at the Sandia Laboratories. The air cleaned by a HEPA filter was used to create a clean environment over the top of the work surface. Clean room technology was developed during the 1960s in the aerospace industry because of the need to fabricate precision machinery. The technology was applied to a number of industries requiring a highly controlled environment, such as the manufacture of semiconductors, precision machinery, and pharmaceuticals.

The use of membranes as a filter medium dates back to the late 1880s. The early membranes were made by gelling and drying colloidal solutions of cellulose esters. These early membranes were used only for ultrafiltration of liquids because the material was unstable when dry. During World War II, the Germans made advances in the technology. This advanced technology then was adapted to many civilian areas following the war. (The application of membranes to air filtration was reported by Goetz in 1953.[6]) Today membranes are made from a wide range of materials and are widely used both in the manufacture of point-of-use filters and also for the sampling of particles.

1.4 Chapter Organization

This chapter will cover the filtration of both air and gases. Three general areas will be covered: the fundamentals of filter behavior (Section 2); filter test methods (Section 3); and then the applications of filters (Section 4). Understanding the fundamentals of filtration is important to allow the user to understand and interpret the filter operational data. Filter test methods, to a great extent, should provide an indication of the performance of the filters. The demand for clean manufacturing conditions has resulted in higher efficiency filters, thereby prompting the development of a number of new filter test methods. Finally, the applications of filtration in air and gas filtration reflect the wide variety of filter media and an overall idea of the many uses of filters.

2. FILTRATION FUNDAMENTALS

The goal of filtration theory is to be able to satisfactorily predict filter collection efficiency, resistance (pressure drop/face velocity) and lifetime from the independent variables describing its structure and operating conditions. The review presented here is of both fibrous filter theory and the capillary tube model theory traditionally invoked to predict membrane filter performance. Based on these theories, equations for predicting filter collection efficiency and resistance for both fibrous and membrane filters will be presented, from which particle size of maximum penetration and filter media figure of merit can then be derived. Filter lifetime cannot yet be predicted theoretically for real filters. While the effects of filter loading upon pressure drop and penetration are being studied theoretically using model filters[7] and lifetime prediction in terms of pressure drop increase may eventually be possible, only empirical results can be presented now. In addition, all the theoretical expressions to be presented in the following sections strictly apply to only clean, virgin fibers and filters.

2.1 Fibrous Filter Theory

The classical collection membranes for particle capture from a fluid flowing past a fiber are illustrated in Figure 5. These five mechanisms are the major particle removal mechanisms over the entire particle size spectrum at ambient conditions. Following are brief descriptions of each mechanism.

(a) Gravity. Large particles (>5 μm) will settle out of the gas stream. This action occurs independently of the fiber and depends on the gas velocity and the particle mass.

(b) Direct Interception. A particle, following the upstream flow streamline, collides with a fiber and is collected. This mechanism can be important over the full particle size spectrum, depending on the ratio of the particle diameter to the fiber diameter.

(c) Inertial Impaction. Inertial impaction occurs when the inertia of the particle causes it to deviate from its initial streamline and collide with the collector (a fiber of the filter). Inertial impaction depends on the gas's viscosity and velocity and also on the particles's diameter and density. Normally, inertial impaction is an important collection mechanism for particles greater than about 0.5 μm in diameter.

(d) Diffusion. Small particles ($<$0.5 μm) have a low enough mass to have their trajectories altered by collision with gas molecules (Brownian motion). Smaller particles have higher diffusion rates. On an average, the particles follow their streamlines. However, the random deviations from streamlines (due to diffusion) lead to a finite probability of collection by the fiber.

(e) Electrostatic Mechanisms. Electrical charges on either the particle or the fiber, or both, create attractive electrostatic forces between the fiber and the particle. This results in collection of the particle.

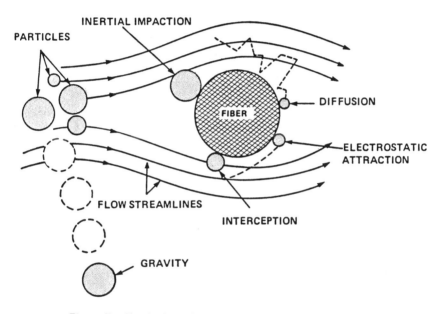

Figure 5: Classical mechanisms: aerosol capture by a fiber.

These mechanisms are considered to act independently of one another, but their contributions are not necessarily additive. Most of the filter modeling work considers three of the mechanisms to be the most important—direct interception, inertial impaction, and diffusion. The effect of gravity is often neglected when discussing small particles because of its minimal effect. Electrical forces have often been neglected because they were considered as only second-order effects. Models which do not include the electrical forces have proven successful in predicting performance. Therefore, this assumption appears justified for uncharged particles without externally applied fields. However, if either the particles are charged or external electric fields exist (in an effort to modify filter performance), or if both conditions exist, then the electrical forces may dominate. Section 2.3 discusses these electrical forces which will be ignored until then.

The relative importance of diffusion, interception, and inertial impaction can be judged from Figure 6 (where the single-fiber efficiencies resulting from each mechanism are calculated).[5] These calculations show that: diffusion dominates for those particles with diameters smaller than about 0.2 μm; inertial impaction and interception dominate for particles with diameters larger than approximately 1 μm; and all the mechanisms may be significant for particles with diameters between 0.2 and 1 μm. An interaction term was omitted from the plot for clarity. Thus, the three mechanisms' efficiencies do not sum to the overall efficiency. Other models would provide slightly different estimates of single-fiber efficiency, but the overall trends remain the same for all the models.

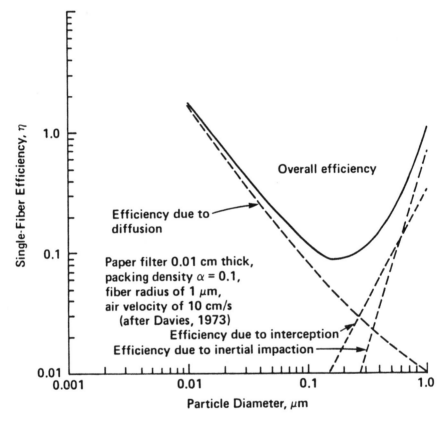

Figure 6: Single-fiber efficiencies (due to diffusion, interception, and inertial impaction).

2.1.1 Particle Trajectory Modeling. In order to fully understand how the three most important mechanisms operate, it is necessary to consider the various models which describe them. A popular approach to modeling fibrous filtration has been particle trajectory analysis. In this approach, the single-fiber collection efficiency is estimated by calculating the limiting trajectory allowing particle-fiber contact. Figure 7 displays schematically the geometry of a trajectory model. The particle trajectories are calculated by solving the equation of motion of the particle. The limiting particle trajectory is the trajectory which just brings the particle in contact with the fiber. The limiting trajectory, at a sufficient distance upstream of the fiber, is parallel to the main flow direction at a distance (Y_c) from the centerline of the fiber. The single-fiber efficiency is the ratio of Y_c to the fiber radius. The single-fiber efficiency can have a value greater than 1. This approach is particularly suitable for the gravitational, inertial impaction, and interception mechanisms. Albrecht[8] and Sell[9] used this approach in their pioneering work, and the subject has been thoroughly reviewed by Davies[5] and Pich.[10]

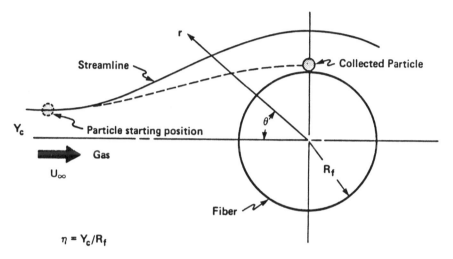

Figure 7: Particle geometry and fiber geometry in a trajectory model.

Trajectory modeling has usually been approached by modeling one, or perhaps two, of the fundamental mechanisms to obtain an expression predicting single-fiber collection efficiency for the mechanism(s). Different researchers arrive at different results, primarily because of the different assumptions concerning the flow field near the fiber and also the simplifications required to develop solutions. The results can be presented in the form of: plots of the results of numerical solutions; semi-empirical expressions; or analytical solutions for the simplest cases. The separate efficiency mechanisms are then combined to give an overall single-fiber efficiency. This overall efficiency can be related (through filter properties) to the overall filter efficiency.

The study of filtration involves applying classical fluid mechanics to the flow around obstacles. In doing so, it is assumed that: the gas stream is a continuum and the Navier-Stokes expressions are applicable. This assumption should be examined to model high-efficiency filters. The fiber Knudsen number is a parameter that relates the fiber size to the distance traveled by gas molecules between collisions (the mean free path). It is given by:

(1) $Kn = l/R_f$

where

  Kn = Knudsen number
  l = mean free path of gas molecule
  R_f = fiber radius.

The mean free path for air is given by:

(2) $l = \dfrac{\mu}{\varrho} \left[\dfrac{\pi M}{2 \, kT} \right]^{1/2}$

where

  μ = viscosity of gas
  ρ = gas density
  M = molecular weight
  k = Boltzmann's constant
  T = absolute temperature
  l = 0.065 μm at normal temperature and pressure in air.

Devienne[11] classified the flow for which the Knudsen number is between 0 and 0.001 as the region in which continuum fluid mechanics is appropriate, and the Navier-Stokes equations can be used to describe the flow. The region bounded by 0.001 < Kn < 0.25 is characterized as the slip-flow regime. Flow within this regime can be modeled theoretically by using the classical hydrodynamics expressions with modified boundary conditions. The region with 0.25 < Kn < 10 is known as the transition regime, and flow for the region for which the Knudsen number is greater than 10 is known as the free molecular gas flow regime. The latter two regimes are not easily dealt with theoretically. For conditions typical in a clean room (70°F and atmospheric pressure), the continuum range is for fiber diameters greater than 130 μm; slip flow is for fibers between 130 to 0.52 μm; and the transition is from 0.52 to 0.013 μm. Typically, the filters of interest include many sub-10 μm fibers. Therefore, the flow regime of interest falls in the slip-flow range.

The isolated cylinder model, where the filter is treated as an array of non-interacting cylinders, is the least complicated of all the models. Contemporary models treat the filter matrix as: parallel cylinders at random spacing; regular arrays of cylinders; or randomly-rotated layers of parallel cylinders. We will limit discussion to the Kuwabara theory which is based on the cell model of randomly spaced cylinders and is discussed next.

The cell model has provided the most successful description of fibrous filters. The filter is treated as a system of parallel, randomly-spaced fibers which are oriented perpendicular to the flow. Each fiber is surrounded by an independent body of fluid (Figure 8a). The polygonal cells are difficult to treat mathematically, and the treatment is simplified to regard the cells as identical concentric cylinders with the dimension shown in Figure 8b. The packing density, α, of the cell is required to equal the fiber volume divided by the total volume. For this particular case:

U_∞ = velocity of gas in the filter
ρ = gas density
μ = gas viscosity.

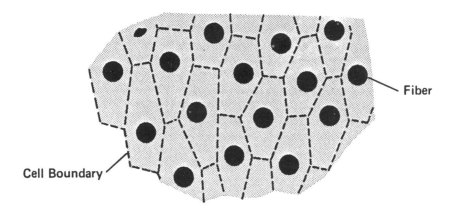

a. Schematic representation of parallel cylinders with the same radius (randomly placed and homogeneously distributed).

Total cell volume = filter volume
fiber volume/cell volume = R_f^2 / R_c^2 = packing density, α

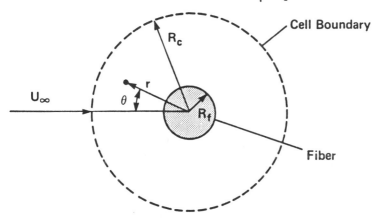

b. Simplified cell used for analysis.

Figure 8: Cell model of filter, after Kuwabara.[12]

(3) α = (fiber radius/cell radius)^2

This model produces a flow field that is valid for a system of cylinders.

The stream function of the Kuwabara[12] continuum hydrodynamics model is given by:

(4) $$\psi = \frac{U_\infty r \sin \Theta}{2 K} \left[2 \ln \left(\frac{r}{R_f} \right) - 1 + \alpha + \frac{R_f^2}{r^2} (1 - \frac{\alpha}{2}) - \frac{\alpha}{2} \frac{r^2}{R_f^2} \right]$$

where

K = $-0.75 - 0.5 \ln \alpha + \alpha - 0.25\alpha^2$
r = radial coordinate
R_f = radius of fiber
U = filter face velocity
U_∞ = fluid velocity within filter = $U/(1 - \alpha)$
α = packing density
θ = angular coordinate.

K is called the Kuwabara hydrodynamics factor. This description of the flow field has been validated experimentally by Kirsch and Stechkina.[13] This formulation is the basis of the descriptions which follow.

2.1.2 Pressure Drop. The relationship between pressure drop and flow is of great practical significance because the amount of energy expended in collecting particles is often as important as the efficiency with which they are collected. Davies[5] defines filter flow resistance by restating Darcy's law to read:

(5) $w = \Delta P/Q$

where

w = resistance
Δp = pressure drop
Q = volumetric flow rate.

Darcy's law which states that the pressure drop is proportional to flow rate has been found to hold for many good-quality air filters. Thus the $\Delta p/Q$ term should be constant and the filter resistance constant. Actually, many lower efficiency filters display an increase in flow resistance as the flow rate is increased. Davies[5] cites compression of the filter and/or a change in the nature of the flow from Stokes's flow to laminar, inertial flow in the range $0.5 < Re < 20$ as possible explanations, where Re is the Reynolds number based on the fiber diameter and is given by the following:

$$Re = \frac{2 R_f \varrho U_\infty}{\mu}$$

where

R_f = fiber radius

Filter flow resistance has been modeled theoretically on the basis of either the flow around fibers (isolated or in arrays) or as the flow through channels. These theoretical results have been extended to give an estimate of filter pressure drop by relating the fiber (channel) resistance and length to the packing density, thickness, and other filter properties. There are also a number of empirical or semi-empirical expressions for pressure drop as a function of filter parameters. The following section presents an expression for fiber drag derived from models of the flow patterns in filters.

2.1.2.1 *Fiber Drag Expressions.* The drag force that a flowing fluid exerts on a body immersed in the fluid is often expressed as:

(6)
$$F_d = (a)\,(KE)\,(Cd)$$

where

F_d = drag force
a = characteristic area
KE = kinetic energy term
Cd = drag coefficient.

For a cylinder transverse to the direction of the fluid flow, the area term is taken as the projected area. This is $2(R_f)(l_f)$, where R_f is the fiber radius and l_f is the fiber length. The kinetic energy term is taken as $1/2\,\rho\,U_\infty$, where ρ is the fluid density and U_∞ is the approach velocity of the fluid to the fiber within the filter. Combining these expressions with Equation 6, the following expression for the drag force is obtained:

(7)
$$F_d = R_f\,l_f\,\varrho\,U_\infty^2\,Cd$$

A dimensionless drag force per unit length of fiber can be defined as:

(8)
$$F^* = \frac{F_d}{l_f}\,\mu\,U_\infty$$

where

F^* = dimensionless drag force and
μ = fluid viscosity.

The dimensionless fiber drag is used because it is convenient mathematically. It can be calculated by integrating the tangential stress on the fiber over the surface of the fiber, using the flow field presented above to provide an expression for the tangential stress. The relationship between F^* and Cd can be determined by combining Equations 7 and 8 to obtain:

(9)
$$F^* = \tfrac{1}{2}\,Re\,Cd$$

2.1.2.2 *Filter Pressure Drop.* The fiber drag can be related to filter pressure drop in filters if certain parameters are known. In the development that follows, the filter is assumed to be homogeneous. The total fiber length, t_{lf}, for a filter of uniform thickness or depth L and with a unit area perpendicular to the flow is:

$$(10) \qquad t_{lf} = \frac{\alpha L}{\pi R_f^2}$$

where

t_{lf} = total fiber length per filter area.

The pressure drop across a filter equals the product of the drag force per unit length of fiber and the total length of fiber per unit area. The drag force per length of fiber can be obtained from F^*, the dimensionless drag per unit length of fiber. Combining Equations 8 and 10 as suggested leads to an expression for pressure drop:

$$(11) \qquad \Delta P = \frac{F^* \alpha L \mu U_\infty}{\pi R_f^2}$$

where

Δp = filter pressure drop

If the Kuwabara[12] flow field is used for the dimensionless drag, the pressure drop is given by:

$$(12) \qquad \Delta P = \frac{\mu U_\infty L}{R_f^2} \frac{4 \alpha}{K}$$

2.1.3 Particle Collection Efficiency. The efficiency of a filter is related to penetration by:

$$(13) \qquad E = 1 - N/N_o = 1 - P_t$$

where

N   is the concentration leaving the filter
N_o is the concentration entering the filter
P_t is penetration ($\equiv N/N_o$).

Penetration as a function of the properties of the filter is given by:

$$(14) \qquad P_t = \exp\left[\frac{-2 \eta \alpha L}{\pi R_f}\right]$$

where

R_f = mean fiber radius
α = packing density of the filter
L   = thickness of the filter
η = single fiber efficiency.

Part of a fundamental evaluation of fibrous filtration consists of quantifying the contribution of various mechanisms to the single fiber efficiency. Of particular interest is the single fiber efficiency as a function of particle diameter. By determining either the maximum penetrating particle diameter or the minimum efficiency, one can develop a conservative estimate of the filter's protection. This section examines each primary collection mechanism and its contribution to filter efficiency and then shows the effects of combining different mechanisms.

2.1.3.1 Particle Collection by Inertial Impaction. Inertial impaction is an important collection mechanism only for the largest particles of concern, typically those larger than 1 μm. The true collection which results solely from inertial impaction (without an interception component) requires that particles (to have their proper mass) be treated as if collision did not occur until the center of the particle reached the fiber. This is not physically true. Therefore, pure inertial impaction is simply a theoretical construct.

Numerical methods are normally used to solve the equation of motion to estimate efficiency. Thus, the results are often presented in the form of graphs or tables. As seen in Figure 9, the Stokes number Stk is the key parameter in particle collection by inertial impaction. Stk is defined[5] as:

$$(15) \qquad\qquad Stk = \frac{\rho_p \, d_p^2 \, U_\infty \, C}{18 \, \mu \, R_f}$$

where

ρ_p = particle density
d_p = particle diameter
U_∞ = velocity of fluid undisturbed by fiber
C = $1 + 2.468 \, l/d_p + 0.826 \, l/d_p \, \exp(-0.452 \, d_p/l)$
l = gas mean free path.

The Stokes number is the ratio of the distance a particle would travel against fluid drag (having an initial velocity U) to the fiber radius. Other authors (i.e., Pich[10]) have defined the Stokes number with respect to the fiber diameter.

For potential flow, the single-fiber collection efficiency resulting from impaction is a function of the Stokes number only. In a viscous flow field, this collection is a function of both the Reynolds number and the Stokes number for an isolated cylinder (single fiber) in a system of cylinders. The packing density is also important.

The single fiber efficiency curves in Figure 9 (η_{st} = single fiber efficiency due to inertial impaction) were originally developed for the interpretation of measurement of droplet sizes in supercooled clouds. A dimensionless group was used as a parameter to include particle diameter effects over a wide range of flow rates:

$$P = Re_p^2/Stk = \frac{18 \, \rho^2 \, U_\infty \, d_p}{\mu \, \rho_p}$$

where

ρ = gas density

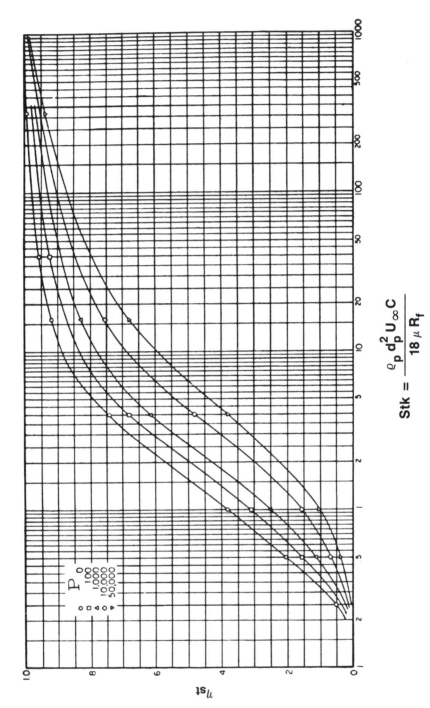

$$Stk = \frac{\varrho_p \, d_p^2 \, U_\infty \, C}{18 \, \mu \, R_f}$$

Figure 9: Collection efficiency for cylinders in an inviscid flow with point particles. (Courtesy, John Wiley and Sons)

U_∞ = gas velocity
d_p = particle diameter
μ = gas viscosity
ρ_p = density of particles
Re_p = Reynolds number of the particle
Stk = Stokes number.

2.1.3.2 Particle Collection by Interception. When applied to aerosol capture by an isolated single fiber, interception describes the collection of those aerosols that follow the fluid streamlines and make contact with the fiber. Strictly speaking, pure interception occurs only when the particles have size but no mass. Thus, the idea of collection resulting solely from interception is conceptually useful but physically unattainable.

An expression for single fiber efficiency based on particles interception in Kuwabara flow was reported by Stechkina et al.[14] and is given by:

$$(16) \qquad \eta_R = \frac{(1+R)^{-1} - (1+R) + 2(1-R)\ln(1+R)}{2K}$$

where

R = the interception parameter, $d_p/2R_f$
K = Kuwabara hydrodynamic factor defined earlier.

The interception parameter, R, is the ratio of the particle diameter to the fiber diameter. Potential flow collection by interception is a function of R only; for viscous flow, the Reynolds number becomes important. Again, in a system of cylinders (fibers), the packing density is also important. For a fixed fiber size, the interception mechanism becomes more significant as the particle size increases.

2.1.3.3 Particle Collection by Diffusion. Diffusion is significant only for small (<0.5 μm in diameter) particles. For these particles, the fluid streamlines represent only the average path of the particle. Their motion actually consists of many random, zigzag deviations from the path of the streamlines (both toward and away from the fiber). These large deviations from the streamlines mean that contact with the collection surface can occur even though the streamline passing through the upstream position of the particle passes the fiber surface at a distance greater than the particle radius. As the particle size decreases, the importance of Brownian motion relative to the fluid flow increases until the distinction between aerosol particle and gas molecule becomes insignificant.

Diffusion modeling is based on the distance the particle can diffuse in a given time. The distance equals the square root of the product of the particle diffusion coefficient and time. For collection on fibers, the residence time in the filter is important. The Peclet number, Pe, relates fluid motion to the diffusion coefficient and hence is important to modeling this form of collection. The Peclet number is defined as:

$$(17) \qquad Pe = \frac{2\,R_f\,U_\infty}{D}$$

where

D = particle diffusion coefficient = $k\,T\,C/\,3\,\pi\,\mu\,d_p$
k = Boltzmann's constant
T = absolute temperature
d_p = particle diameter.

An example of the collection efficiency due to diffusion is given by Lee and Liu:[15]

(18) $$\eta_D = 2.6\left(\frac{K}{1-\alpha}\right)^{-0.333} Pe^{-0.667}$$

This expression shows that diffusive collection is a function of the Peclet number and a hydrodynamics factor. In terms of filter and aerosol properties, collection efficiency which results from diffusion increases with decreasing particle diameter, gas velocity, and fiber size.

2.1.3.4 Particle Collection by Diffusion and Interception. The efficiency expression for the combined mechanisms of diffusion and interception[15] is:

(19) $$\eta_{D,R} = 1.6\left(\frac{K}{1-\alpha}\right)^{-0.333} Pe^{-0.667} + 0.6\left(\frac{1-\alpha}{K}\right)\left(\frac{R^2}{1+R}\right)$$

This combination of mechanisms should be adequate to describe the collection of particles of diameters less than 0.5 μm. Lee and Liu[16] and Rubow[4] have shown that a model based only on these two mechanisms can successfully predict the particle size of maximum penetration and adequately predict filter efficiencies.

2.1.3.5 Particle Collection by Diffusion, Interception, and Inertia. The fan model of Stechkina et al.[14] has been shown to fit both particle collection and pressure drop experimental data fairly well in the range of the most penetrating particle size. The single fiber collection efficiency for the fan model is given by:

(20) $$\eta = \eta_D + \eta_{D,R} + \eta_R + \eta_{St,R}$$

Diffusion:

$$\eta_D = 2.9\,K^{-0.333}\,Pe^{-0.667} + 0.624\,Pe^{-1}$$

Diffusion-Interception:

$$\eta_{D,R} = 1.24\,K^{-0.5}\,Pe^{-0.5}\,R^{0.667}$$

Interception:

$$\eta_R = (2K)^{-1}\,[2(1+R)\,\ln(1+R) - (1+R) + (1+R)^{-1}]$$

Impaction-Interception:

$$\eta_{St,R} = \text{Computed numerically}$$

These expressions developed by Stechkina et al.[14] are similar but not identical to those presented earlier. Stechkina et al. use them to compute the efficiency of each layer in the model filters. The fan model assumes that the fibers have random orientation in layers. Each layer can be rotated, creating an illusion of the spreading of a lady's fan.

These filter models are often applied by including an inhomogeneity factor, ϵ. This correction is included in the overall efficiency equations as follows:

(21)
$$P_t = \exp\left[\frac{-2\,\eta\,\alpha\,L}{\pi\,R_f\,\epsilon}\right]$$

The inhomogeneity factor is an estimation of the nonuniformity of a real filter and is defined as the ratio of the drag forces acting per unit fiber length or pressure drop and model filter or:

(22)
$$\epsilon = \frac{\eta}{\eta_{st}} \quad \text{or} \quad \epsilon = \frac{\Delta P_{theoretical}}{\Delta P_{experimental}}$$

where:

η = theoretical single particle efficiency
η_{st} = actual single particle efficiency.

To summarize, the theoretical approach to predicting filter performance consists of combining fundamental expressions and empirical corrections. The complexity of the filter material makes an exact prediction based on *a priori* assumptions and measures of the filter properties unlikely. However, understanding the mechanisms of filter behavior is very useful in gauging the limitations of various media.

2.1.3.6 Correlation of Actual Performance with Models. Many experimental investigations of fibrous filter efficiency have been reported. Only those models that make use of flow fields based on systems of cylinders using the Kuwabara flow equations will be considered here.

Figure 10 compares the theories of Stechkina et al.[14] and Yeh and Liu[17] with experimental data collected by Lee and Liu.[16] The proper constants accounting for lack of uniformity, random fiber orientation, and similar nonideal conditions were incorporated into the models. The inhomogeneity factor is about 1.67 for these polyester filters. Overall, both theories fit the data well. The theory of Stechkina et al. is seen to agree closely at high velocities while deviating at low velocities. Stechkina et al.[14] have stated that some of the assumptions which they made while developing the impaction model (described in their earlier paper) were inadequate. On the other hand, the theory of Yeh and Liu is seen to agree best with the data at low filtration velocities.

A semi-empirical method based on earlier work by Friedlander[18] for generalizing filter efficiency data was reported by Lee and Liu.[16] Starting with the single fiber efficiency equation 20 for combined diffusion and interception and using empirical coefficients

$$\eta = \beta_1\left(\frac{1-\alpha}{K}\right)^{1/3} Pe^{-2/3} + \beta_2\left(\frac{1-\alpha}{K}\right)\frac{R^2}{1+R}$$

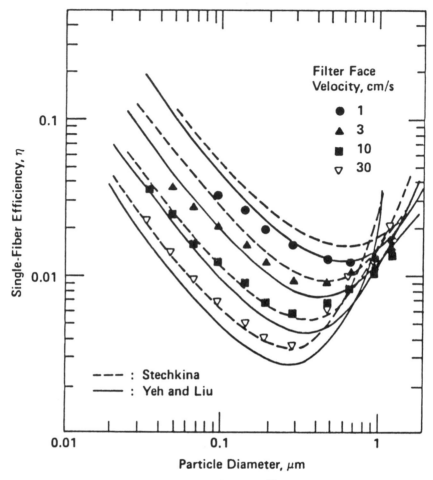

Figure 10: Comparison of data of Lee and Liu[16] with theories of Stechkina et al.[14] and Yeh and Liu.[17] Fiber diameter of 11 μm; packing density of 0.151.

if both sides are multiplied by

$$Pe\ R/\ (1+R)^{1/2}$$

then the above equation can be reduced to

$$(23) \quad \eta\ Pe\ \frac{R}{\sqrt{1+R}} = \beta_1\left(\frac{1-\alpha}{K}\right)^{1/3} Pe^{1/3}\ \frac{R}{\sqrt{1+R}} + \beta_2\left[\left(\frac{1-\alpha}{K}\right)^{1/3} Pe^{1/3}\ \frac{R}{\sqrt{1+R}}\right]^3$$

Therefore, the coefficients can be determined by plotting

$$\eta\ Pe\ R/\sqrt{1+R}$$

and

$$\left(\frac{1-\alpha}{K}\right)^{1/3} Pe^{1/3} R/\sqrt{1+R}$$

where

η = the single fiber efficiency
R = interception parameter
α = the fiber volume fraction or the filter solidity
K = the hydrodynamic factor
Pe = Peclet number

This correlation (Figure 11) was found to apply over a wide range of experimental conditions. Filter performance data can be plotted on a log-log graph with two dimensionless parameters over a wide range of filter parameters and particle diameters. The importance of this approach is that a semi-empirical correlation can be used in applications of high-efficiency collection of sub-micron-sized particles. Thus, the correlation can serve as a guide in the design of filter media.

2.1.3.7 Most Penetrating Particle Fiber. As shown in the description of collection mechanisms, fibrous filters function such that there is a particle diameter which has the greatest penetration and thus the lowest efficiency. The most penetrating particle diameter depends on the filter fiber diameter, volume fraction, and the gas velocity. Lee and Liu[19] differentiated Equation 20 for combined diffusion and interception mechanisms with appropriate corrections for slip and obtained:

(24) $d_{p\,min} = 0.885 \left(\frac{K}{1-\alpha}\right)\left(\frac{l^{1/2}kT}{\mu}\right)\left(\frac{D_f^2}{U}\right)^{2/9}$ for $0.075 < l/d_p < 1.3$

where

d_{pmin} = most penetrating particle diameter
d_p = particle diameter
K = hydrodynamic factor
l = mean free path
k = Boltzmann's constant
μ = viscosity of gas
D_f = fiber diameter, $2R_f$
U = gas velocity
α = fiber volume fraction (packing density).

This equation is shown in Figure 12. Increasing the velocity through the filter and reducing the fiber diameter will reduce the value of the most penetrating particle diameter. Emi and Kanaoka[20] compared the predictions of Equation 24 with data obtained with HEPA media and found that the trends were correctly predicted.

The particle size dependent penetrations have been measured for filter media typically used in both HEPA and ULPA filters. All of the mechanisms expected from the theoretical development are evident in the results. The penetration data reported by Liu et al.[21] are shown in Figures 13 and 14. The most penetrating particle diameter and its dependence on flow velocity are apparent.

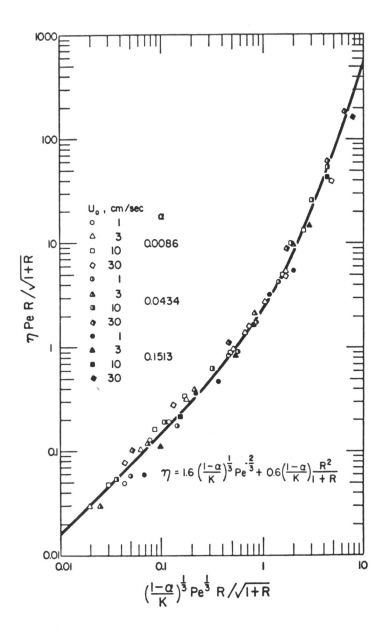

Figure 11: Correlation of filtration data. (Reprinted by permission of the publisher from "Theoretical Study of Aerosol Filtration by Fibrous Filters," by K.W. Lee and B.Y.H. Liu, *Aerosol Science and Technology,* 1, pp. 156. Copyright 1982 by Elsevier Science Publishing Co., Inc.)

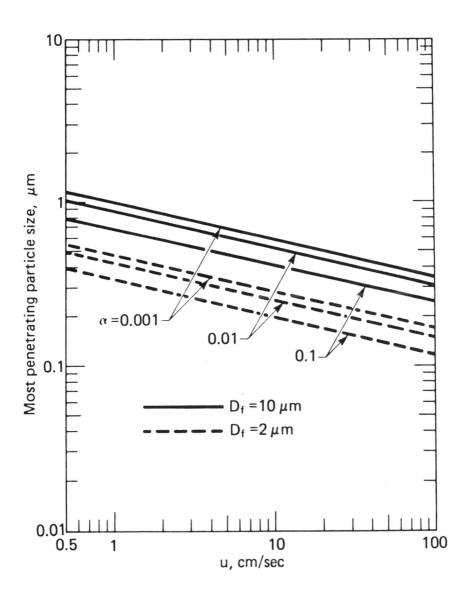

Figure 12: Calculated most penetrating particle size as a function of velocity, fiber volume fraction and fiber diameter. (Courtesy, Journal of the Air Pollution Control Association)

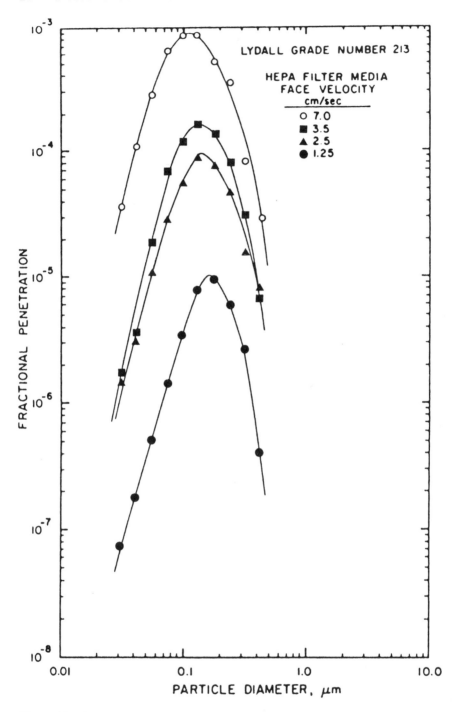

Figure 13: Aerosol penetration through HEPA filter. (Courtesy, Institute of Environmental Sciences)

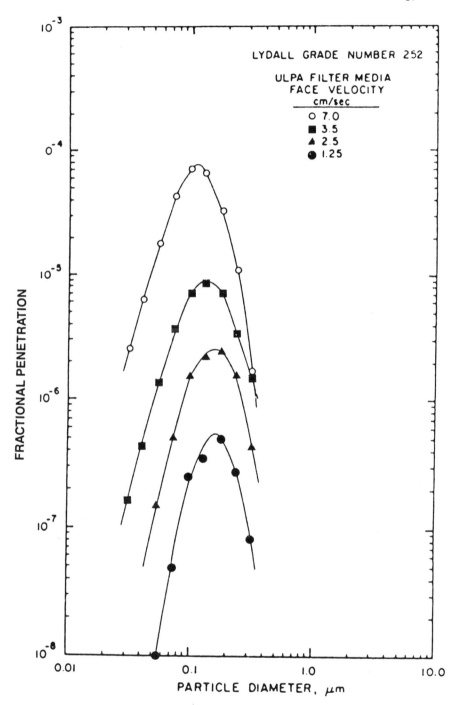

Figure 14: Aerosol penetration through ULPA filters. (Courtesy, Institute of Environmental Sciences)

2.1.3.8 Figure of Merit. Another criterion of filter performance is the relative efficiency (or penetration) as a function of pressure drop. The figure of merit is given by:

(25) $f = -\log P_t/\Delta P$

where

P_t = penetration (outlet concentration/inlet concentration)
ΔP = pressure drop.

The figure of merit is a function of particle diameter and flow velocity through the filter and provides a good way of comparing filters tested under identical conditions. The larger the figure of merit is, the better the performance, assuming both P_t and ΔP individually fall within acceptable limits.

2.2 Membrane Filters

There are four types of membrane filters:

(1) Solvent cast filters: filters are formed from one or more types of cellulose esters or other materials such as polysulfone.

(2) Nuclepore filter: a film of polycarbonate is exposed to radiation and the tracks are etched to form a porous sheet.

(3) Teflon: polymer particles are solvent cast sintered and then stretched to form a weblike structure. Often a support membrane is used for strength.

(4) Silver membranes: silver particles are sintered into a porous sheet.

The most appropriate performance equations vary among the different filter types, the fibrous filter equations of Section 2.1 being most applicable to the solvent cast and Teflon filters and the equations of a capillary pore model best describing the performance of the Nuclepore filter. Before presenting these equations, the structure and composition of these first three listed membrane filter types will be reviewed.

Solvent Cast Filters—The cellulosic material can be esters of cellulosic acetate, nitrate or triacetate. The cellulose esters are in solution with two or more solvents and are cast on a flat surface. The evaporation of the solvents leaves a very porous structure. The choice of the filter material is dictated by the application. The surface of a typical filter material with a pore size of 0.45 μm was shown in Figure 2.

The performance of the filter is usually specified by the pore size. The smaller the pore size is, the greater the particle collection efficiency and the gas pressure drop. The bubble point is usually specified as an indication of the integrity of the membrane. The procedure to determine bubble point is as follows. For a serial stack of filters or a single membrane, nitrogen is used to force the liquid through the membrane holder. Tubing is attached downstream of the holder and immersed in a beaker of water. Pressure is increased to displace all liquid so that air impinges on the wet membrane. The bubble point pressure is defined as the pressure in kg/cm^2 or psi where air passes through a wet membrane and is viewed downstream as bubbles. The performance of a typical cellulosic filter media is shown in Table 2.

Table 2: Performance of Selected Cellulosic Nitrate Filters

Pore Size (μm)	Porosity %	Weight (g/m²)	Nominal Thickness (μm)	Nitrogen Flow Rate (l/min/cm²)	Bubble Point (psi)
3.00	80	44	150	28.3	13
1.2	80	45	145	20.4	16
0.80	79	46	145	15	19
0.65	76	50	135	10.2	18.5
0.45	74	54	148	5.8	26
0.30	75	55	140	3.7	45
0.20	72	57	135	2.4	60
0.10	67	61	110	0.67	41

Asymmetric filters are cast of polysulfone by a unique process. It produces a highly asymmetric membrane whose pore size decreases gradually from an average of 10-20 μm on the upstream side to 0.45, 0.2, or 0.1 μm on the downstream side (Figure 15).[23]

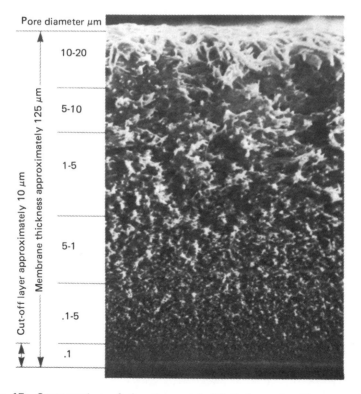

Pore diameter μm

10-20

5-10

1-5

5-1

.1-5

.1

Membrane thickness approximately 125 μm

Cut-off layer approximately 10 μm

Figure 15: Cross-section of the Brunswick SE Series polysulfone membrane. (Courtesy of Filterite, Brunswick Technetics)

Other microporous membranes cast in a conventional manner have pores that are comparable in size on the upstream and downstream sides, and throughout their cross section. The advantage of the large pores on the upstream side is that the membrane can retain a considerably greater quantity of particles before becoming clogged.

The "cutoff" layer is only about 10 μm thick. Other membrane media, which are fairly uniform throughout, have a "cut off layer" equal to their thickness (approximately 100-125 μm). A thin cut off layer gives the media significantly higher flows with much less pressure drop.

Nuclepore Membranes–The manufacture of the Nuclepore filter is a two-step process. A polycarbonate film is exposed to collimated, charged particles in a nuclear reactor. Particles pass through the material, leaving behind sensitized tracks. The pore density (pores/cm^2) is controlled by the residence time in the reactor. The exposed film is then immersed in an etch bath where the tracks are preferentially etched into uniform, cylindrical pores. A specific pore size is produced by controlling the length of the etching process. The development of this kind of filter is described by Price and Walker[24] and Fleischer et al.[25] The performance of several filters is shown in Table 3.

Table 3: Characteristics of Nuclepore Filters

Pore Size (μm)	Nominal pore density (pores/cm^2)	Nominal thickness (μm)	Flow Rate air (1/min/cm^2)	Bubble point (psi)
12.0	1 x 10E5	6	65	<1
10.0	1 x 10E5	8	50	<1
8.0	1 x 10E5	10	45	2
5.0	4 x 10E5	10	42	4
3.0	2 x 10E6	10	42	6
2.0	2 x 10E6	10	36	10
1.0	2 x 10E7	10	36	15
0.8	2 x 10E7	10	36	18
0.6	3 x 10E7	10	16	25
0.4	1 x 10E8	10	16	33
0.2	3 x 10E8	10	6	60
0.1	3 x 10E8	5	2.2	>100
0.08	6 x 10E8	5	2.2	>100
0.05	6 x 10E8	5	0.25	>100
0.03	6 x 10E8	5	0.03	>100

Teflon Filters–Teflon membranes (hydrophobic filters) are manufactured in a unique process where polytetrafluoroethylene (PTFE) is expanded in both uniaxial and biaxial manners. The resulting membrane is thin, extremely porous and behaves as an absolutely retentive membrane. Teflon membranes are asymmetric in that a polypropylene web support is laminated to one side (to improve handling characteristics). Table 4 summarizes the performance of several Teflon filters.

Teflon membranes prove inert to most solvents which are chemically aggressive; strong acids; and bases. The backing material imposes certain chemical and thermal limitations. These membranes can be sterilized by steam at 121°C, 15 psi and can be used in temperatures up to 140°C. Beyond 140°C, the laminate begins to soften.

Table 4: Properties of Teflon Membrane Filters

| Polymer | Pore size (μm) | Porosity (%) | Flow rate | | Bubble point[3] | | Water breakthrough | | Temperature maximum °C |
			Methanol[1] mL/min/cm^2	Air[2] L/min/cm^2	kg/cm^2	PSI	kg/cm^2	PSE	
Teflon	1.0	91	90	7	0.21	3	0.49	7	145
	0.5	84	40	3	0.49	7	1.41	20	145
	0.2	78	15	2	0.91	13	2.81	40	145

1. Flow rates are determined by filtering 500 ml of methanol through a 9.6 cm^2 filter area under constant vacuum 0.7 kg/cm^2 (10 PSE).

2. Initial air flow rates are determined under constant vacuum 0.7 kg/cm^2 (10 PSI).

3. Bubble point is defined as the minimum pressure in kg/cm^2 or PSI required to force air through a membrane prewet with methanol.

Source: Based on MicroFiltration Systems Catalog. (22)

2.2.1 Performance Prediction. The prediction of the pressure drop of a cast membrane filter, according to Rubow and Liu,[26] can be accomplished best by using the fibrous filter model described in Section 2.1. Rubow and Liu[26] found that, by using the diffusion and interception mechanisms, the particle size dependent penetration for Millipore AA filters, with a pore diameter of 0.8 μm, could be matched with a fiber diameter of 0.555 μm, an inhomogeneity factor of 1.67 and actual values of filter thickness and porosity. This work demonstrates that applying the fibrous filter model to membranes is not ideal yet certainly can develop an understanding of the effect of the filter parameters on performance. This is especially useful since the complex structure of cast membrane filters is not well described by simple pore models. The Nuclepore-type filter's performance can be predicted best by using the capillary pore model. The flow pressure drop model assumes that the membrane is traversed by parallel capillaries. This model assumes a system of equidistant, parallel, circular capillaries of diameter D_f perpendicular to the filter surface and with a length equal to the filter thickness L.

The flow in a capillary in the continuum region is described by the Hagen-Poiseuille law which, for a system of capillaries, is:

$$(26) \qquad \Delta P = \frac{8 \, \mu \, L \, U}{\pi \, a^4 \, N}$$

In the range of small Knudsen numbers, the slip correction must be taken into account. Modifying the Hagen-Poiseuille law by including the slip correction to the capillary model gives:

$$(27) \qquad \Delta P = \frac{8 \, \mu \, L \, U}{\pi \, a^4 N (1 + 3.992 \, Kn_f)}$$

where

μ	=	viscosity
a	=	pore radius = $D_p/2$
L	=	length of pore
N	=	number of pores
U	=	superficial velocity Q/A
A	=	filter area
Kn_f	=	l/a
l	=	mean free path.

Diffusion reflection of the gas molecules from the pore surface has been assumed.

A relatively good agreement of these equations with experimental data has been claimed by Pich.[10]

2.2.2 Particle Collection. The model of particle deposition of particles on a membrane filter assumes the deposition of particles on the surface by inertial deposition and interception, while the pores collect particles by diffusion. Membrane filters can collect particles much smaller than the pore diameters.

Surface collection is described by interception and sieving and impaction.

If d_p is the particle diameter, then the interception is described by the parameter:

(28)
$$N_r = D_p/d_p$$

for $N_r < 1$. The particle is captured when it approaches the wall of the pore at a distance equal to its radius.

Sieving is a special case of direct interception and is described by the same parameter N_r, where $N_r > 1$. Obviously, for $N_r > 1$, all particles are captured by the filter which acts as a sieve. Spurny et al.[27] reported an equation for the efficiency of a filter from interception.

(29)
$$\eta_R = 2\,N_r - N_r^2$$

The impaction of particles around a pore on a filter surface was investigated by Kanaoka et al.[28] The equations of motion were solved numerically with an orifice flow field. The results have the familiar Sigmoid shape when the efficiency is plotted as a function of the Stokes number.

The collection of particles by inertial impaction is reported by Pich[10] as:

(30)
$$\eta_{Stk} = \frac{2\,E_I}{1+\beta} - \frac{E_I^2}{(1+\beta)^2}$$

where

$$E_I = 2\,Stk\,\sqrt{\beta} + 2\,Stk^2\,\beta\,\exp\left(-\frac{1}{Stk\,\sqrt{\beta}}\right) - 2\,Stk^2$$

$$\beta = \frac{\sqrt{1-\alpha}}{1 - \sqrt{1-\alpha}}$$

$$Stk = \frac{C\,D_p^2\,\rho_p\,U}{9\,\mu\,d_p}$$

Particle removal by diffusion in the pores is described by equations developed from diffusion batteries. The entrance effects are neglected, and the problem is reduced to diffusion of particles to a wall in laminar flow at very small Reynolds numbers. The mechanism is described by a dimensionless parameter:

(31)
$$N_d = \frac{4\,L\,D}{d_p^2\,U}$$

where D is the diffusion coefficient of the particles. The efficiency of collection of particles in a tube is described by Gormley and Kennedy[29] as:

(32)
$$\eta_D = 2.57\,N_d^{2/3} - 1.2\,N_d - 0.177\,N_d^{2/3}$$

This equation is valid for $N_d < 0.03$. The equation derived by Twomey[30] can be applied for $N_d > 0.03$. It is:

$$\eta_D = 1 - 0.81904 \exp(-3.6568\ N_d)$$
$$- 0.09752 \exp(-22.3045\ N_d)$$
(33)
$$- 0.03248 \exp(-56.95\ N_d)$$
$$- 0.0157 \exp(-107.6\ N_d)$$

The overall filter efficiency can be obtained by combining the expressions for the various particle collection mechanisms. It is assumed that each mechanism is independent and the pores are not interactive. The total collection efficiency is given by an expression developed by Spurny et al.:[27]

(34) $$E_T = \eta_{Stk} + \eta_D + 0.15\ \eta_R - \eta_{Stk}\ \eta_D - 0.15\ \eta_{Stk}\ \eta_R$$

where 0.15 was empirically determined for Nuclepore filters. This equation tends to underestimate interception effects because of the construction of flow near the pore.

Membrane filters demonstrate the same general particle size dependent shape of the penetration curve as do fibrous filters (Figure 16). Filters with very high efficiencies or low penetrations are difficult to measure and result in incomplete curves.

Nuclepore filters have a similar kind of behavior except that the "window" between diffusional and inertial interception collection is broader (Figure 17). As the pore size of the filter decreases, the particle size corresponding to the minimum in collection efficiency becomes smaller. The curves reported by Spurny et al.[27] exhibited good agreement between theory and the experimental data.

2.3 Electrostatics

Electrical forces can play an important role in improving both fibrous and membrane filter performance by increasing filter collection efficiency and, in some cases, decreasing pressure drop. In addition to being an important consideration in laboratory investigations, electrostatic enhancement of filters has been suggested as a way of obtaining cleaner working spaces.

There are three major electrical forces between fibers and particles:

1.   image force (charged particles only)

2.   dielectrophoretic force (charged collector only)

3.   coulomb force (charged particles and collector)

The image force is the attractive force created between a charged body and a charge induced in an adjacent, electrically neutral surface (i.e., a fiber or a membrane collector). The charged particle polarizes the fiber by causing a charge redistribution within the fiber. The resulting force can be calculated by treating the induced fiber charge as a fictional charge located at a point within the fiber. This computational approach is the classical method of image analysis. The image force is always attractive.

The dielectrophoretic force also results from polarization. This force results when charge separation in an electrically neutral body takes place in a nonuniform electric field. The charge of the sign displaced toward the region of higher electric field creates a larger coulomb force on the body than the equal charge of the opposite sign displaced toward the region of lower electric field. Polarization in a nonuniform electric field always creates a net electrical force in the direction of increasing field. Since the electric field surrounding electrically charged fibers always decreases in magnitude with radial distance from the fiber, a particle will always experience an attractive dielectrophoretic force pulling it towards the fiber.

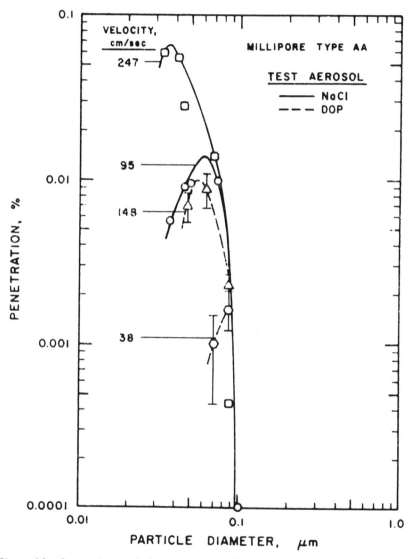

Figure 16: Comparison of the experimentally determined aerosol penetration data for the AA filter observed in this study and that obtained by Rubow.[4] (Courtesy, Institute of Environmental Sciences)

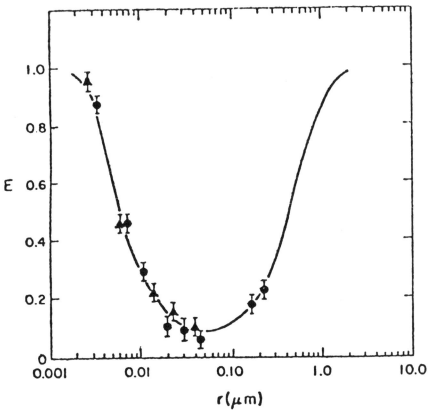

Figure 17: Comparison of the collection efficiency as theoretically predicted and experimentally determined for a 5 μm pore diameter Nuclepore filter at a face velocity of 5 cm/sec. (Courtesy, ASTM)

The coulomb force is the familiar repulsion or attraction between charges. A special case of the coulomb interaction exists under concentrated particle conditions from the net charge of the particle cloud or space charge. When the particle cloud is uniformly charged, mutually repulsive coulomb forces drive the particle apart and toward any grounded or uncharged surfaces (i.e., a fiber). The space charge field precipitates or deposits charged particles on surrounding surfaces until the space charge electric field decreases to insignificant values. Only then can the charge on an individual particle dominate particle behavior through the image force. Several approaches have been developed over the years to utilize these forces in various combinations. The particle filter under various conditions of charge and fields has been reviewed by Pich.[32]

The mathematical expressions to describe filter electrical effects were reported by Shapiro et al.[33] and are summarized in Table 5. The coulombic forces are the most effective for the filtration of large and intermediate-sized particles. The space charge effect and image force may be important for highly charged, concentrated submicron particles.

Table 5: General Expressions for Electrostatic Forces Between Particles and a Single Cylindrical Collector

Type of interaction	Radial force
1. Image force	$-\dfrac{a_c}{16\pi\,\epsilon_0}\dfrac{q^2}{(R_f-r)^2}$
2. Dielectrophoresis due to collector charge	$-\dfrac{Q_e^2}{4\,\pi\,\epsilon_0}\,a_p\,\dfrac{r_p^3}{r^3}$
3. Coulombic force	$\dfrac{Q_e\,q}{2\pi\,\epsilon_0\,r}$
4. Space charge effect	$-\dfrac{q^2\,N_p\,R_f^2}{3\,\epsilon_0\,r}$
5. Coulombic force due	$E_0 q\left[1+a_c\dfrac{R_f^2}{r^2}\right]$
6. Dielectrophoresis due to external electric field	$-8\pi\epsilon_0\,a_p a_c\,E_0^2 p\,\dfrac{r_p^3}{R_C}\left(\dfrac{R}{r}\right)^3\left[a_c\left(\dfrac{R_f}{r}\right)^2+1\right]$

Source: Shapiro et al. (33).

Nomenclature:

$a_p=\dfrac{\epsilon_p-\epsilon_0}{\epsilon_p+2\epsilon_0}$ - constant.

$a_c=\dfrac{\epsilon_c-\epsilon_0}{\epsilon_c+\epsilon_0}$ - constant.

E_0 = Electrostatic field.

N_p = Number density of aerosol particles.

 q = Aerosol particle electrostatic charge.

Q_e = Charge per unit length of the cylindrical collector.

 r = Radial coordinate.

r_p = Aerosol particle radius.

R_f = Collector radius.

ϵ_0 = Free space dielectric constant.

ϵ_c = Collector dielectric constant.

ϵ_p = Aerosol particle dielectric constant.

2.3.1 Passive Electrostatic Systems. The first application of permanently-charged fibers is the Hansen filter.[5] This filter is made by dusting natural wool with a fine powder of colophony resin. The mixture is then mechanically blended until the resin particles develop a permanent charge. The resin particles (approximately 1 μm in size) are small when compared with the 20 μm wool fibers. Because of the resins's high resistivity, the filters would retain their charge up to several years when kept in sealed containers. These filters were used in military gas masks by the Danish, Dutch, French, and Italian armies in the 1930s and in civilian gas masks in England beginning in the 1930s. The Hansen filter is still widely used in industrial respirators.

In recent years a material suggested for passive electrostatic filters is fibers with permanent dipoles, such as electrets. One material is formed by passing a polymer sheet across a corona discharge which imparts positive and negative charges on opposite sides of the material. The sheet is then split into small fibers which are then formed into a filter mat. This material is called Filtrete and is sold by the N.V. Verto Co., Rotterdam, Netherlands.[34] The 3M Co. of St. Paul, Minnesota, has obtained rights to the process and uses the filter material for respirators under applying high voltage to a polymer melt during a spinning process to form a fiber. This fiber is commercially available from the Carl Freudenberg Co., West Germany.[35] The permanently-charged filters have a superior performance over purely mechanical filters with the same fiber diameter and packing. In particular, the efficiency is improved in the 0.1 to 1 μm size range. However, the permanent electrical field and the particle collection enhancement may be reduced in an unpredictable way by material collected in the filter.[36]

2.3.2 Active Electrostatic Systems. A number of combinations of particle charging and electrostatic fields have been developed to enhance filtration. A typical hardware configuration is a corona precharger followed by a HEPA filter which has an applied electric field.[37,38] It is claimed that these systems have lower pressure drop and greater particle collection than mechanical filters as shown in Figure 18.

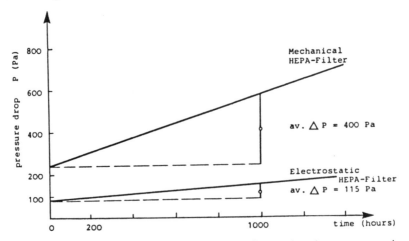

Figure 18: Pressure drop as a function of time. Comparison between a mechanical filter and the Delbag electrostatic HEPA Filter. (Courtesy, Uplands Press Ltd.)

Masuda and Sugita[39] reported the test data shown in Figure 19. The effects of corona precharging and applied electric field were shown to be additive. An improvement in efficiency of more than two orders of magnitude was reported. Masuda and Sugita attributed this effect to the prefiltering of the corona discharge unit.

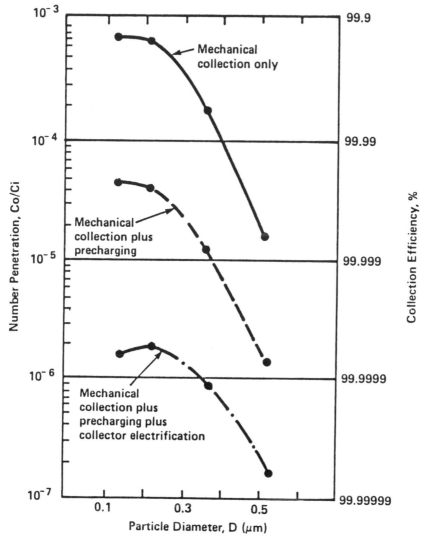

Figure 19: Electrical enhancement of HEPA filter performance.[39]

Electrostatic filtration has been demonstrated in the laboratory as an approach to increasing the efficiency of the filter while lowering the pressure drop. However, the concept has not been proven in the manufacturing environment. Currently it is difficult to gauge the impact of electrostatic filtration on clean rooms.

3. FILTER TESTING FUNDAMENTALS

The rapidly evolving field of contamination control is creating a need for new test methods. Contemporary products manufactured in clean areas require greater protection than what was needed a few years ago. Some of the test procedures use DOP (dioctyl phthalate), a substance which may actually be detrimental to products. In addition, many standard methods were developed over 30 years ago using the technologies of that time. Filters are tested to insure that the filter has the desired efficiency at the rated gas flow and pressure drop. This section will review the fundamentals of filter testing and describe various standard tests as well as some research and development tests.

The components of a filter test apparatus include an aerosol generator, a test filter holder, and particle detection instruments. The development of a testing methodology requires matching the aerosol generation technique to the instrumentation and obtaining the desired filter data. For example, if a polydisperse test aerosol is used, a very high resolution particle size distribution-measuring instrument is required for the upstream and downstream measurements. If a monodisperse aerosol is used, the particle size (once determined) does not need to be measured during the filter efficiency test. Also, simple total particle detection instruments can be used. Additional trade-offs exist between testing the media or the entire filter assembly and the time required for the test.

3.1 Particle Generation

Aerosols can be classified as either dispersion aerosols or condensation aerosols.[40] Dispersion aerosols are formed by mechanically breaking up a solid or liquid through processes such as grinding or atomization or by redispersing a powder. Condensation aerosols are formed either when vapors are condensed or when a gas phase reaction produces an aerosol product. In general, dispersion aerosols are larger than condensation aerosols and tend to be more polydisperse. Techniques have been developed that allow the use of either kind of aerosol as a test aerosol for testing a filter.

3.1.1 Atmospheric Aerosol. One approach with the availability of sensitive aerosol monitoring instrument is the use of ambient aerosol as a challenge aerosol. An atmospheric aerosol is made up of both a condensation aerosol and a dispersion aerosol. The condensation aerosol from combustion sources and photochemical atmospheric reactions is found in a mode at about 0.3 μm, and the dispersion aerosol from dust, sea salt, and pollen is at 10 to 20 μm. The concentration of an atmospheric aerosol can vary widely, exhibiting diurnal variation (as well as longer term variations), depending on large scale meterological activity. However, the general shape of the particle size distribution is remarkably similar in urban locations.[41] The use of an atmospheric aerosol as a challenge aerosol is limited by its variability in concentration. However, it can be used to advantage when the rapid screening of a filter is required and single particle counting aerosol instrumentation is available.

3.1.2 Dispersion Test Aerosol. Dispersion aerosols may be either liquids or solids. Solid dispersion aerosols are produced by grinding processes, powder resuspension, and dispersion of a solid in liquid particles, followed by evaporation of the liquid. Liquid dispersion aerosols are produced by nebulizers, nozzle sprayers, and other processes which produce particles through liquid shear.

3.1.2.1 Grinding Processes and Powder Resuspension. The aerosols produced by both of these processes are polydisperse. The process of grinding will naturally produce a wide distribution of particle sizes. Resuspension of an originally monodisperse submicron aerosol to produce a monodisperse aerosol is very difficult. This is because the particle-to-particle adhesion forces are too large to be overcome completely by the dispersion process. A survey of various powder suspension techniques was reported by Corn and Esmen[42] in *The Handbook on Aerosols.* The usual solids-dispersing device consists of a feeder to deliver the dust to a high-energy section, such as a compressed air venturi, for dispersal.

3.1.2.2 Dispersion of Solid Particles in Liquids. Solid particles in liquid suspension can be aerosolized by producing liquid particles and evaporating the liquid. If the solid particles were monodisperse originally, and only one particle was entrained in each liquid droplet, the solid aerosol (remaining after evaporating the liquid) will be monodisperse, even though the liquid droplets were not monodisperse. This is the approach commonly used with monodisperse polystyrene latex particles. Particles dispersed in this way have a large residual electrical charge and must be neutralized soon after formation to prevent coagulation or loss on the wall. The procedure can be used to produce a well-characterized test aerosol. The concentration of particles in the liquid must be kept low, insuring that only one solid particle is likely to be in any given liquid particle. Otherwise, doublet and triplet particles will be formed, and the monodispersity of the aerosol will be degraded. This low particle concentration limit makes this technique difficult to use in testing high efficiency filters because the number of particles penetrating would be few and difficult to detect.

3.1.2.3 Liquid Particle Dispersion. Polydisperse aerosols are produced by most processes that break up liquids with shear force. Nebulization or atomization is a widely-used technique, and various types of compressed air and ultrasonic nebulizers are commercially available. The performances of a number of these are summarized in Table 6.

Two types of monodisperse liquid particle generators are widely cited in the literature. The spinning disk generator produces the particle by throwing a ribbon of liquid from a disk spinning at high speed.[42] The vibrating orifice generator injects the liquid from a orifice driven at ultrasonic frequencies.[44] Both instruments produce a large (20 to 30 μm) monodisperse aerosol and can produce particles in the range of 1 μm in size by evaporating the solvent. The major limitation of these devices for filter testing is in being able to obtain a challenge aerosol which is concentrated enough to test high efficiency filters.

3.1.3 Condensation Test Aerosols. The condensation of vapors on solid nuclei to produce a test aerosol is an important technique for the production of well-defined aerosols in the laboratory. The size of the aerosol depends on the number of nuclei available and the super saturation in the vapor-gas mixture. By controlling these parameters, a wide range of nearly monodisperse particles from about 0.05 to 5 μm can be generated. An early laboratory device was described by Sinclair and La Mer.[45] The hot-DOP generator used in MIL-STD 282 is a condensation generator on a large scale.

Table 6: Typical Operating Characteristics of Selected Nebulizers

Nebulizer	P (psig)	A (µL/L)	W (µL/L)	Q (L/min)	VMD (µm)	σ_g	Fluid
Dautrebande D-30	10	1.6	9.6	17.9	1.7	1.7	Water
	20	2.3	8.6	25.4	1.4	1.7	Water
	30	2.4	8.2	32.7	1.3	1.7	Water
Lauterbach	10	3.9		(1.7)	3.8	2.0	Water
	20	5.7	(12)	(2.4)	2.4	2.0	Water
	30	5.9		(3.2)	2.4	2.0	Water
Collison	20	7.7	12.7	7.1	(2.0)	(2.0)	Water
(3-jet model)	25	6.7	12.6	8.2	2.0	2.0	Water
	30	5.9	12.6	9.4	--	--	Water
	40	5.0	12.6	11.4	--	--	Water
DeVilbis #40	10	16	(10)	10.8	4.2	1.8	Water
	15	15.5	8.6	13.5	3.5	1.8	Water
	20	14	7.0	15.8	3.2	1.8	Water
	30	12	7.2	20.5	2.8	1.8	Water
Lovelace Water	15	27	(10)	1.3			
(Baffle screw set for	20	40	10	1.5	5.8	1.8	Water
optimum operation at	30	31	11	1.6	4.7	1.9	Water
20 psig)	40	21	9	2.0	3.1	2.2	Water
	50	27	11	2.3	2.6	2.3	Water
Retec X-70/N	20	56	20	5.0	5.1	2.0	Water
	20	53	12	5.4	5.7	1.8	Water
	30	54	11	7.4	3.6	2.0	Water
	40	53	7	8.6	3.7	2.1	Water
	50	49	9	10.1	3.2	2.2	Water
DeVilbiss Ultrasonic (setting #4 Somerset, PA)	--	150	33	41	6.9	1.6	Water
Wright Nebulizer	24	29	--	12	2.6	2.0	Water
TSI Models 9302A	25	(4.2)	--	6.5	2.0	<2.0	Water
(from product literature)		(0.3)	--	6.5	0.8	<2.0	DOP
Laskin Nozzle (Flanders Filters, Inc., product literature, 1984)	20-25	14-28	--	(10)	0.7	1.7	DOP
Hydrosphere Nebulizer	20	24.5	--	110	4	2.3	Water

Notes: (1)   Values in parentheses are estimates.

(2)   Conditions are approximately ambient.

Key:

P = Gauge pressure at air inlet.
A = Output particle concentration.
W = Evaporation rate into dry air.

Q = Volumetric flow rate of air.
VMD = Volumetric mean diameter.
σ_g = Geometric mean standard deviation.

A modern example of this kind of generator as applied to laboratory tests of filters was described by Lee and Liu.[16] DOP is dissolved in a solvent (i.e., alcohol), and the solution is atomized. The polydisperse aerosol initially produced is then heated and vaporized. The vapor subsequently cools and condenses on the nuclei from the original drops to form a monodisperse aerosol. The particle size is easily varied by controlling the concentration of DOP in the solution.

3.2 Instrumentation

Generally, two types of measurement are required to characterize an aerosol—particle size distribution and concentration. The range of size measurements required for research is broad, and a variety of instruments is necessary to span the entire range.

3.2.1 Particle Concentration. The methods used for particle concentration measurement include:

(1) electrical detection

(2) fluorometric detection

(3) condensation nuclei counters

(4) light scattering.

When these methods are used for filter evaluations, the size of the challenge particles must be known and nearly monodisperse to obtain reproducible results. The light scattering photometer has been used widely for filter efficiency and leak testing for over 30 years. The condensation nuclei counter has recently proven very useful for filter testing (largely because of its sensitivity and wide dynamic range). The other two methods are less popular and will be described briefly.

The electrical detector simply measures the charge transferred by the particles. With an aerosol of known size and charge, the concentration can be determined. This method is most successful when the aerosol size distribution has been preseparated by a mobility separator. It is also useful in applications to determine the relative efficiency of a filter with an aerosol having a controlled narrow particle size distribution because the absolute concentration of the particles is not needed.

Fluorometric detection requires that a challenge aerosol be generated from a dye (i.e., uranine or methylene blue). The particles are collected on filters. The filters are extracted, and the relative concentration is measured with a calibrated fluorometer. The fluorometric approach requires careful calibration and attention to background fluorescence when detecting low concentrations. The method is much more time-consuming than methods which rely on particle counters using light scattering or electrical current detection. However, some situations (i.e., the need to measure particles larger than 5 μm or deposition on a surface) develop where this is the preferable choice.

3.2.1.1 Condensation Nuclei Counter. The Condensation Nuclei Counter (CNC) can count single particles which are smaller than 0.1 μm. However, the counting efficiency tails off significantly when the particles are smaller than 0.02 μm.[47] The CNC does not have the ability to discriminate by size and operates by growing large, easily detected particles from all the aerosol particles which enter the counter. The counter grows particles by causing vapor to condense on all of the entrained particles and then counts the large (10 to 12 μm) particles optically.

There are two kinds of CNCs: (1) those which induce condensation by adiabatic expansion; (2) and those which induce condensation by reducing the temperature.[48] The adiabatic instruments, usually of earlier design, operate in a cyclical manner where the chamber containing the aerosol is sealed off from its source and then adiabatically expanded to induce condensation. This action interrupts the aerosol flow. Only the CNCs based on condensation by temperature reduction prove compatible with continuous flow sampling. The continuous-flow CNC (Figure 20) has a thermoelectric cooling element which reduces the temperature of the alcohol-saturated aerosol immediately upstream of the optical counting stage. This action makes it possible for the aerosol to flow continuously. However, the aerosol transit time of this CNC (designed for 0.01 cfm aerosol flow) results in a response time of 30 to 60 s to changes in the input aerosol. The literature provides discussions of modified custom designs which reduce the response time.[49,50] TSI, Inc. markets a version of the continuous flow CNC.[51]

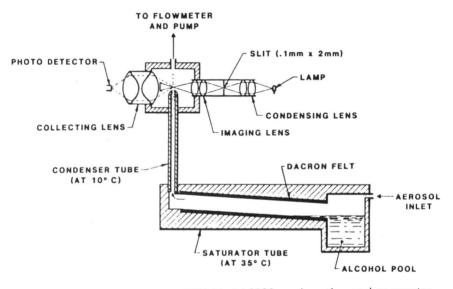

Figure 20: Schematic diagram of TSI Model 3020 condensation nucleus counter. (Courtesy, Institute of Environmental Sciences)

3.2.1.2 Light-Scattering Photometer. The light-scattering photometer has been widely used for filter efficiency and leak testing.[52] A diagram of the optical arrangement is shown in Figure 21. The sample of air is passed through the focal plane of a lens system. Clean air is passed near the lenses to prevent soiling. The optical system consists of a high intensity white light source, condensing lens, and a collecting lens to collimate the light into the detector. A light stop on the second condensing lens forms a shadow on the collecting lens. Light is scattered by the particles, first into the shadow and into the collecting lens. The light stop shields the detector from direct view of the light source.

The Mie light-scattering theory applies when addressing the issue of light-scattering from particles which are between 0.05 and 10 μm in size. The Mie intensity parameters are complicated functions of the index of refraction, size, and angle.

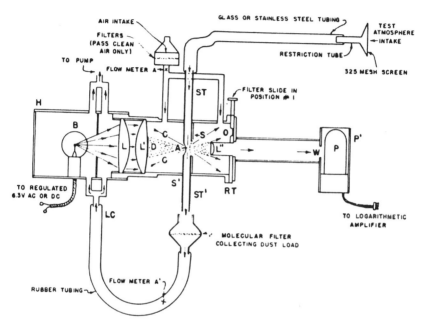

H – Sealed light source housing.	A – Stray-light-limiting aperture.	W – Photomultiplier tube window: admits scattered light only.
B – Lamp.	S – Stray-light-limiting diaphragm.	P – Photomultiplier tube.
L – Condenser Lens.	ST – Test atmosphere inlet tube.	P' – Photomultiplier tube housing.
L' – Condenser Lens.	ST' – Outlet tube.	S' – Sealed housing.
L" – Scattered - light collecting lens.	O – Optical calibration filters.	LC – Lamp-cooling air inlet.
D – Diaphragm stop: produces conical shadow.	RT – Rotable flange.	
C – Cone of light.		

Figure 21: Light scattering photometer. (Courtesy, Phoenix Precision Instrument)

Figure 22 provides plots of the light scattering flux from Mie intensity parameters for various values of the size parameter, α:

(35)
$$\alpha = \frac{\pi \, d_p}{\lambda}$$

For $d_p < \lambda$, the intensity parameters are well-behaved. However, as d_p approaches and exceeds λ ($\alpha \cong 3$), marked oscillations build up. This relationship bodes ill for the design of an optical scattering instrument in that it implies a nonunique relationship between light scattering at a fixed angle and particle size. The collection angle and the size parameter, α, and refractive index of the particles determine the instrumental response. A computer response is shown in Figure 22: an effective collection angle of interest is about 20 degrees. This instrument is most sensitive to particles in the 0.1 to 1 μm range. Light scattering instruments can be calibrated to provide mass concentrations for aerosols if the particle size distribution and particle refractive index are constant.

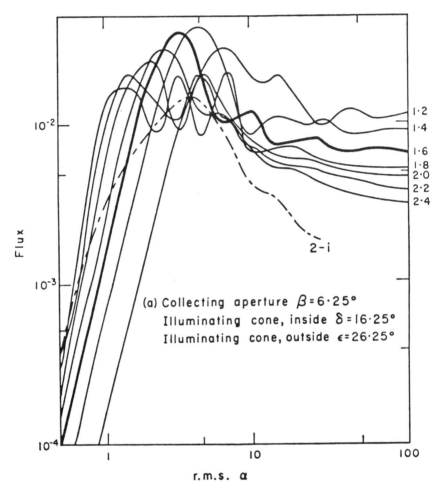

Figure 22: Response curves of Sinclair aerosol photometer. (Courtesy, Academic Press)

3.2.2 Particle Size Distribution Measurement. The development of a variety of instruments in the last few years has increased the options available for filter testing. Understanding the limitations of the instruments will help in interpreting filter test data.

3.2.2.1 Optical Particle Counters. The major variables which distinguish one OPC from another are: (a) the light source (laser versus white light); (b) the scattering volume definition/flow rate capability (aerodynamic versus optical focus; 1 cfm units versus low flow units); (c) the scattering angle (forward scattering versus 90° scattering; wide angle versus narrow angle); and (d) the collection optics and photodetector. Each variable is discussed in further detail below. An example of an optical counter is shown in Figure 23.

While the first optical particle counters were built with incandescent light sources, more recent designs call for helium-neon lasers (λ = 633 nm). This kind of design achieves higher values of incident light intensity in practical instrument housing than white light detectors do. Therefore, in comparison with instruments which use white light, laser instruments can detect smaller particles. Indeed, white light instruments suffer severe losses in their counting efficiency for particles smaller than 0.5 μm.[53]

Aerodynamic focusing refers to the feature in a design whereby the aerosol stream which is being monitored is confined to the desired sensing volume with particle-free sheath air. Most designs of optical particle counters employ this technique, both to minimize the wall effects and also to confine the aerosol particles to the volume of uniform illumination (without creating turbulent currents that could recirculate particles back through the sensing volume). As mentioned earlier, the molecular scattering from the illuminated volume sets a limit on its size. Thus, to detect particles smaller than 0.1 μm in size, the viewing volume must be small enough so that the molecular scattering does not exceed the single particle scattering. A small viewing volume dictates low sample flow rates through the instrument.

In order to monitor the outlet of a filter where the aerosol particle concentration is low, there must be a high sample flow rate (to obtain statistically significant counts in a practical amount of time). Clean room air which has 10 countable aerosol particles/ft^3 generates 10 counts in 1 min at a sample flow of 1 cfm. At lower sample flows, the time needed to generate this number of counts increases proportionately.

The particle counts can be modeled by the Poisson distribution:

$$(36) \qquad P(N) = \frac{C^N \exp(-C)}{N!}$$

where

    $P(N)$ = the probability of occurrence of count N
    C = the mean long-term value of the count.

The distribution has the property that the standard deviation of C is:

$$(37) \qquad \sigma = C^{1/2}$$

The particle count is given by:

$$(38) \qquad C = N_t Q_s t$$

where

Q_s is the volumetric sample flow
t   is the time of sampling
N_t is the true concentration.

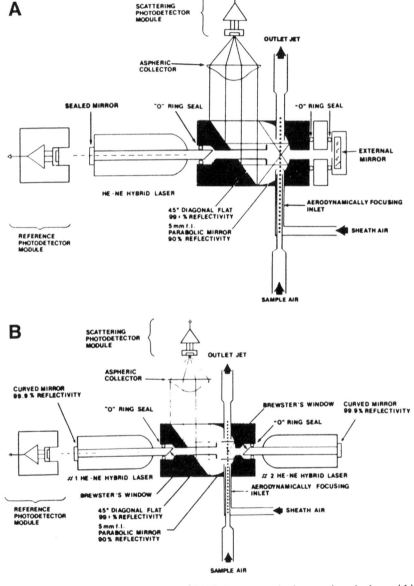

Figure 23: Optical systems used in PMS Laser resolution cavity devices. (A) Above. Used in LAS-X instrument series. (B) Below. Used in highest sensitivity instrument model LAS-HS. (Courtesy, Institute of Environmental Sciences)

In Poisson statistics, the standard deviation varies as the square root of the sample size, favoring the highest flow rate instrument available. Currently the highest flow rate instruments sample at a flow rate of 1 cfm. This sort of instrument does not achieve as low a size sensitivity as various low-flow instruments do, currently being limited to approximately 0.1 μm particle detection when using laser light. Nonetheless, this combination of flow rate and size sensitivity proves the best combination for monitoring a clean room.

The viewing volume affects coincident counting, where two or more particles are counted as one. It also affects smear-out, where a single particle is counted more than once. Optical counters which count single particles incorporate the assumption that the signal received by the pulse height analyzer originates from only one particle. When more than one particle is in the viewing volume simultaneously, a coincidence error occurs. The two particles are classified as one large particle. The result is that a distortion occurs in the measured particle size distribution. The coincidence error is given by:

(39)
$$\frac{N_i}{N_t} = e^{-N_t V}$$

where

N_i is the indicated number
N_t is the true number
V   is the detection volume in the sampler.

The instrument manual will provide guidelines on the levels of particles where this error will occur. This kind of error is negligible as long as the concentration-measuring volume product is less than 0.1.[54] Coincident counting has relatively little importance at an aerosol particle concentration which is characteristic of a high efficiency filter. However, the challenge aerosol may have a high enough concentration to cause coincidence error without dilution. Particles which traverse the viewing volume slowly or recirculate in and out of the viewing volume can be counted twice. Therefore, maintaining the flow rates within the design limits is also important.

The viewing volume contributes to an optical counter's noise level (signal level generated by the instrument when no particle is actually in the viewing volume). Two sources of noise are molecular Rayleigh scattering and stray light scattering from imperfect optical surfaces. Noise levels in a 1 cfm optical particle counter should correspond to less than an equivalent concentration of 0.1 particles/ft^3. In a lower flow instrument, this equivalent concentration (attributed to noise) rises.

To translate a scattering event into a particle count, the scattered light must be collected and focused onto a photoelectric transducer (either a photomultiplier tube or, more likely, a photodiode). The quantum efficiency of a photodiode exceeds a photomultiplier tube's quantum efficiency, and the photodiode itself costs less in both dollars and operating power. The optical system for collecting the scattered light uses both lenses and mirrors.

An arrangement which is particularly attractive is that which uses an ellipsoidal mirror. The scattering takes place at one focal point with the photosensing surface positioned at the other focal point. This design collects light scattered over a wide angle. Other designs rely on either parabolic mirrors to direct the scattered light to a lens for focusing on the photodetector or on a coaxial lens behind the light trap to collect the small angle forward scattering.

3.2.2.2 *Aerodynamic Particle Sizer.* The Aerodynamic Particle Sizer (APS) sizes aerosol particles by measuring their speed in an air flow that is rapidly accelerated.[55] Inertia causes the particle motion to lag behind the gas motion (in proportion to the square of particle size). The APS pulls the aerosol particles to be measured through a small orifice in order to accelerate the gas flow to a controlled, high velocity. Aerosol particles which are suspended in the gas also accelerate with the gas. However, they do not achieve the final gas velocity because of their relaxation time delays (time required to equilibrate with step function changes in the surroundings). This relaxation time, τ, depends on the particle size and density, along with the gas viscosity.[40]

(40) $$\tau = d_p^2 \, \varrho_p / 18 \, \mu$$

where

ρ_p = particle density
μ = gas viscosity.

The relation times for submicron-sized particles are on the order of microseconds (in air at ambient temperature).

The APS measures the time of flight between two legs of a split laser beam. This amount of time translates into aerodynamic particle diameter (the diameter of that unit density sphere having the same settling velocity as the particle) through an empirically-developed curve. The detected size range is between approximately 0.5 and 15 μm for which times of flight vary between 800 and 3000 ns. The resolution in size at the smaller size end of this range deteriorates as the particle moves with velocities very close to that of the gas.

3.2.2.3 *Electrical Mobility Analyzers.* Electrical mobility analyzers separate aerosol particles by controlling the particle electrical charge and then drifting those particles in controlled electric fields. When coupled with a Condensation Nuclei Counter (CNC), or an electrical detector, this kind of analyzer measures particles less than 0.1 μm in diameter. Particles in this size range cannot be detected by either an optical particle counter or the APS.

The two basic designs commercially available are the integral mobility analyzer and the differential mobility analyzer. Each design incorporates particle-free sheath air to define the aerosol flow paths through the field region. These air flow paths are independent of the electric field. Each has a separation chamber which consists of a polished metal rod placed coaxially within a cylindrical metal tube whose walls are electrically grounded. The metal rod is electrically biased, thereby creating a radial electric field which removes electrically-charged particles from the aerosol stream flowing along the tube's outer walls. When no electric field is applied in the integral analyzer, all of the particles will exit the tube and then be counted. As the voltage on the center rod is increased, more of the electrically-charged particles are attracted to the rod, thus removing them from the effluent air.

The differential mobility analyzer has two exit ports, a sampling port immediately below the center rod into which particles of a narrow mobility band can be deflected, and an excess flow port through which all other particles leave the separation chamber. With no applied voltage, clean sheath air which is adjacent to the center rod flows through the sampling port, and all aerosol (along with some excess clean air) flows through the overboard port. It is assumed that these air flow patterns do not change with applied electric field. Instead, the aerosol particle trajectories change. The radial electric field attracts charged particles of the opposite sign, resulting in an entrainment of those particles with just the right electrical mobility in the air stream exiting the sampling port. Particles with a higher mobility are captured by the center rod and removed from both air streams. The particles with lower mobility remain in the excess air and go overboard.

Both kinds of mobility analyzers relate the aerosol particle electrical mobility to particle size by the relationship:

$$(41) \qquad Z = \frac{n_p\, eC}{3\,\pi\mu\, d_p}$$

where

Z = aerosol particle electrical mobility
n_p = number of unit charges on the aerosol particle
e = unit electrical charges
C = Cunningham slip correction factor, a dimensionless number dependent on l/d_p where l = the mean free path of an air molecule (0.065 μm at 1 atm)
μ = gas viscosity
d_p = particle diameter.

The cuts in mobility are determined by the values of v needed to deflect particles with various limiting trajectories:

$$(42) \qquad v = EZ$$

where

v = aerosol particle drift velocity
E = applied electric field.

Reproducible operation depends on control of both the electric field (a straightforward matter) and the particle charge (not as straightforward a matter). The integral mobility analyzer positively charges all aerosol particles by making them pass through a zone of low electric field which contains a high density of positive ions. Ion diffusion is the dominant particle-charging mechanism. By maintaining a fixed aerosol particle residence time and a fixed ion density, a reproducible (though empirical) relationship between particle size and particle charge can be maintained and a sensitive relationship between particle mobility and diameter established.[56,57]

The differential mobility analyzer controls the particle charge in a different manner. In order to reduce the charge distribution on the aerosol particles to a Boltzmann distribution, it passes the particles through a Kr 85 neutralizer. With the particle size cutoff (none larger than between 0.5 and 1.0 μm) known, the data reduction incorporates the assumption that all particles in the largest size cut (lowest mobility increment) are singly charged. The mobility of the multiply-charged fractions of these same-size particles and their concentration are calculated from the assumed Boltzmann charge distribution. This concentration value is then subtracted from the appropriate higher mobility channel. When all of the multiply-charged fractions are subtracted from each higher mobility cut, it is assumed that the remainder is the single-charged component of that mobility cut which is related to diameter.

This procedure makes it possible to determine the single-charged particle population of each mobility channel and the total population of each size channel computer from the assumed Boltzmann charge distribution. The particle size decade over which the mobility measurements work well falls between 0.02 μm and 0.2 μm.

The differential mobility analyzer can also be used as a separator to generate a monodisperse challenge aerosol. The output of a polydisperse aerosol generator can be separated by the device into a narrow band of mobility corresponding to a nearly monodisperse size distribution.

3.3 Sampling

The ability to obtain an accurate sample from the filter duct to the sensor requires that a volume of particles be removed from the free stream and transported to the instrument. Two considerations are: the need for isokinetic sampling and losses from tubing transport.

Isokinetic sampling requires that the velocity in the sampling nozzle match the velocity in the air stream. Subisokinetic conditions tend to cause oversampling of the large particles while super-isokinetic conditions tend to cause over-sampling of small particles. If the particles sampled are submicron in size, isokinetic sampling may not be a large error. The errors associated with deviations from isokinetic sampling are summarized by Hinds.[58]

The mechanisms for particle losses in sampling lines include: gravitational sedimentation; inertial impaction at the bends and elbows; and diffusional collection. Also electrostatic attraction can affect particle loss at the walls. The larger particles are affected by internal and gravitational effects. Submicron-sized particles are removed by gravitational effects. Similar to filtration, the 0.1 to 0.3 μm particles are the most penetrating. An empirical analysis of tubing loss was reported by Bergin.[59]

Filters with penetrations of 10^{-6} or lower are difficult to test in terms of performance. This issue was discussed by Rivers and Engleman.[60] The very low particle counts downstream of the filter require an evaluation of the background counts and instrumental noise to obtain sufficient particle counts for a statistically valid test as discussed in the section on optical particle counters. The sampling times may be quite long. Also, most instruments have a limited dynamic range, thereby preventing the use of the same sensor for both the downstream and upstream measurements. The generation of concentrated challenge aerosol to produce adequate downstream particle counts causes problems in two areas: alteration of the filter properties through particle loading and saturation of the particle counter used upstream. An approach often used to allow sampling of the challenge aerosol is to reduce the concentration transported to the instrument by dilution with clean air. All diluters have size-dependent losses. For accurate measurements with dilution, the diluters should be calibrated.

The need for accurate evaluation of high efficiency filters has pushed particle measurement technology to its limits with current instrumentation.

3.4 Standard Tests

A number of filter performance tests have been developed and used to verify filter protection in various applications. As mentioned earlier, the fraction of particles penetrating most filters is highest for particles between about 0.1 and 0.3 μm in diameter. Filters developed for protection against submicron aerosols have therefore been tested with aerosols in this size range. This was done so that the test would be as severe as possible. A summary of standard filter tests appears in Table 7. The challenge materials are aerosols of dioctyl phthalate, NaCl, methylene blue, uranine, and several test dusts. The DOP aerosol is monodisperse, and the others are polydisperse. In all cases, overall penetration—rather than penetration as a function of particle size—is measured.

The United States standards for testing high efficiency filters involve the "hot DOP" test and the "cold DOP" test. The equipment used for hot DOP filter testing (Figure 24) consists of a vaporization/condensation generator. The output of the generator is adjusted with the Owl, a multi-angle light scattering photometer which measures the light-scattering intensity at selected angles. This provides an indication of particle size and polydispersity. The concentration of aerosol at the inlet and outlet of the filter is measured with another instrument, the Sinclair-Phoenix light photometer described earlier. In recent years, this approach has been criticized because the aerosol is not really monodisperse and does not fall into the most penetrating particle diameter range for most modern media.[36,68,69] The conclusion from Bergman et al.[36] was that these limitations are mutually correcting. DOP has been recently classified as a suspected carcinogen, and corn oil has been recommended as an alternative material.

The performance of high-efficiency filters depends critically on the integrity of the filters. Leaks will dramatically degrade the performance of the filter. Thus, filters are usually tested at the manufacturer's plant on a fixture designed to identify any areas which require repairs (ANSI/ASMR N510). A typical unit is shown in Figure 25. A nebulizer, called the Laskin nozzle, is used to generate an aerosol which is then blown to the filter mounted in a fixture. An operator scans the exposed face of the filter in the open air with a tube connected to a photometer.

3.5 Developmental Tests

A number of filter-testing methodologies have been reported in the literature, and a few are summarized in Table 8.

3.5.1 Flat Media Testing. Specialized filter efficiency tests for the 0.03 to 1.0 μm diameter range were developed by Liu and Lee[70] and used the condensation aerosol generator. This method was found to be well-suited to filter testing, as the generator was stable and the particle size could be easily and reproducibly changed. The apparatus can be used to develop the entire penetration versus particle size relationship within the particle generator's range.

Table 7: Summary of Methods for Filter Efficiency Measurement: Standard Tests

Method	Aerosol generation and property					Aerosol detection		Reference
	Material	Dispersity	Generation technique	Diameter (MMD) (mm)	Concentration (mg/m³)	Detector	Penetration measured (%)	
STANDARD TEST:								
1. DOP Test Hot (U.S. standard)	Dioctyl phthalate	Monodisperse	Vaporization-condensation	0.3	100	Light-scattering photometer	0.001 and up	DLB-76-2-639 (1960) (61) MIL-STD-282
2. Sodium flame test (British standard)	NaCl	Polydisperse	Atomizing 2% aqueous soln.	0.6 (0.01-1.7)	--	Flame photometer	0.001-0.1	British Standard 3928 (1969) (62)
3. Methylene blue test (British standard)	Methylene blue	Polydisperse	Atomizing 1% aqueous soln.	(0.01-1.6)	--	Stain density	0.01-60	British Standard 2831 (1971) (63)
4. Uranine test (French standard)	Uranine	Polydisperse	Atomizing 1% aqueous soln.	0.3	10	Fluorometer	0.001 and up	STDX 44-011 (1978) (64)
5. ASHRAE test--Prev. NBS and AFI tests (U.S. standard) include three parts:								
1. Weight arrestance	72% fine dust 23% Molocco black 5% 7 cotton linters	Polydisperse	AFI dust feeder	--	>1,000	Gravimetric	2-80	ASHRAE Standard 52-76 (1976) (65)
2. Dust spot efficiency	Atmospheric aerosol	Polydisperse	--	>0.3	>0.1	Opacity meter	2-80	ASHRAE Standard 52-76 (1976) (65)
3. Dust holding capacity	Same as 5(1)	Polydisperse	AFI dust feeder	--	>1,000	Gravimetric	2-80	ASHRAE Standard 52-76 (1976) (65)
6. SAE test	AC-fine AC-coarse	Polydisperse Polydisperse	Sonic dust Feeder and others	8 32	>1,000	Gravimetric	--	SAE J726b (1979) (67)
7. Cold DOP	Dioctyl phthalate	Polydisperse	Nebulizer	0.7	100	Light-scattering	100	ANSI N 101.N (1972) (68) ANSI N 510

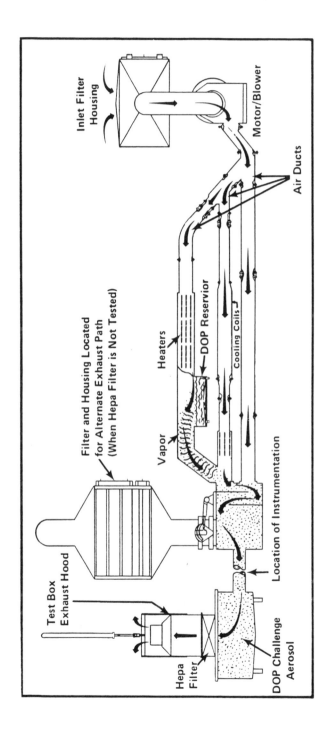

Figure 24: Penetrometer (instrumentation not shown). (Courtesy, Flanders Filters)

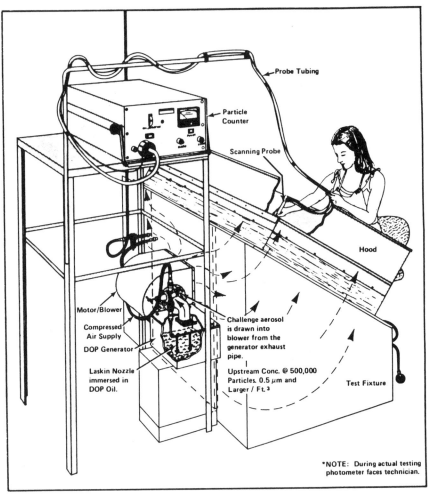

Figure 25: The "cold" DOP test. (Courtesy, Flanders Filters)

3.5.2 Laser OPC Modification of Standard Tests. With the development of improved filter media, lack of sensitivity has proven to be a distinct limitation of the standard methods. Cadwell,[71] and Rivers and Engleman,[60] reported the substitution of Laser Spectrometers for the light scattering photometer in standard test arrangements. The upstream measurement required a 1000:1 dilution to avoid saturating or coincidence errors in the particle counter. This approach now is widely used by filter manufacturers and is being evaluated as a test method in DOE laboratories.[72,73]

3.5.3 Point-of-Use Filters. Testing point-of-use filters has two important obstacles. The filters have very low penetrations, and the pressure drop across the filter may be large. These obstacles require the construction of a special apparatus. The approach reported by Rubow and Liu[83] is shown in Figure 25.

Table 8: Summary of Methods for Filter Efficiency Measurement: Specialized Tests

Specialized test method	Aerosol generation and property					Aerosol detection		Reference
	Material	Dispersity	Generation technique	Diameter (MMD) (μm) or as indicated	Concentration (mg/m³)	Detector	Penetration measured (%)	
1. Monodisperse aerosol-electrical detection	Dioctyl phthalate	Monodisperse	Vaporization condensation	0.3-1.0	100	Electrical aerosol detector	0.001 and up	(70)
2. Polydisperse aerosol-CN detection	Ambient aerosol	Polydisperse	Ambient	0.04 nmd	$N=10^5/cm^3$	CN counter	0.1 and up	(92)
3. Monodisperse and polydisperse aerosol-optical detection	Gold, silver atomizer	Polydisperse Polydisperse	Vapor-condense Spray-drying	.005-.03 nmd 1 and below	$10^7/cm^3$ Up to 10^7	CN counter CN counter	0.001 and up 0.01 and up	(74) (75)
	Latex spheres	Monodisperse	Spray-drying	0.09-2	Up to $10^4/cm^3$	Photometer	0.01 and up	(76) (77)
	Ambient aerosol	Polydisperse	Ambient	>0.3 nmd	$N=10^3/cm^3$	Optical counter	0.01 and up	(73)
	PSL, oleic acid, uranine	Both	Spray drying Vapor-condense	1-3	0.02-4	Photometer	0.01 and up	(78)
4. Polydisperse aerosol-fluorometric detection	Uranine	Polydisperse	Atomizer	0.025, 0.27	--	Fluorometric	0.01 and up	(79)
5. Airborne radioactivity-alpha detection	Ambient aerosol with decay product	Polydisperse	Ambient	0.04 nmd (0.001-0.04)	$N=10^5/cm^3$	Alpha detector	1.0 and up	(80)
6. Miscellaneous	NaCl, uranine oil, fly ash	Polydisperse	Atomizer and spinning disk	0.05-6.0	Varied	Fluorometric, gravimetric, titrametric, atomic absorption, microscopic, etc.	0.01 and up	(81)
7. Polydisperse aerosol - CN detection	NaCl	$\sigma_g = 1.63$	Atomized	0.22	$N=10^7/cm^3$	CN counter	0.0001 and up	(82)

MMD = mass mean diameter.
nmd = number mean diameter.

It consisted of an atomization aerosol generator to produce a concentrated aerosol of NaCl. This was followed by a drying and charge neutralization chamber and a particle detector to measure the upstream and downstream particle concentration across the filter.

This generator was capable of generating highly concentrated aerosols of up to 7 particles/cc. The mean particle diameter was varied by varying the concentration of the saline concentrations among 0.001, 0.01, 0.1, 1.0 and 10%; corresponding mean particle diameters of 0.033, 0.041, 0.046, 0.067 and 0.089 μm were obtained based on measurements made with an electrical aerosol analyzer. The geometrical standard deviation of the aerosol was found to range from 1.7 to 2.2.

The aerosol concentration upstream and downstream of the filter was measured with a Condensation Nucleus Counter, and occasionally, a laser optical particle counter was used to check for the penetration of large particles above 0.1 μm.

The extremely low particle penetration through the filter required precautions to establish the "zero-count" baseline and to measure the true particle concentration downstream of the filter. Needle valves, toggle valves and 3-way valves were all found to be particle producers. Particle concentrations as high as 1 particle/cc could be measured downstream of the valves when placed directly in the clean air stream for flow control. As a consequence, the system in Figure 26 was designed such that the flow could be controlled by means of valves placed either upstream of the filter or in paths of high aerosol concentration.

To eliminate particle generation due to mechanical movement of the system, the three lines leading from the test filter holder, the expansion system and the pure air supply were held rigidly in place. The exit of these three lines were placed inside an inverted 500 ml glass flask or, in the case of a complete filtration unit of high flow capacity, inside a 50 gallon drum. The sampling lines to the CNC and OPC were also inserted into these containers for sampling air from the vessel for particle counting.

To establish a clean air baseline for the particle counters, the aerosol generator was turned off, and the particle count downstream of the test filter was measured. Concentrations as low as 10^{-5} particles/cc could be obtained. Next, the aerosol generator was turned on, and the count downstream of the filter was again measured (valve 6 open and valve 7 closed). This provided the downstream particle concentration value. To establish the upstream particle concentration, valve 6 was closed and valve 7 opened. The upstream particle concentration was then similarly measured by the CNC in the "photometric mode."

Nawakowski and de Pinillos[84] discussed sampling in systems requiring pressure reduction upstream of the optical particle counter. Also they observed that sampling through a tee gave higher counts than isokinetic sampling.

Another issue is the shedding of particles from filters. Accomazzo and Grant[85] described mechanical and hydrodynamic shocking filters to stimulate particle release. They found that fibrous filters shed more particles than membrane filters. Thorogood and Schwarz[86] described a test rig used to expose filters to clean gas and continuously measure the downstream concentration of particles.

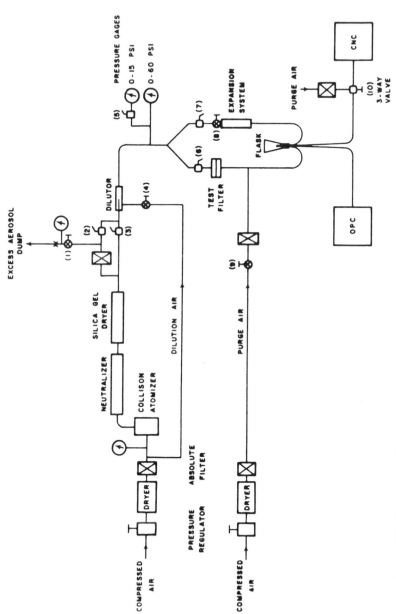

Figure 26: Schematic diagram of the apparatus for testing ultra-high efficiency membrane filters. (Courtesy, Institute of Environmental Sciences)

4. FILTER APPLICATIONS

Filters are applied in two distinct ways: in ventilation systems to clean the atmosphere and as point-of-use filters in air and gas piping. In a ventilation system, large volumetric flow rates of air are filtered. These filters must have a low pressure drop with a moderate to high efficiency, depending on the requirements of the facility. Ventilation system filters are usually fibrous and have low particle-shedding characteristics. The filters themselves are usually disposable.

Point-of-use filters are packaged for installation in process piping. While the pressure drop may not be a primary concern, the high efficiency is. The stream is not recycled through the media as it is in a ventilation system. These filter media can be either fibrous or membrane.

The design of either kind of filter depends on the trade-off between pressure drop and collection efficiency, filter area, and cost. The design must also address the question of compatability between the chemical properties of the gas stream and the filter's resistance. Accomazzo and Grant[85] described applications of membrane filters in process gases.

4.1 Ventilation System

The ventilation system in a clean area must do two things: deliver clean air to the work area (to protect critical components); and remove any contamination which is generated in the process equipment. A simplified ventilation system is shown in Figure 27.

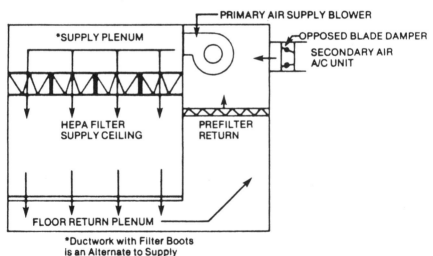

Figure 27: Diagram of a clean room. (Courtesy, ASHRAE)

Several different kinds of filters are used to maintain a clean area. Low efficiency filters prefilter the air, and the HEPA filters are used as final filters. The fibrous filter can be manufactured by a variety of methods and is available in a wide range of efficiencies. In Table 8, the most common types of filters are described. For critical applications, the filters are usually retested after installation. Table 9 shows how ventilation filters (by media type) perform according to various MIL-STD and ASHRAE standards.

Table 9: Performance of Ventilation Filters

Media type	ASHRAE weight arrestance %	ASHRAE spot test efficiency %	MIL-STD 282 DOP efficiency %	ASHRAE dust holding capacity g/1000 cfm
Open cell foam	70-80	15-30	0	180-425
Thin mats 5 to 10 μm	80-90	20-45	0	90-180
fibers 1/4 to 1/2"	85-90	25-40	5-10	90-180
thick	90-95	40-60	15-25	270-540
Mats of 3 to 5 μm fibers 1/4 to 1/2" thickness	>95	60-80	35-40	180-450
Mats of 1 to 4 μm fibers	>95	80-90	50-55	180-360
mats of 0.5 to 2 μm fibers	-	90-98	75-90	90-270
HEPA	-	-	95-99.999	500-1000

Source: ASHRAE handbook, 1982 (87).

A HEPA-type of filter is usually made of glass fiber paper packaged in a square fixture. The flat filter paper is folded with separators between the sheets. The filter assembly then is bonded into a frame with glue to prevent leaks. The design of the paper separators and the exact configuration of the filters depend on the specific manufacturer. A filter with a 2 by 2 ft opening typically will have up to 200 ft^2 of paper. The types and grades of filters depend on the application and thoroughness of the testing.

HEPA filters are manufactured in two grades: Grade 1 (Fire Resistant, which is recommended for both cleanroom and containment applications) and Grade 2 (Semi-Combustible, which may be used in cases where fire is not a hazard). The different types are shown in Table 10.

Filters can be manufactured in a number of physical dimensions and shapes. Entire walls or ceilings of clean rooms can be formed from filters. The sizes of high efficiency filters are shown in Table 11. The grades largely differ in the degree of testing and, to a lesser extent, some details in construction.

4.2 Point-of-Use Filters

There is a wide variety of forms of construction and media for point-of-use filters. The filter package may be constructed for installation in a wide range of piping systems. The media may consist of porous metal, screens, fibrous media, filament media, flat paper, or membrane filters. The choice for filtering process gases is a membrane filter in a cartridge.

Table 10: Performance Classification of HEPA Filters

Type	Applications	Testing and performance
A	Industrial noncritical	MIL-STD-282 at rated flow Hot DOP $>$ 99.97% at 0.3 μm
B	Nuclear containment	MIL-STD-282 at rated and 20% flow, Retested at Dept. of Energy Facilities MIL-F-51068
C	Laminar flow	MIL-STD-282, Pin hole scan with cold DOP ANSI/ASME N510)
D	Ultra low penetration air	Pin hole scan with cold DOP, 99.9995% at 0.12 μm determined by laser optical particle counters
E	Toxic, nuclear, biological hazard containment	MIL-F-51477 or MIL-F-51068

Source: IES Recommended Practice: HEPA Filters (1986) (88).

Table 11: Filter Size and Related Airflow

Size (in.)			Size (mm)			Rate Airflow/Min[1,2,3]	
H	W	D	H	W	D	(ft^3/min)	(m^3)
8	8	5 7/8	203	203	150	35	1.0
12	12	5 7/8	305	305	150	90	2.5
24	24	5 7/8	610	610	150	500	14.2
24	30	5 7/8	610	762	150	625	17.7
24	36	5 7/8	610	915	150	750	21.3
24	48	5 7/8	610	1219	150	1000	28.3
24	72	5 7/8	610	1829	150	1500	42.5
30	30	5 7/8	762	762	150	780	22.1
30	36	5 7/8	762	915	150	925	26.2
30	48	5 7/8	762	1219	150	1250	35.4
30	60	5 7/8	762	1524	150	1560	44.2
30	72	5 7/8	762	1829	150	1875	53.1
36	36	5 7/8	915	915	150	1125	31.9
36	72	5 7/8	915	1829	150	2250	63.7

1. The rated airflow for a 76-mm (3-in.) nominal deep filter is 1/2 the volume flow for a 149-mm (5.875-in.) deep filter for the separator type and the separatorless style. The rated volume flow for 292-mm (11.5-in.) deep filter is double the rated flow for a 149-mm (5.875-in.) deep filter.

2. The maximum pressure drop is 25.4 mm (1 in.) w.g. at rated flow except for ULPA filters and filters constructed in accordance with MIL-F-51477 and MIL-F-51068.

3. 76 mm (3 in.) nominal deep separatorless style filters should have the same pressure drop and flow rating as a 149-mm (5.875-in.) separator type filter.

Source: IES Recommended Practice: HEPA Filters (9186) (88).

4.3 Sampling Filters

Another use for filters is obtaining samples of particles, with weighable collections of particles usually filtered by glass fiber filters. The collection of particles within the matrix of fibers and irregular surfaces makes identifying individual particles impractical with this substrate. Also, the very low concentrations found in a clean room make mass sampling infeasible.

Membrane filters are the method of choice for sampling particles for microscopic identification, with open face filters usually used. For more information on the microscopic analysis of particle deposits, see McCrone et al.,[89] McCrone and Delly,[90] and Bloss.[91]

5. SUMMARY

Filtration provides the principal barrier to particle contamination in clean spaces and gases. Fiber filters are used primarily because of low pressure drop and cost. However, a wide range of other specialized filters is used in contamination control for sampling and as point-of-use filters.

The demand for cleaner gas has encouraged researchers to examine the fundamentals of filtration and the development of efficient media. An overview of the fundamentals of filtration was presented in this chapter to provide insight into filter behavior. Although filtration theory cannot accurately predict filter performance solely from structural characteristics, it does provide a good relative prediction of the effects of either media or operational changes. A fundamental limitation of the theory is the difficulty in describing the complex structure of the media. With the use of an empirically determined inhomogeneity factor, the prediction of performance can be quite good. Also filtration theory, supported by laboratory data, shows that the particle size dependent penetration does not monotonically increase with reduced particle size. Instead, the influence of diffusion produces a maximum in penetration. The particle diameter at the maximum penetration depends on the air velocity and the physical properties of the filter.

The development of the ULPA media and the attention now being paid to the measurement of performance of membrane filters has resulted in close examination of filter evaluation methodology. The long-standing standard methods are inadequate for current needs. Modern aerosol instrumentation (i.e., laser optical particle counters, laser aerodynamic particle sizers and continuous flow condensation nuclei counters) are being applied to filter testing. The testing of high pressure drop, low penetration point-of-use filters presents several difficult complications such as filter loading during the test and particle generation in the pressure reduction apparatus used to reduce upstream pressure (to allow particle measurement at atmospheric pressure). It is expected that developments in testing will continue over the next few years.

REFERENCES

1. Jahreis, C.A., *Filtration and Separation* 18 (4), pp. 310-311 (1981).
2. Locke, B.R. et al., *J. Env. Sci.* 28 (6), pp. 26-29 (1985).

3. Osburn, C.M., "Aerosol Control in Semiconductor Manufacturing." In: *Aerosols.* Edited by B.Y.H. Liu, D.Y.H. Pui and H.J. Fissan. Elsevier Science Pubishing Co., Inc., pp. 673-675 (1984).

4. Rubow, K.L., "Submicron Aerosol Filtration Characteristics of Membrane Filters." Ph.D. Dissertation, University of Minnesota (1981).

5. Davies, C.N., *Air Filtration,* New York, NY, Academic Press (1973).

6. Goetz, A., *Am. J. Public Health* 43, pp. 150-159 (1953).

7. Emi, H., Wang, C.S. and Tien, C., *AIChE Journal* 28 (3), pp. 397-405 (1982).

8. Albrecht, F., *Physik Zeits* 32, p. 48 (1931).

9. Sell, W., *V.D.I. Forschung. Heft* 347, p. 1 (1931).

10. Pich, J., "Gas Filtration Theory." pp. 1-117 In: *Filtration Principles and Practices,* Second Edition. Edited by M.J. Matteson and C. Orr. New York and Basel, Marcel Dekker (1987).

11. Devienne, M., *Frottement et Echanges Termiquer dans les Gas Rarifies.* Paris: Gauthier-Villars (1958).

12. Kuwabara, S., *J. Physics Society of Japan* 14 (4), pp. 527-532 (1959).

13. Kirsch, A.A. and Stechkina, I.B., *J. Aerosol Science* (8), pp. 301-307 (1972).

14. Stechkina, I.B., Kirsch, A.A. and Fuchs, N.A., *Ann. Occup. Hyg.* 12, pp. 1-8 (1969).

15. Lee, K.W. and Liu, B.Y.H., *Aerosol Science and Technology* 1, pp. 147-161 (1982b).

16. Lee, K.W. and Liu, B.Y.H., *Aerosol Science and Technology* 1, pp. 35-46 (1982a).

17. Yeh, H. and Liu, B.Y.H., *J. Aerosol Sci.* 5, pp. 191-204 (1974).

18. Friedlander, S.K., *Ind. Eng. Chem.* 50, pp. 1161-1164 (1958).

19. Lee, K.W. and Liu, B.Y.H., *JAPCA* 30 (4), pp. 377-381 (1980).

20. Emi, H. and Kanaoka, C., "Collection Efficiency of High Efficiency Particulate Air Filter." Proceedings: International Symposium on Powder Technology, pp. 517-524 (1981).

21. Liu, B.Y.H., Rubow, K.L. and Pui, D.Y.H., "Performance of HEPA and ULPA Filters." Proceedings: Institute of Environmental Sciences, Las Vegas, NV, April 30-May 2 (1985).

22. MicroFiltration Systems, 6800 Sierra Court, Dublin, CA (1985).

23. Brunswick Technetics, 4116 Sorrento Valley Boulevard, San Diego, CA (1986).

24. Price, P.B. and Walker, R.M., *J. Appl. Phys.* 33, pp. 3407-3412 (1962).

25. Fleischer, R.L., Price, P.B. and Walker, R.M., *Science* 149, pp. 383-393 (1965).

26. Rubow, K.L. and Liu, B.Y.H., "Characteristics of Membrane Filters for Particle Collection." In: *Fluid Filtration: Gas, Volume I, ASTM STP 975.* Edited by R.R. Raber. Philadelphia, PA, American Society for Testing and Materials (1986).

27. Spurny, K.R. et al., *Envir. Sci. and Tech.* 3, pp. 453-464 (1969).

28. Kanaoka, C., Emi, H. and Aikura, T., *J. Aerosol Sci.* 10, pp. 29-41 (1979).

29. Gormley, P.G. and Kennedy, M., *Proc. Roy. Irish Acad.* 52A, pp. 163-169 (1949).

30. Twomey, S., *Bull. Obs. Puy de Dome* 10, pp. 173-180 (1962).

31. Nuclepore, 7035 Commerce Circle, Pleasanton, CA (1985).

32. Pich, J., "Theory of Electrostatic Mechanism of Aerosol Filtration." In: *Fundamentals of Aerosol Science.* Edited by D.T. Shaw. New York, NY, John Wiley and Sons (1978).
33. Shapiro, M., Laufer, G. and Gutfinger, C., *Atmospheric Environment* 17 (3), pp. 477-484 (1983).
34. Van Turnhout, J. et al., "Electric Filters for High Efficiency Air Cleaning." *Proceedings: Second World Filtration Congress,* Sept. 18-20, London, England. Croyden, England, Uplands Press (1979).
35. Brown, R.C., "Electrical Effects in Dust Filters." Proceedings: Second World Filtration Congress, Sept. 18-20, London, England. Croyden, England, Uplands Press (1979).
36. Bergman, W. et al., "Electric Air Filtration: Theory, Laboratory Studies, Hardware Development, and Field Evaluations." UCID-19952. Lawrence Livermore National Laboratory (1984).
37. Cucu, D.D., Stiehl, H.H. and Lippold, H.J., *Filtration and Separation.* 21, (2), pp. 96-99 (1984).
38. Cucu, D.D., "The Homogeneity of the Particle Charging Within the Ionizer of an Electro-ULPA-Filter." Proceedings: Annual Technical Meeting of the Institute of Environmental Sciences, Mt. Prospect, Illinois, Institute of Environmental Sciences, pp. 74-76 (1985).
39. Masuda, S. and Sugita, N., "Electrically Augmented Air Filter for Producing Ultra-Clean Air and Its Collection Performance." Proceedings: 6th International Symposium on Contamination Control, pp. 243-246 (1982).
40. Fuchs, N.A., *The Mechanics of Aerosols.* (Revised and Enlarged Edition). New York, MacMillan Publishing Company (1964).
41. Willeke, K. and Whitby, K.T., *J. Air Poll. Cont. Assoc.* 25, pp. 529-534 (1975).
42. Corn, M. and Esmen, N.A., "Aerosol Generation." In: *Handbook on Aerosols.* Edited by R. Dennis. NTIS No. TID-26608 (1976).
43. Raabe, O.G., "The Generation of Aerosols of Fine Particles." In: *Fine Particles.* Edited by B.Y.H. Liu. New York, NY, Academic Press (1976).
44. Bergland, R.N. and Liu, B.Y.H., *Environ. Sci. Technol.* 7, pp. 147-153 (1973).
45. Sinclair, D. and LaMer, V., *Chem. Rev.* 44, p. 245 (1949).
46. Whitby, K.T., Lundgren, D.A. and Peterson, C.M., *Int. J. Air Water Pollut.* 9, p. 270 (1965).
47. Agarwal, J.K. and Sem, G.J., *J. Aerosol Sci.* 11, pp. 343-357 (1980).
48. Ensor, D.S. and Donovan, R.P., *J. Env. Sci.* 28 (2), pp. 34-36 (1985).
49. Stolzenburg, M.R. and McMurray, P.H., "A Theoretical Model for an Ultrafine Aerosol Condensation Nucleus Counter." In: *Aerosols.* Edited by B.Y.H. Liu, D.Y.H. Pui and H.J. Fissan. New York, NY, Elsevier Sci. Publishing (1984).
50. Bartz, H. et al, "Response Characteristics of Four Condensation Nuclei Counters to Aerosol Particles in the 3-50 nm Diameter Range." *J. Aerosol Sci.* (in press).
51. TSI, Inc., *Particle Technology Instruments.* Form No. TSI 3000-R681 8M-2MBRI (1986).
52. Sinclair, D., *J. Air Pollution Control Assoc.* 17, p. 106 (1967).

53. Gebhart, J., "Counting Efficiency and Sizing Characteristics of Optical Particle Counters." Proceedings: Annual Tech. Meeting of the Inst. Env. Sci., pp. 102-105 (1985).
54. Raasch, J. and Umhauer, H., *Particle Characterization* 1 (2), pp. 53-58 (1984).
55. Agarwal, J.K. et al., *J. Aerosol Sci.* 13, pp. 222-223 (1982).
56. Liu, B.Y.H., Pui, D.Y.H. and Kapadia, A., "Electrical Aerosol Analyzer: History, Principle, and Data Reduction," In: *Aerosol Measurement.* Edited by D.A. Lundgren et al., University Presses of Florida (1976).
57. Pui, D.Y.H. and Liu, B.Y.H., "Electrical Aerosol Analyzer: Calibration and Performance." In: *Aerosol Measurement.* Edited by D.A. Lundgren et al., University Presses of Florida (1976).
58. Hinds, W.C., *Aerosol Technology,* New York, NY, Wiley (1982).
59. Bergin, M.H., *Microcontamination* 5 (2), pp. 22-28 (1987).
60. Rivers, R.D. and Engleman, D.S., *Journal of Environmental Sciences* 27 (5), pp. 31-36 (1984).
61. *Penetrometer, Filter Testing DOP Q-107.* 150. DLB-76-2-639. U.S. Army Armament Research and Development, Attn: DRDAR-TSC-S, Aberdeen Proving Ground, MD.
62. *Method for Sodium Flame Test for Air Filters,* BS3928. British Standards Institution, 2 Park St., London, W1A 2BS, England (1969).
63. *Methods of Test for Air Filters Used in Air Conditioning and General Ventilation,* BS2831. British Standards Institution, 2 Park St., London, W1A 2BS, England (1971).
64. *Method to Measure Average Filtration Efficiency with Uranine Aerosol,* STD X44-011. L'Afnor, Tour Europe, 92 Cour Bevote, France (1978).
65. *Method of Testing Air-Cleaning Devices Used in General Ventilation for Removing Particulate Matter,* ASHRAE Std. 52-76. American Society of Heating, Refrigerating, and Air-Conditioning Engineers, 1791 Tullie Circle, NE, Atlanta, GA (1976).
66. *Efficiency Testing of Air-Cleaning Systems Containing Devices for Removal of Particles,* ANSI N 101.1. American National Standards Institute, 1430 Broadway, New York, NY (1972).
67. SAE J7266, Society of Automotive Engineers. Dept. 716, 400 Commonwealth Dr., Warrendale, PA (1979).
68. Kapoor, J.C., Subramanian, K.G. and Khan, A.A., *Filtration and Separation,* pp. 133-134, March/April (1977).
69. Stafford, R.G. and Ettinger, H.J., *Atmospheric Environment, 6 (5), pp. 353-362 (1972).*
70. Liu, B.Y.H. and Lee, K.W., *Environ. Sci. Technol.* 10, p. 345 (1976).
71. Cadwell, G.H., "A Brief History of the Modern ULPA Filter." Proceedings: Institute of Environmental Sciences, Las Vegas, NV, April 30-May 2 (1985).
72. Salzman, G.C. et al., "Potential Application of a Single Particle Aerosol Spectrometer for Monitoring Aerosol Size at the DOE Filter Test Facilities." Proceedings: 17th DOE Nuclear Air Cleaning Conference (1982).
73. Scripsick, R.C., "New Filter Efficiency Tests Being Developed for the Department of Energy." In: *Fluid Filtration: Gas Volume I, ASTM 975.* Edited by R.R. Raber. Philadelphia, PA, American Society for Testing and Materials (1986).

74. Sinclair, D., *J. Aerosol Science,* 7, pp. 75-79 (1976).
75. Silverman, L. and McGreevy, G., *Atmos. Environment* 1, pp. 1-10 (1967).
76. Rimberg, D., *Am. Ind. Hyg. Assoc. J.* 30, pp. 394-401 (1969).
77. Lindeken, C.L., Gorgin, R.L. and Petrock, K.S., *Health Physics* 9, pp. 305-308 (1963).
78. Shleien, B., Cochran, J.A. and Friend, A.G., *Am. Ind. Hyg. Assoc. J.* 27, pp. 353-359 (1966).
79. Posner, S., "Air Sampling Filter Retention Studies Using Solid Particles." Proceedings: Seventh AEC Air Cleaning Conference, TID-7627 (1961).
80. Lindeken, C.L., *Am. Ind. Hyg. Assoc. J.* 22, pp. 232-237 (1961).
81. Lundgren, D.A. and Gunderson, T.C., *Am. Ind. Hyg. Assoc. J.* 35, p. 866 (1975).
82. Remiarz, R.J., Johnson, B.R. and Agarwal, J.A., "Automated Systems for Filter Efficiency Measurement." In: *Fluid Filtration: Gas, Volume I, ASTM 975.* Edited by R.R. Raber, Philadelphia, Pa, American Society for Testing and Materials (1986).
83. Rubow, K.L. and Liu, B.Y.H., Evaluation of Ultra-High Efficiency Membrane Filters. Proceedings: Institute of Environmental Sciences, Las Vegas, NV (1984).
84. Nawakowski, C.E. and Martinez de Pinillos, J.V., "Monitoring of Particles in Gases with a Laser Counter." In: *Microelectronics Processing: Inorganic Materials Characterization.* Edited by L.A. Casper, Washington, D.C., American Chemical Society (1986).
85. Accomazzo, M.A. and Grant, D.C., "Mechanisms and Devices for Filtration of Critical Process Gases." In: *Fluid Filtration: Gas, Volume I, ASTM STP975.* Edited by R.R. Haber, Philadelphia, PA, American Society for Testing and Materials (1986).
86. Thorogood, R.M. and Schwarz, A., "Performance Measurement of Gas Ultrafiltration Cartridges." Proceedings: Institute of Environmental Sciences Annual, Dallas, TX, May 6-8 (1986).
87. *ASHRAE Handbook: 1982 Applications.* Atlanta: The American Society of Heating, Refrigerating and Air-Conditioning Engineers, Inc.
88. *IES Recommended Practice: HEPA Filters.* IES-RP-CC-001-86. Institute of Environmental Sciences (1986).
89. McCrone, W.C., McCrone, L.B. and Delly, J.G., *Polarized Light Microscopy.* Ann Arbor, Michigan: Ann Arbor Science Publishing (1978).
90. McCrone, W.C. and Delly, J.G., *The Particle Atlas,* 2nd Edition. Ann Arbor, Michigan: Ann Arbor Science Publishing (1973).
91. Bloss, F.D., *An Introducction to the Methods of Optical Crystallography.* Saunder College Publishing (1961).
92. Appel, B.R. and Wesolowski, J.J., "Selection of Filter Media for Particulate Sampling with a Lundgren Impactor." AIHL Report No. 125. Air and Industrial Hygiene Laboratory, Berkeley, CA (1972).

2

Instrumentation for Aerosol Measurement

Benjamin Y.H. Liu and David Y.H. Pui

INTRODUCTION

Particulate contamination is a major problem in the semiconductor and electronic industry. Deposition of particles on the wafer surface and on other sensitive electrical or mechanical parts can lead to the failure of the device, resulting in a loss of yield. Control of particulate contamination is of vital importance in the fabrication of integrated circuits and other precision electrical or mechanical components.

Solid or liquid particles suspended in a gas are usually referred to as aerosols. The suspending medium can be air, or some other gas, such as argon, nitrogen, helium, etc., the latter being important as processing gases for semiconductor manufacturing. The branch of science dealing with such gas-borne particles is known as Aerosol Science.

This chapter provides an overview of the instrumentation for aerosol measurement. The characteristics of airborne particles, their deposition rate on wafer and other surfaces and the performance of air filters will also be briefly discussed.

SIZE DISTRIBUTION OF AEROSOLS

Aerosol particles may consist of microscopic bits of material of only a few nanometers in diameter containing a few thousand molecules each to particles of a near macroscopic size of a fraction of a millimeter in diameter. The nominal size range of interest in aerosol science is 0.002 to 100 μm. The lower size limit corresponds to the smallest particles that can be detected by an aerosol measuring instrument, viz. the condensation nucleus counter. The largest particles are those that would begin to fall out from air too quickly to be considered an aerosol.

In normal ambient air, particles are found over this entire size range of 0.002 to 100 μm. Figures 1 and 2 show some sample size distribution of atmospheric aerosols presented in the form of a number and a volume (or mass) distribution.

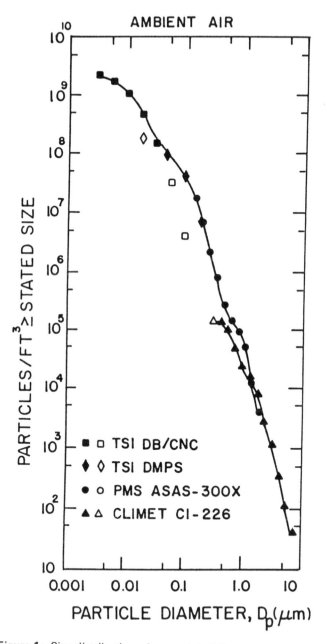

Figure 1: Size distribution of aerosols in Minneapolis ambient air.

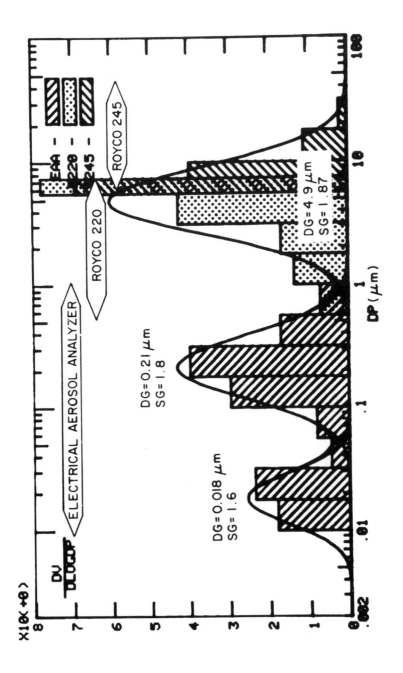

Figure 2: Trimodal size distribution of atmospheric aerosols according to Whitby (1978).

Figure 1 is a cumulative number distribution curve showing the number of particles per unit volume of air—the number concentration—larger than the stated particle size plotted as a function of particle size. The resulting curve is one with a sharply declining slope as the particle size is increased. Above a particle size of approximately 0.1 μm, the slope becomes approximately constant with a value of -3 on the log-log plot. Such a size distribution is known as a Junge distribution,[1] in honor of the original discoverer of the size distribution. For an aerosol with a Junge distribution, the particle concentration decreases by a factor of eight for a factor of two increase in size and by a factor of thousand for a factor of ten increase in size. For instance, if the concentration of atmospheric particles larger than 0.1 μm is 1×10^7 particles/ft^3, then the concentration of particles larger than 0.2 μm would be 0.125×10^7 particles/ft^3, and that larger than 1.0 μm would be 10^3 particles/ft^3, if the Junge distribution is followed.

In the case of the size distribution plot shown in Figure 2,[2] the quantity $dV/d\log D_p$ is plotted against the particle size in a semilog plot. Here dV is the volume concentration of particles (volume of particles per unit volume of air) in the log size interval $d\log D_p$. In such a plot, the area under the curve between any two particle sizes is proportional to the volume (or mass) concentration of particles in that size interval. On such a plot, the atmospheric aerosol size distribution often has a bimodal or trimodal shape as shown. The coarse particle above 2 μm correspond to mechanically generated particles, such as particles generated by wind-blown dust, dust on the roadway made airborne by motor vehicle traffic, etc. Particles in this mode tend to contain elements that are characteristic of the crustal composition of the earth, such as silicon, calcium, iron, aluminum, etc. In contrast, particles in the two lower modes, the so-called nuclei and accumulation modes, are the result of combustion and gas-to-particle conversion processes. They usually contain products of combustion, such as elemental and organic carbon, and products of gas-to-particle conversion processes in the atmosphere, such as sulfates and nitrates, etc. The relative amounts of material in the different modes depend on the locality and weather and on the pollution by local industry and motor vehicle traffic.

Aerosols in the clean room contain both atmospheric particles that have penetrated through the clean room filter and particles generated internally in the clean room. Figure 3 shows the size distribution of a clean room aerosol measured at the University of Minnesota. Also shown for comparison is the clean room aerosol size distribution according to Federal Standard 209b. Because of the high efficiency of the clean room filter for small particles, very few particles below 0.05 μm are found in the clean room. In the large particle size range, particles generated internally in the clean room are added to those penetrating through the filter. This causes the slope of the distribution curve to be less steep than that in the ambient atmosphere. The size distribution of aerosols above 0.1 μm appear to be well approximately by a straight line with a negative slope of 2.2 as described in Federal Standard 209b used for clean room classification. Similar size distribution measurement has been reported by Donovan et al.[3] and Locke et al.[4]

Although the high efficiency particulate air (HEPA) filter used for clean room air filtration is quite efficient (>99.97% for 0.3 μm diameter particles), it does not provide a perfect barrier between the inside and the outside environment. Figure 4 shows the concentration of atmospheric particles and particles in the clean room simultaneously measured by two condensation nucleus counters (CNC).[5] The indoor concentration is seen to be some five orders of magnitude lower than that in the ambient. However, the correspondence of the peaks in the curves shows that a finite fraction of atmospheric particles have penetrated through the filter, and that particles in the clean room detected by the CNC likely have their origin in the ambient atmosphere.

Figure 5, adapted from Reference 6, is a comparison of the size distribution of atmospheric aerosols, aerosols in the clean room according to Federal Standard 209b and the size distribution of other "particulate systems" in the solar system. Figure 6, adapted from Reference 7, shows a comparison of the size of particulate air contaminants, the feature size of the very large scale integrated (VLSI) circuits and the size of some common particulate systems.

The usual "design rule" for semiconductor manufacturing states that particles down to 1/10 of the characteristic feature size of the device must be controlled. Thus for a one megabit chip with a characteristic feature size of only 1.0 μm, particles as small as 0.1 μm must be controlled. However, many semiconductor devices have thin oxide layers that are only 0.02 μm thick. Smaller particles may also be important. These particles may not produce "killer" defects. However, they may produce hidden defects that may impair the long term performance and reliability of the device. Contamination control in semiconductor manufacturing, therefore, should be approached from the point of view of *total particulate control*, i.e., the control of airborne particles of all sizes.

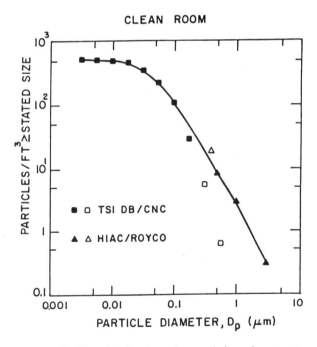

Figure 3: Size distribution of aerosols in a clean room.

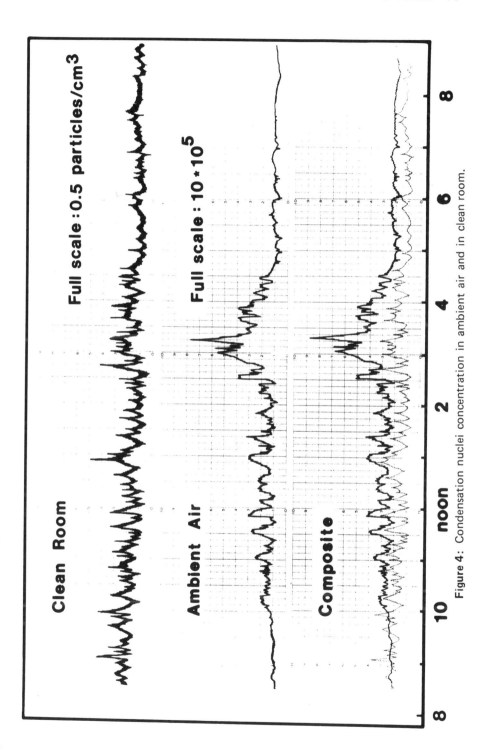

Figure 4: Condensation nuclei concentration in ambient air and in clean room.

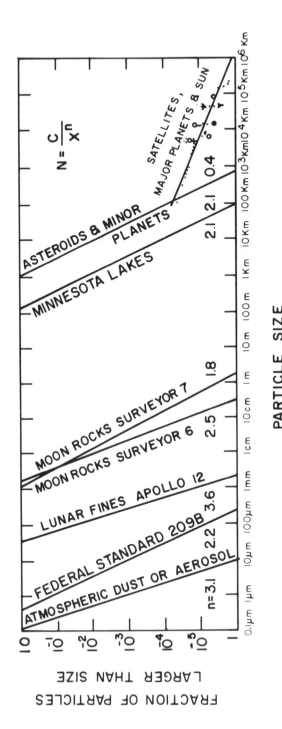

Figure 5: Size distribution of particulate systems in solar system (adapted from Heywood, 1970).

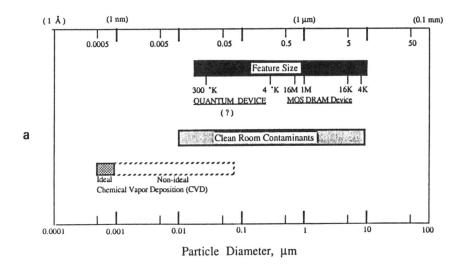

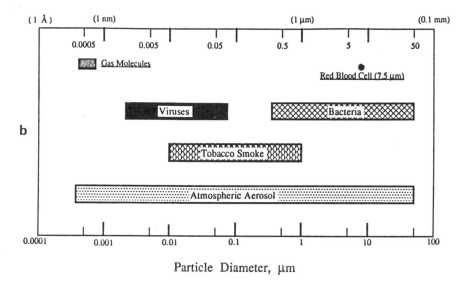

Figure 6: Size of particulate contaminants, (a) feature size of VLSI circuits and (b) some common particulate systems.

PARTICLE DEPOSITION ON SURFACES

The rate of deposition of aerosol particles on wafer or some other surface governs the rate with which the surface becomes contaminated by airborne particles. The two major mechanisms responsible for particle deposition on surfaces are sedimentation and diffusion. Sedimentation is important for large particles, viz. those above 1.0 μm in diameter, while diffusion is important for small particles below 0.1 μm. In the intermediate size range, both sedimentation and diffusion are important and must be considered. When the particle and/or the surface are electrically charged, enhanced deposition due to electrostatic effects can also occur and must also be considered.

Figure 7 shows the settling speed of unit density spheres in air at a temperature of 20°C and 1 atmosphere pressure calculated from the Stokes law with the correction for particle slip:

$$(1) \qquad V_s = \rho_p D_p^2 C g / 18 \mu$$

where

V_s = settling speed, cm/s

ρ_p = particle density, g/cm^3

D_p = particle diameter, cm

μ = gas viscosity, poise

g = acceleration of gravity

In Equation (1), C is the slip correction given by

$$(2) \qquad C = 1 + 2.592\lambda/D_p + 0.84\lambda/D_p + e^{-0.435\ D_p/\lambda}$$

The slip correction is important when the particle diameter is of the same order of magnitude as the mean free path, λ, of the surrounding air. For air under normal temperature and pressure conditions, λ is equal to 0.0652×10^{-4} cm.

The rate of diffusion of small aerosol particle to a surface is governed by its diffusion coefficient, which is given by

$$(3) \qquad D = kTC/3\pi\mu D_p$$

where k is the Boltzmann's constant and T is the absolute temperature. The diffusion coefficient for aerosol particles is shown in Figure 8 together with the electrical mobility of singly charged particles and the particle relaxation time, the latter being important when considering the motion of charged particles and the dynamic behavior of airborne particles.

The theoretical calculation of the particle deposition rate on wafer surfaces is complicated by the airflow pattern and the existence of a boundary layer over the wafer surface. Such a calculation has been reported recently by Liu and Ahn.[8] They considered the three cases shown in Figure 9. Case (a) corresponds to a free standing wafer in a vertical laminar flow (VLF) clean room; Case (b), a wafer in a horizontal-flow clean hood, and Case (c), vertical wafers in a wafer carrier. Particle deposition occurs by both sedimentation and diffusion in cases

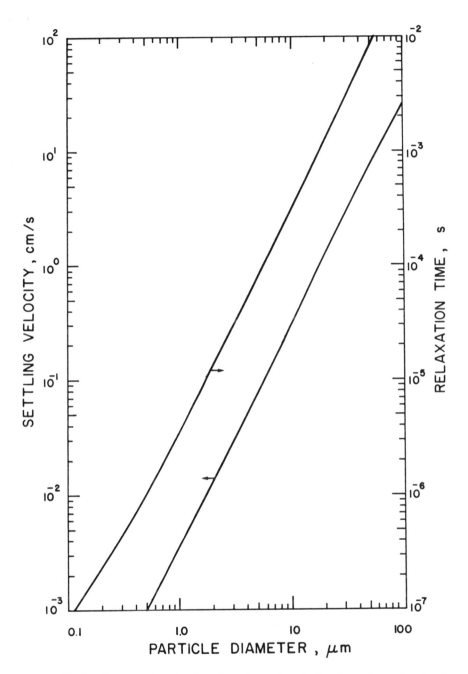

Figure 7: Settling speed and relaxation time of unit density spheres in air at 20°C and 1.0 atmosphere.

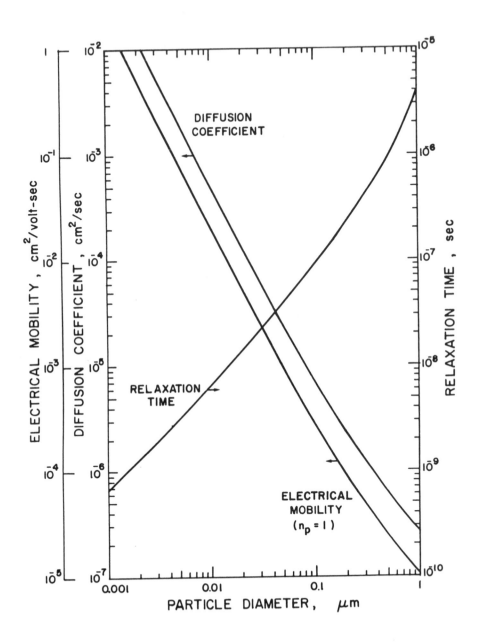

Figure 8: Diffusion coefficient, electrical mobility of singly charged particles and relaxation time.

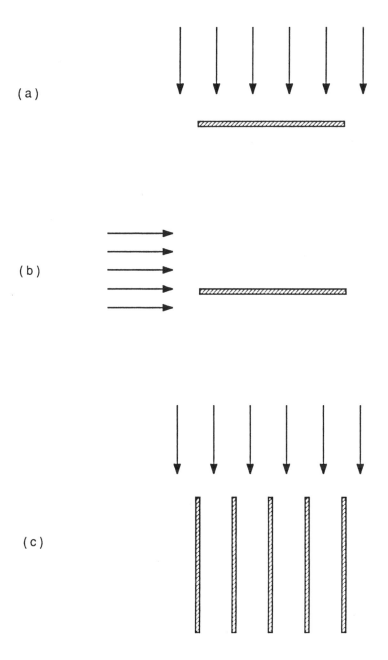

Figure 9: Schematic diagram showing (a) horizontal wafer in a vertical laminar flow (VLF) clean room, (b) horizontal wafer in a horizontal clean hood and (c) vertical wafers in a wafer carrier.

(a) and (b), while in case (c) the deposition is by diffusion only. Figures 10 through 12 show the calculated deposition velocity, V, for aerosol particles where by definition,

(4) $V = j/N$

j is the number of particles depositing on a unit wafer surface area in a unit time interval and N is the airborne particle concentration. The deposition velocity for sedimentation is the same as the settling speed of the particle, whereas for diffusion, the deposition velocity can be interpreted as an equivalent migration velocity due to Brownian motion. It is interesting to note in Figure 10 that the migrating velocity of a 0.01 μm particle, for instance, due to diffusion is the same as that of a 2.5 μm particle due to sedimentation.

The result for a horizontal wafer in Figures 10 and 11 shows that the deposition velocity first decreases with increasing particle size for small particles, then increases with increasing particle size for large particles. There is a minimum in the deposition velocity for particles in the vicinity of 0.2 μm and this minimum deposition velocity is of the order of 0.5×10^{-4} cm/s. It should be noted that a deposition velocity of 0.5×10^{-4} cm/s translates to a deposition rate of 1.87 particles per day for a 125 mm diameter wafer and an airborne particle concentration of 100 particles/ft^3.

The influence of electrical charge on particle deposition on semiconductor wafers has not been studied in detail. Figure 13 shows the deposition rate constant[9] for uncharged particles and particles carrying 1, 2 and 3 elementary units of charge in a spherical vessel of 250 liters in volume. While not exactly comparable to the case of particle deposition in clean rooms, the figure does show the importance of electrical charge on particle deposition on surfaces.

INSTRUMENTATION FOR AEROSOL MEASUREMENT

Because of the low contaminant levels in clean rooms and clean processing gases, only single particle counting instruments, viz. those capable of counting single individual particles, are suitable for particle measurement in such applications. Instruments such as transmissometers, photometers, the electrical aerosol analyzer, etc. which are used for air pollution, industrial hygiene and related studies involving high particulate concentration levels, are generally unsuited for use in clean rooms. Figure 14 summarizes the measuring range of different aerosol measurement devices, both for clean room and other applications.

Optical Particle Counter

Single particle optical counters, or OPC's, are undoubtedly the most widely used instrument for particle measurement in the clean room and in clean process gases. These instruments operate by passing single, individual particles through an illuminated viewing volume in the instrument and counting the pulses of light scattered by each particle as it passes through the viewing volume. The amplitude of the light pulse, as measured by a photomultiplier tube or a solid state photodetector is taken as a measure of the particle size. The pulse signal is then processed electronically to yield a pulse-height histogram, which is then con-

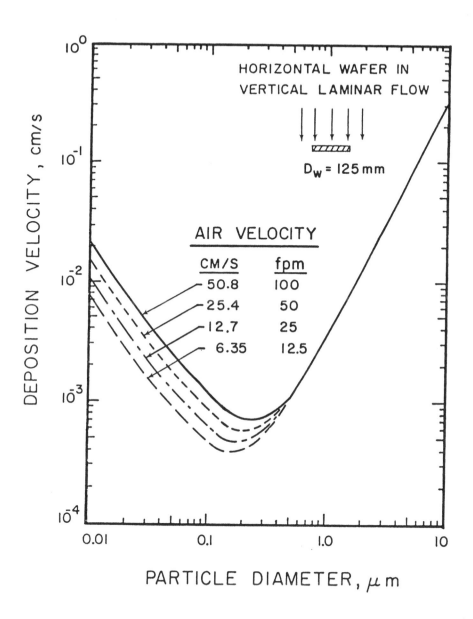

Figure 10: Deposition velocity of aerosol particles for a 125 mm diameter wafer in vertical laminar flow.

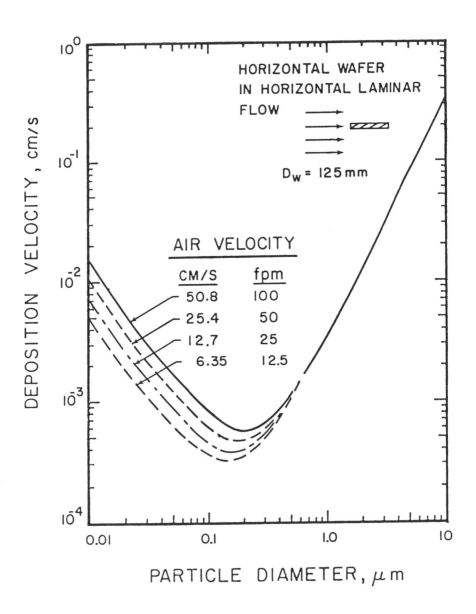

Figure 11: Deposition velocity of aerosol particles for a 125 mm diameter wafer in horizontal laminar flow.

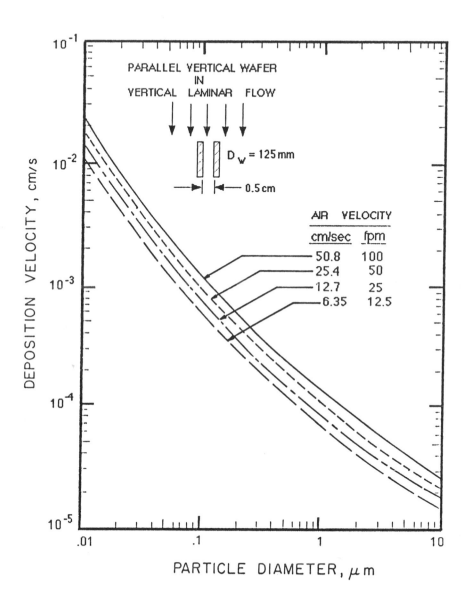

Figure 12: Deposition velocity of aerosol particles for 125 mm diameter, vertical wafers in a wafer carrier.

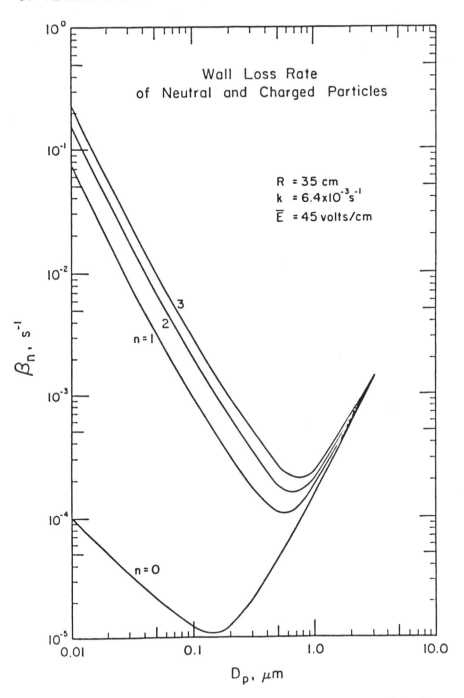

Figure 13: Deposition rate constant for charged and uncharged particles in a 250 liter spherical vessel.

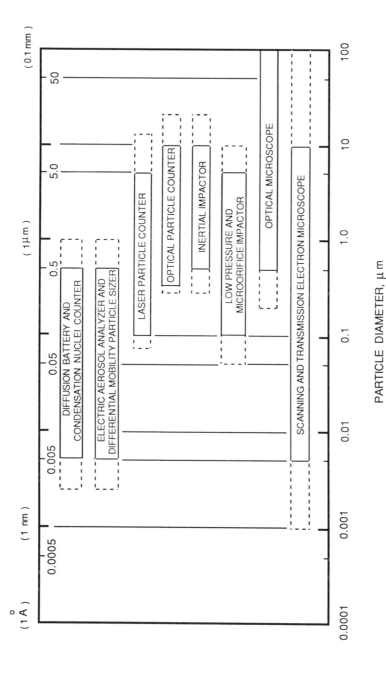

Figure 14: Particle size range of aerosol measuring instruments.

verted to a histogram for particle size distribution using an appropriate calibration curve. Figure 15 shows the schematic operating principle of an optical particle counter.

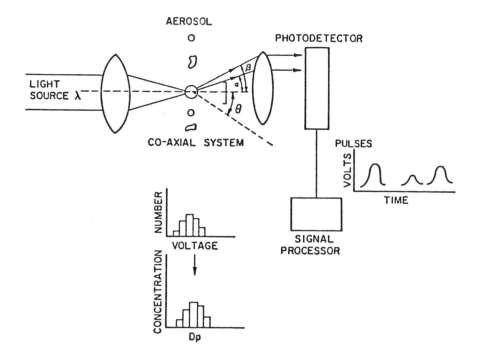

Figure 15: Schematic diagram showing the operating principle of a single particle optical counter.

There are a large number of commercial optical particle counters available. They can be differentiated by the light source (laser or incandescent white light) they use, the sampling flow rate of the instrument, the number of channel of data the instrument provides, and other distinguishing characteristics, such as portability and ability to be interfaced with computers, etc. In general, instrument using a laser light source can detect smaller particles than a corresponding instrument using an incandescent light source because of the higher illuminating intensity of the laser, and a higher flow rate instrument can count more particles in a given time period than an instrument of a lower sampling flow rate. The latter property is important for particle counting in clean rooms. The lower detection limit of the white-light counter is usually 0.3 μm whereas laser counters can detect particles as small as 0.1 μm. A sampling flow rate of 1 cubic feet per minute (cfm) is usually considered high and a flow rate of 0.01 cfm is usually considered low. Instruments with intermediate flow rates of 0.1 and 0.25 cfm are also quite common. Table 1 summarizes various particle counting instruments available at the time of writing of this chapter.

Table 1: Selected Aerosol Measuring Instruments for Clean Rooms

Manufacturer	Model No.	Flowrate, cfm	Size Range, μm	Number of Size Channels	Illumination Source	Remarks
Climet Instrument Co.	CI-208C	0.25 or 1	0.3->20	16	white light	
P. O. Box 151	CI-208P	0.25	0.3->10	6	white light	high pressure counter
Redlands, CA 92373	CI-226	0.25 or 1	0.3->20	-	white light	
	CI-6300	1.0	0.19->5.0	6	laser	
	CI-6400	0.1	0.1->0.5	6	laser	
	CI-8060	1.0	0.3->10	8	white light	
Misc/Royco	5300	1.0	0.5->15	8	white light	
141 Jefferson Drive	4101	1.0	0.5-	6	white light	
Menlo Park, CA 94025	4102	0.1	0.5-	6	white light	
	4130	1.0	0.3-	6	laser	
	5100	1.0	0.25->10	6	laser	
	5000	0.01-0.05	0.3->10	6	white light	disk drive counter
Met One	205	1.0	0.16->10	6	laser	
481 California Ave.	200	1.0	0.3->10	6	white light	
Grants Pass, OR 97526						
Particle Measuring	LPC-101	0.1	0.1->1.0	4	laser	
System, Inc.	LPC-525	1.0	0.2->5.0	5	laser	
1855 S. 57th Ct.	LPC-555	1.0	0.3->5.0	4	laser	
Boulder, CO 80301	LPC-101-HP	0.1	0.1-			high pressure counter
	HPLAS		0.3-			
Status	4000	1.0	0.3->5.0	5	white light	
(Faley International Corp.)	5000	0.0177	0.3->5.0	5	white light	
P. O. Box 669	2100	0.01	0.5->5.0	2	white light	
El Toro, CA 92630-0669)	2100P	0.01	0.5->5.0	2	white light	
TSI, Inc.	3755	0.1	0.5->5.0	2	laser	multipoint application
500 Cardigan Road	3760	0.05	0.01->1.0	1	laser	condensation nucleus counter
P. O. Box 64394						
St. Paul, MN 55164						

Optical particle counters differ widely in their design and performance characteristics. Figures 16a, b and c show the optical systems used in two PMS laser counters and a TSI laser counter. Pictures of some commercial instruments are shown in Figures 17a and b. Some laser counters make use of the "active" scattering principle (Figures 16a, b),[10,11] in which the particles are passed through the resonant cavity of a helium-neon laser whereas others use the "passive" scattering principle, in which the light beam is first focussed and the aerosol passed through this focused light beam for optical detection. In other instruments (Figure 16c) a solid-state laser is used as the light source to obtain a small, light-weight portable sensor.

The response of an optical particle counter (the amplitude of the output pulse) is a function of the optical design of the instrument. For a specific instrument, the response is a function of the size and refractive index of the particle. In the case of nonspherical particles, the instrument response also depends on the particle shape and the orientation of the particle in the instrument viewing volume. In general, the instrument response can be determined by feeding particles of a known particle size and refractive index into the instrument and measuring the response experimentally.

Detailed calibration studies of optical particle counters have been reported by various investigators including the authors of this chapter. Some sample results[12-14] are shown in Figures 18 through 23. These figures pertain to an "active scattering" laser counter of the narrow angle type manufactured by PMS, Inc. Figure 18 shows the response of the instrument when sampling monodisperse polystyrene latex (PSL) particles of various particle sizes. The response has been plotted in the form of cumulative distribution curves in which the fraction of particles indicated by the instrument is plotted against the apparent size of the particles indicated by the instrument for various monodisperse particle sizes. In Figures 19 through 25 the median output pulse amplitude of the photodetector is plotted against the particle size for four test aerosols: polystyrene latex (PSL), dioctylphthalate (DOP), sodium chloride and carbon black. The result shows that the response of the instrument is a complicated function of particle size, shape and refractive index. For accurate particle size measurement, therefore, the response of the instrument must be taken into account. Similar response curves for an active scattering, wide angle instrument manufactured by PMS, Inc. is shown in Figure 23.

Aerodynamic Particle Sizer

An instrument that differs from the above is the Aerodynamic Particle Sizer[15] manufactured by the TSI, Inc. In this instrument (Figure 24), the particles are accelerated through a small nozzle to different speeds. The larger the particle size, the lower the speed of the particle due to particle inertia. The particle velocity at the nozzle exit is then measured to provide a measure of particle size. This principle enables the "aerodynamic size" of the particles to be measured. The aerodynamic size is defined as the diameter of a unit density sphere having the same settling speed as the particle in question. Because of this ability to measure the aerodynamic particle size, the instrument is capable of measuring the settling speed of the particle, which is directly related to rate of settling of particles on surfaces. In general, the conventional light-scattering optical particle

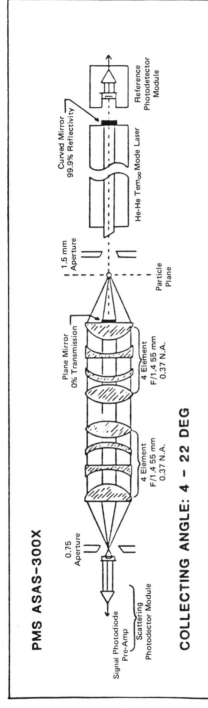

Figure 16: Schematic diagram of (a) a narrow angle, active scattering, laser OPC (PMS ASAS -300X), (b) a wide angle, active scattering, laser OPC (PMS LAS-X) and (c) a laser OPC using a solid-state laser (TSI LPC).

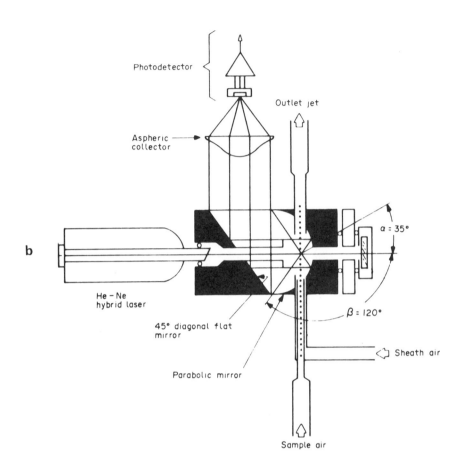

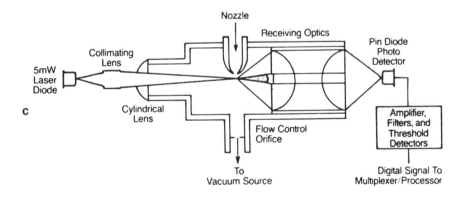

Figure 16: (Continued)

a

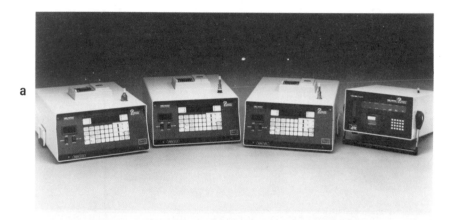

b

Figure 17: Pictures of (a) Hiac/Royco laser OPC's and (b) TSI laser OPC.

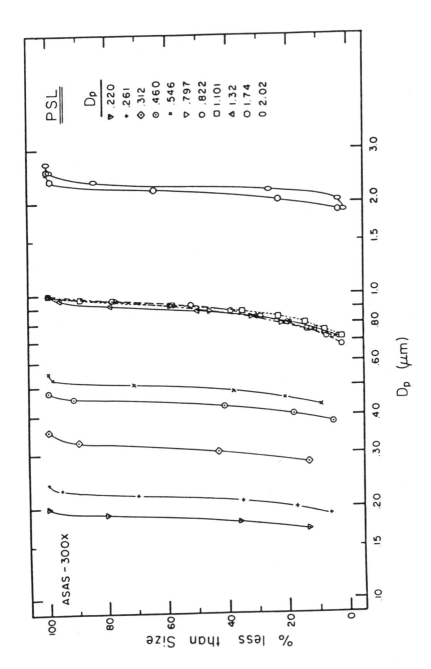

Figure 18: Response of PMS ASAS-300X counter to monodisperse PSL particles.

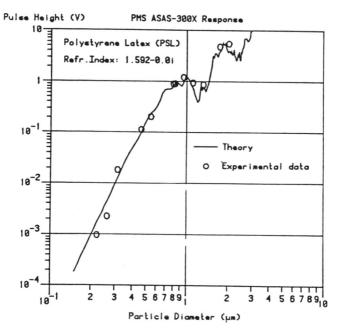

Figure 19: Response of PMS ASAS-300X counter to PSL particles.

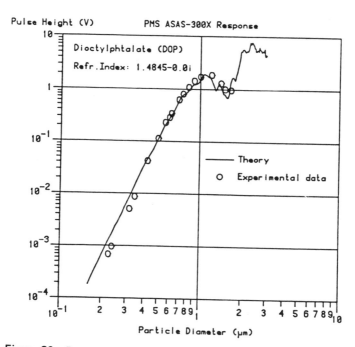

Figure 20: Response of PMS ASAS-300X counter to DOP particles.

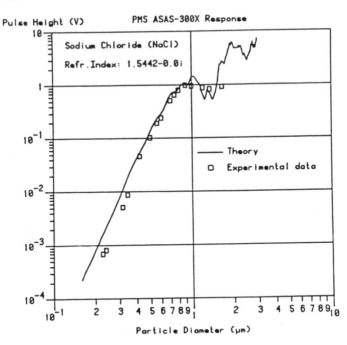

Figure 21: Response of PMS ASAS-300X counter to NaCl particles.

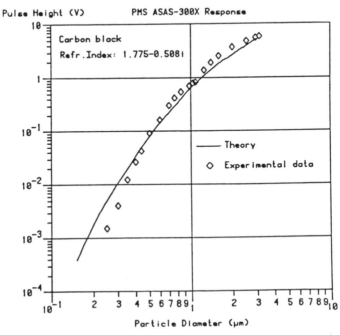

Figure 22: Response of PMS ASAS-300X counter to carbon black particles.

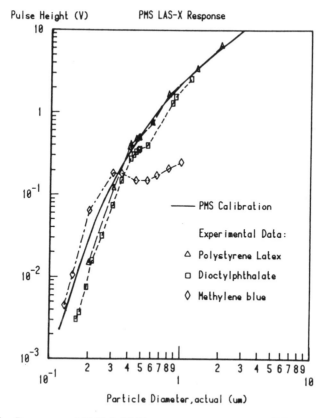

Figure 23: Response of PMS LAS-X counter to transparent (PSL and DOP) and absorbing (methylene blue) particles.

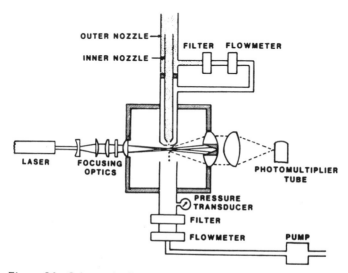

Figure 24: Schematic diagram of TSI Aerodynamic Particle Sizer.

counters measure an optical equivalent size which is related to the geometrical size of the particle in a complicated manner that is not easily related to its geometrical size or dynamic property, such as settling speed. Figure 25 shows the response of the APS to the aerodynamic size of aerosol particles.[16]

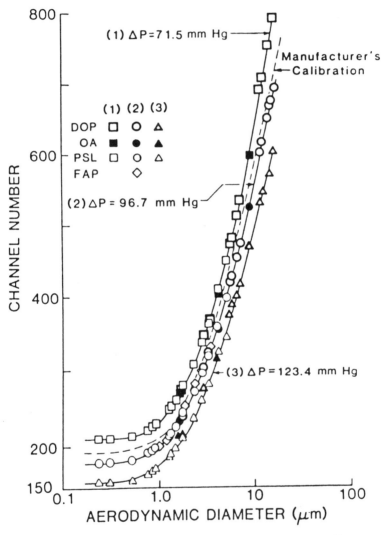

Figure 25: Response of TSI Aerodynamic Particle Sizer.

Condensation Nucleus Counter and Diffusion Battery

For particles below 0.1 μm in diameter, the condensation nucleus counter can be used. The instrument is based on the condensation-growth of small particles in a supersaturated atmosphere of a vapor, such as water or alcohol.

According to the Kelvin equation, a particle of radius, a, will serve as a "nucleus of condensation" in an atmosphere when the saturation ratio of the vapor exceeds the value,

$$(5) \qquad p/p_o = e^{-2\sigma/a\rho_1 RT}$$

where p/p_o is the ratio of vapor pressure to that at saturation, σ is the surface tension of the liquid, ρ_1 is the density of the liquid, R is the gas content and T is the absolute temperature. Thus, by placing small particles in a supersaturated atmosphere, the particles can be made to "grow" to a sufficiently large size for easy detection by light scattering. By this means, particles as small as 0.002 μm can be detected when a suitably high supersaturation is used.

A schematic diagram of a commercially available CNC is shown in Figure 26. The instrument[17] can also be viewed as an optical counter with a particle magnifier. In this instrument, the airstream containing particles that are normally too small to be detected by direct light scattering ($<$0.1 μm) is first passed through a heated alcohol bath at 35°C to saturate the air with alcohol vapor and then through a condenser tube at 10°C to produce the required supersaturation. In the condenser the particles grow to a size of approximately 12 μm. The particles are then detected by light scattering.

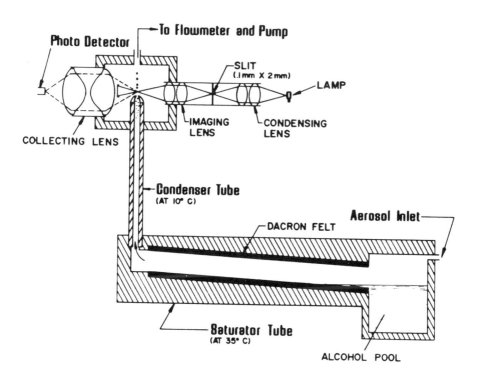

Figure 26: Schematic diagram of TSI condensation nucleus counter.

The lower particle size limit of the CNC has been determined empirically. Figure 27 shows the "counting efficiency" of the instrument as a function of particle size. The counting efficiency is the fraction of particles in the airstream that is actually counted by the instrument and is a result of particle loss in the flow passages in the instrument and the nonuniform temperature and vapor concentration profiles in the condenser. The result shows that the instrument measures particles with near 100% efficiency down to a lower size limit of approximately 0.03 μm, below which the counting efficiency drops off sharply, reaching the 50% level at 0.01 μm. At the lower limit of the instrument of approximately 0.005 μm, only 10% of the particles are counted. An experimental CNC capable of counting particles with high efficiency to 0.002 μm has been developed in the authors' laboratory.

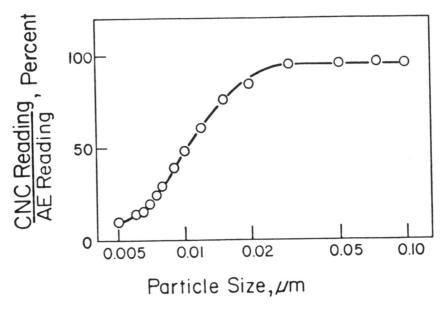

Figure 27: Counting efficiency of TSI CNC.

The condensation nucleus counter is an aerosol concentration measuring device. As such, it does not provide information on particle size. For particle size distribution measurement, the instrument is often used with a diffusion battery. The diffusion battery consists of a number of screens as diffusion collectors for small aerosol particles. By passing the aerosol through the diffusion battery and measuring the loss of particles in each diffusion stage with the CNC, the size distribution of the aerosol can be determined. The DB/CNC combination can measure the size distribution of aerosols in the 0.002 to 0.5 μm diameter range. A schematic diagram of a TSI diffusion battery is shown in Figure 28 which is based on the research of Sinclair and Hoopes.[18] The theory of wire screen diffusion battery has been developed in considerable detail by Yeh et al.[19]

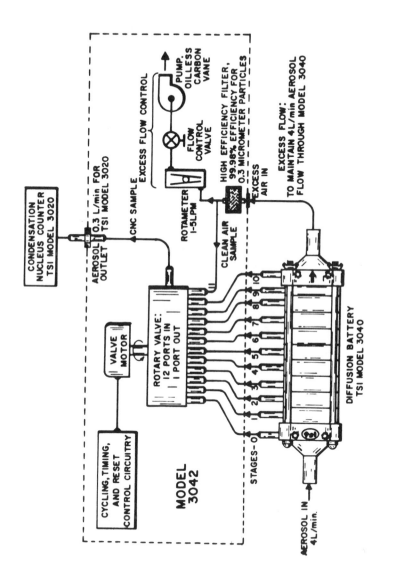

Figure 28:   Schematic diagram of diffusion battery.

AEROSOL SAMPLING INSTRUMENTS

Because of the low particulate contaminant levels in clean rooms, aerosol sampling instruments such as filter samplers, electrostatic samplers, etc. that have been developed for air pollution and industrial hygiene studies are generally not suited for particle sampling in the clean room. More useful, however, are inertial impactors, which can collect and deposit particles on a small sampling area, thus allowing high magnification electron microscopy to be used for particle identification.

Figure 29 is a schematic diagram of a cascade impactor,[20] consisting of a number of impaction stages arranged in series. As air flows through each stage, it is accelerated by a nozzle. The jet of air then impinges on a flat surface causing particles larger than a certain characteristic "cut size" to impact on the surface. The surface is usually coated with an adhesive layer to prevent particle bounce. By arranging several of these impactor stages in series with progressively smaller nozzle diameters, particles with progressively small sizes are collected. The conventional impactor can collect particles in the 0.3 to 20 μm diameter range. However, by operating the impactor at a low pressure,[21] or using very small "micro-orifices,"[22] particles as small as 0.02 μm can be collected. The theory of inertial impactors has been developed in considerable detail by Marple and Liu.[23]

MONODISPERSE AEROSOL GENERATORS

In aerosol studies, it is often necessary to generate a monodisperse aerosol for instrument calibration. Monodisperse aerosols are also needed for various experimental purposes. For optical particle counter calibration, polystyrene latex particles are often used. These are available in the form of uniform spheres suspended in water. The suspension can be atomized to form a monodisperse aerosol. Figure 30 shows monodisperse latex particles of several different sizes that can be used for aerosol instrument calibration.[24]

For generating monodisperse aerosols of different materials, laboratory aerosol generators are needed. Two widely used instruments for monodisperse aerosol generation are the Berglund-Liu vibrating orifice monodisperse aerosol generator[25] and the Liu-Pui electrostatic classifier.[26]

Figure 31 is a schematic diagram of the Berglund-Liu vibrating orifice monodisperse aerosol generator. The instrument operates by forcing a solution through a small (10 and 20 μm) diameter orifice, which is vibrated by a piezoelectric ceramic to cause the liquid jet to break up into uniform size droplets. The volatile solvent is then allowed to evaporate, leaving the nonvolatile solute as small aerosol particles. The system is capable of generating monodisperse aerosols of a known particle size in the 0.5 to 50 μm diameter range. Figure 32 shows some of the monodisperse solid and liquid particles generated by the instrument.

For generating monodisperse aerosols below 0.5 μm in diameter, the electrostatic classifier of Liu and Pui can be used. Figure 33 is a schematic diagram of the instrument. In this device, a liquid solution is first atomized to produce a polydisperse spray of solution droplets. The droplets are then allowed to dry to obtain a polydisperse aerosol. This polydisperse aerosol is then brought a

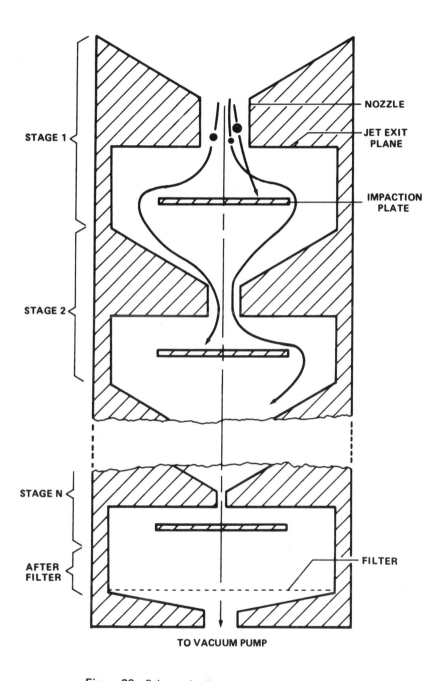

Figure 29: Schematic diagram of cascade impactor.

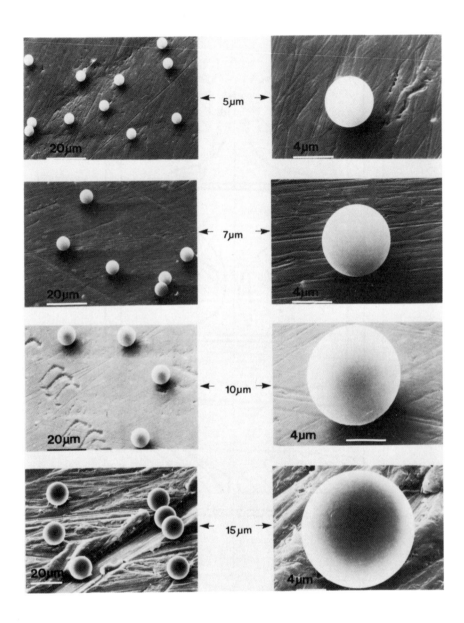

Figure 30: Monodisperse polystyrene latex particles.

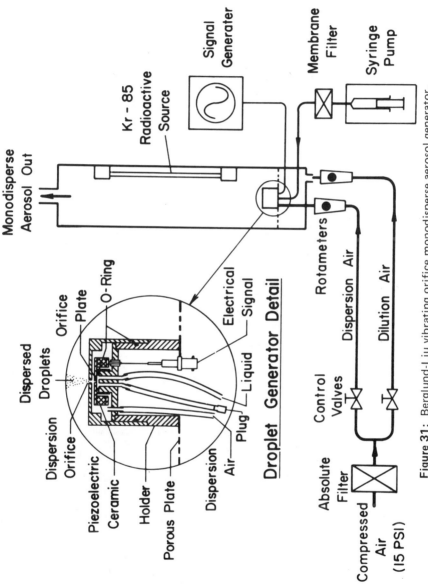

Figure 31: Berglund-Liu vibrating orifice monodisperse aerosol generator.

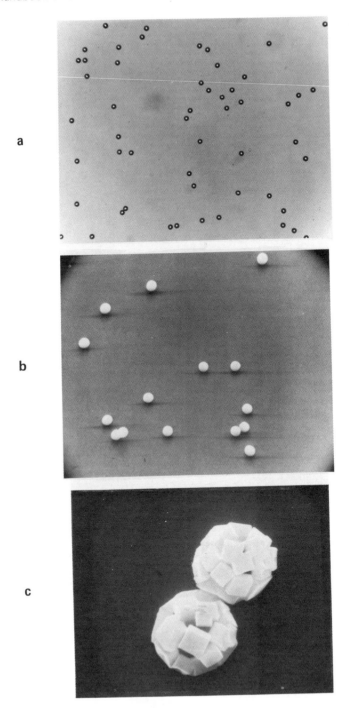

Figure 32: Monodisperse (a) DOP, (b) methylene blue and (c) NaCl particles produced by Berglund-Liu vibrating orifice monodisperse aerosol generator.

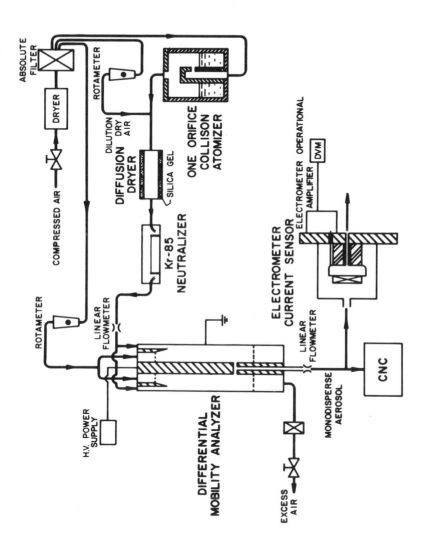

Figure 33: Electrostatic classifier of Liu and Pui.

state of charge equilibrium by exposing to a radioactive ionizer to develop a low level electrical charge on the particles. This charged aerosol is then classified electrostatically in the classifier to obtain a monodisperse aerosol. The instrument is capable of generating monodisperse aerosols in the 0.01 to 0.5 μm diameter range. With some simple modification, particles as large as 3 μm can be generated.

PERFORMANCE OF HEPA AND ULPA FILTERS

HEPA filters have been widely used for air filtration in the clean room. More recently, fibrous filters with higher efficiencies have become available and are used for clean room air filtration. These are sometimes referred to as ultra-low penetration air—or ULPA—filters.

The efficiency of a filter is a function of the size of the aerosol particles and the velocity of air through the filter media. Figure 34 shows some recent measurement results reported on the performance of HEPA and ULPA filters.[27]

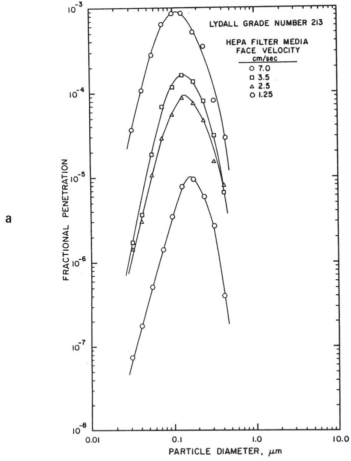

Figure 34: Aerosol penetration through (a) a HEPA and (b) an ULPA filter.

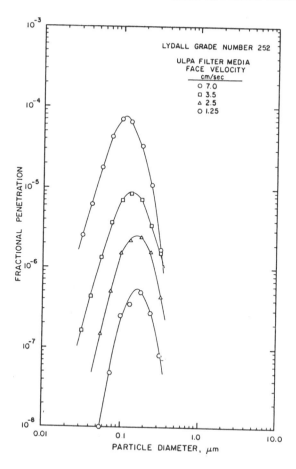

Figure 34: (Continued)

In these plots the aerosol penetration through the filter media is shown as a function of particle size for various filtration velocities. It is interesting to note that in all cases, the penetration is found to first increase with increasing particle size for small particles, then decrease with increasing particle size for large particles. There is a well-defined peak in the penetration curve. This peak penetration generally occurs in the vicinity of 0.15 μm and tends to decrease with increasing flow rate through the filter. In addition, the entire penetration curve tends to become higher with increasing flow velocity through the filter.

The above behavior of HEPA and ULPA filters is consistent with the filtration mechanisms of diffusion and interception. Large particles are collected primarily by the mechanism of interception, which becomes more effective as the particle size is increased. In contrast, small particles are collected by diffusion, which becomes more effective as the particle size becomes smaller. The peak in aerosol penetration corresponds to the transition from diffusion dominated filtration regime to one dominated by interception.

Acknowledgement

Much of the information in this chapter on instrument calibration, filtration, aerosol size distribution and clean room performance has been based on research sponsored by a consortium of companies at the University of Minnesota. Members of this Particulate Contamination Control Research Consortium include Donaldson Company, Liquid Air Corporation, Millipore Corporation, Texas Instruments, Inc., and TSI, Inc. The support of the consortium for this work is gratefully acknowledged.

REFERENCES

1. Junge, C.E., *Air Chemistry and Radioactivity,* Academic Press, New York (1963).
2. Whitby, K.T., *Atmos. Environ.,* Vol. 12, 135-159 (1978).
3. Donovan, R.P., Locke, B.R., Ensor, D.S. and Osburn, C.M., *Microcontamination,* 39-44, December (1984).
4. Locke, B.R., Donovan, R.P., Ensor, D.S. and Caviness, A.L., *J. Environ. Sci.* 28(6), 26-29 (1985).
5. Liu, B.Y.H., Lee, J.W., Pui, D.Y.H., Ahn, K.H. and Gilbert, S.L., "Performance of a Model Clean Room," presented at the 8th International Symposium on Contamination Control, Milan, Italy, September 9-11 (1986).
6. Heywood, H., "The World of Small Particles," Lecture given at the Loughborough University, Great Britain (1970).
7. Larrabee, G.B., *Chem. Eng.* 92(11), 51-59 (1985).
8. Liu, B.Y.H. and Ahn, K.H., "Particle Deposition on Semiconductor Wafers," Submitted to *Aerosol Science and Technol.* (1986).
9. McMurry, P.H. and Rader, D.J., *Aerosol Sci. Technol.* 4, 249-268 (1985).
10. Knollenberg, R.G., *SPIE Vol. 92 Practical Applications of Low Power Lasers* 92, 137-162 (1976).
11. Knollenberg, R.G., "Open-Cavity Laser 'Active' Scattering Particle Spectrometry from 0.05 to 5 microns," in *Fine Particles: Aerosol Generation, Measurement, Sampling and Analysis,* (B.Y.H. Liu, Ed.), Academic Press, New York (1976).
12. Liu, B.Y.H., Szymanski, W.W. and Ahn, K.H., *J. Environ. Sci.* 28(3), 19-24 (1985).
13. Liu, B.Y.H., Szymanski, W.W. and Pui, D.Y.H., *ASHRAE Trans.* 92 (Part 1) (1986).
14. Szymanski, W.W. and Liu, B.Y.H., *Particle Characterization* 3, 1-7 (1986).
15. Remiarz, R.J., Agarwal, J.K., Quant, F.R. and Sem, G.J., "Real Time Aerodynamic Particle Size Analyzer," in *Aerosols in the Mining and Industrial Work Environment,* (V.A. Marple and B.Y.H. Liu, Eds.), 879-895 (1983).
16. Chen, B.T., Cheng, Y.S. and Yeh, H.C., *Aerosol Sci. Technol.* 4, 89-97 (1985).
17. Agarwal, J.K. and Sem, G.J., *J. Aerosol Sci.* 11, 343-357 (1980).
18. Sinclair, D. and Hoopes, G.S., *Am. Ind. Hyg. Assoc. J.* 36, 39-42 (1975).
19. Yeh, H.C., Cheng, Y.S. and Orman, M.M., *J. Colloid and Interface Sci.* 86, 12-16 (1982).

20. Marple, V.A. and Willeke, K., "Inertial-Impact," in *Aerosol Measurement,* (Lundgren et al., Ed.), 90-107, University Press of Florida, Gainsville, Florida (1979).
21. Hering, S.V., Flagan, R.C. and Friedlander, S.K., *Environ. Sci. Technol.* 12, 667-673 (1978).
22. Kuhlmey, G.A., Liu, B.Y.H. and Marple, V.A., *Am. Ind. Hygiene Assoc. J.* 42, 790-795 (1981).
23. Marple, V.A. and Liu, B.Y.H., *Environ. Sci. Technol.* 8, 648-654 (1974).
24. Blackford, D.B., "The Assessment of a New Range of Large Monosized Polystyrene Spheres and Their Potential as Calibration Material for Particle Size Analysis Equipment," to be published in *Aerosol Sci. Technol.* (1986).
25. Berglund, R.N. and Liu, B.Y.H., *Environ. Sci. Technol.* 7, 147-153 (1973).
26. Liu, B.Y.H. and Pui, D.Y.H., *J. Colloid Interface Sci.* 47, 155-171 (1974).
27. Liu, B.Y.H., Rubow, K.L. and Pui, D.Y.H., *Proc. 31th Annual Technical Meeting of the Institute of Environmental Sciences,* 25-28 (1985).

BIBLIOGRAPHY

1. Hinds, W.C., *Aerosol Technology,* Wiley, New York (1982).
2. Liu, B.Y.H., *Fine Particles: Aerosol Generation, Measurement, Sampling and Analysis,* Academic Press, New York (1976).
3. Liu, B.Y.H., Pui, D.Y.H. and Fissan, H., *Aerosols: Science, Technology and Industrial Applications,* Elsevior, New York (1984).
4. Lundgren, D.A. et al., *Aerosol Measurement,* University Press of Florida, Gainesville, Florida (1979).
5. Marple, V.A. and Liu, B.Y.H., *Aerosols in the Mining and Industrial Work Environment,* Ann Arbor Sci. Publishers, Ann Arbor, Michigan (1983).

3

Clean Room Garments and Fabrics

Bennie W. Goodwin

INTRODUCTION

A clean room is a work area which incorporates high standards of environmental control—humidity, temperature and air cleanliness—to minimize all forms of particulate matter and contamination. The heart of the clean room is the HEPA filter system. These filter systems are designed to provide air quality equivalent to the work being performed. A properly designed and equipped clean room will provide the appropriate environment for the class of work intended—from Class 1 or Class 10 manufacturing facilities, to Class 100,000 assembly areas.

However, introduce people into the clean room, and the contamination level rises noticeably. As much as 80% to 95% of the particulate contamination may come from people who work in clean rooms. A good HEPA filter system will maintain an acceptable level of cleanliness, providing the workers are properly attired. As important as the proper design and construction of the clean room, is the selection, use and care of clear room garments and accessory items.

The primary function of clean room garments is to assist in controlling particulate contamination in the clean room. With this in mind, clean room apparel must be treated as a *personnel filter.* The purpose of this *personnel filter system* is to control particulate contaminants from the employee's body and personal clothing. Too often clean room employees are allowed to become sloppy in the way they wear their clean room garments. When this happens, even the best HEPA filter system cannot compensate for this lack of control.

Ergonomics, how employees view themselves in their work place, is often cited as the reason for using clean room garments which may compromise good contamination control procedures. Concerns such as these are voiced by clean room employees:

"The garments just don't fit right."

"These coveralls are big enough for two of us."

"This 'bunny suit' is too hot."

"You can't tell the girls from the boys."

"Don't they make comfortable uniforms?"

"Why do I have to wear this 'plastic' stuff?"

Clean rooms are environmentally controlled work areas designed for the specific purpose of providing a *special clean place* in which to perform certain tasks. Specially designed equipment is used in performing these tasks, including employee work apparel. Clean room garments are *personnel filters*—not fashion items.

Clean room garments must be designed to cover the worker's body and street clothing to keep "human dust" out of the clean room space. Special garment construction techniques are required to allow clean room apparel to function properly. Limited linting fabrics, manufactured specifically for use in clean rooms, must be selected for the application required.

As important as these factors are, these apparel items must be cleaned in a clean room processing facility, and worn as intended by clean room personnel to be effectively utilized in the *contamination control system*. These fundamental requirements for clean room garments will be reviewed and discussed in the following sections.

CLEAN ROOM GARMENT DESIGN/STYLES

Clean room garments play a critical role in controlling particulate contamination in the clean room. Some garments may be inherently "unclean" by virtue of poor design. In selecting proper clean room apparel, careful attention must be given to styling features which may present contamination problems.

Clean room apparel is designed to provide the best possible "personnel filter system" for the environment in which it is used. The design of clean room garments should be simple, with no pockets, pleats, belts or other construction features which may become lint traps.

Basic clean room garments consist of head, body, hand and foot covers, in a size range that provides proper fit for all employees. These apparel items are all critical requirements for clean room attire.

Many design compromises are used in an attempt to please clean room employees. Clean room garments which are full-cut provide comfort for the wearer while maintaining their effectiveness as a "personnel filter." The contamination control of clean room apparel may be compromised by using such styling features as elastic waists, stockinette or knitted cuffs, and pencil holders or badge tabs. These features may make clean room garments more acceptable to employees. However, if quality levels are to be maintained, clean room employees must abide by the rules that govern good design practices for clean room garments.

Head Covers

When considering personnel contamination control for the clean room, head covers are of major importance. One of the "dirtiest" parts of the human body is the head and the openings on it. Hats or caps need to fit snugly around the head, covering all hair to contain loose hairs and skin flakes.

Hoods provide total coverage of the head and neck, and have a full shoulder-drape (long bib) to fit inside the coverall or frock. Hoods always are worn with the shoulder-drape under the coverall or frock. This allows particulates from the head and neck to fall into, and be contained within, the garment.

Hoods may be open-face (Figure 1a), or totally cover the face with the exception of the eyes (Figure 1b). Face masks should be used with open-face hoods and other head coverings as required to maintain established contamination control levels. All facial hair is to be covered.

Employees often complain about not being able to hear well when wearing a hood because the ears are covered. Some hood designs compensate for this by using a more open material across the ears (Figure 1c). "Ear vents" are not recommended, since they provide an escape route for particulate matter above the clean work space, and may contribute to contamination.

Other types of head covers are snoods (Figure 1d), caps (Figure 1e) and hats (Figure 1f). These coverings are designed to cover only the employee's hair. They do not cover the neck, ears or face, and should be used only in work areas where minimum particulate control is required.

In an effort to provide more comfortable head covers, and please employees, some clean room operators use nonwoven cellulosic disposable snoods. These may be used as the only head cover, or as a liner for hoods. Cellulosic head covers should not be worn in electronics clean rooms where particulate control is the primary concern. The cellulosic fibers may shed from the snood, get into the environment and contaminate the clean work place.

Body Covers

Clean room garments should provide the degree of body coverage determined by the needs of the user. Adjustable collars and cuffs *(sleeves and pant legs)* provide a snug fit, and reduce the flow of contaminants through these openings.

Coveralls. Coveralls (Figure 2a) have a full length zipper from the crotch to the base of the military collar, with a covered fly. Grippers (snaps) are recommended for the final fastening of the collar. Coverall pant legs should meet and slightly overlap the employee's shoes, and fit inside the clean room bootie.

The standup military collar (Figure 2b) is adjustable to fit snugly around the employee's neck. This reduces the flow of particulates through the neck opening caused by the bellows effect as air moves inside the coverall. The use of "V-neck" and other collarless design features compromises the contamination control function of coveralls (Figure 2c). These loose fitting neck openings allow contaminants to flow up through the neck, and be expelled through the face openings in the head cover into the clean work space.

One of the major complaints of workers who wear coveralls is the bulkiness. Extra space inside the coverall provides a corridor for air circulation, and adds comfort for the employee. Elastic waists (Figure 2d) and inside tunnel belts are

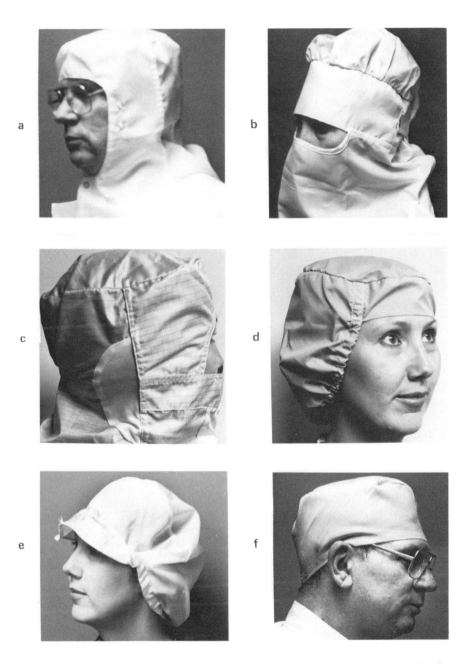

Figure 1: Head covers. (a) Open face hood. (b) Full-coverage hood. (c) Hood with ear-vent. (d) Snood. (e) Woman's cap. (f) Man's hat.

sometimes used to provide a more stylish garment. These features may create a comfort problem by reducing the air flow inside the garment.

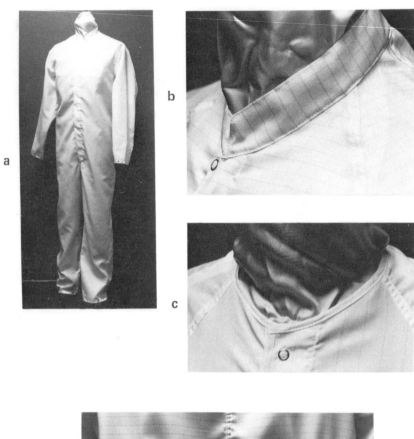

Figure 2: Coverall. (a) Complete view. (b) Standup military collar. (c) V-neck collar. (d) Elastic waist.

Frocks. Frocks should be of simple design with adjustable standup collar and cuffs. The recommended closure is a full length zipper fly front to the base of the collar (Figure 3a). Frocks with gripper (snap) front closure may be used in work areas where minimum particulate control is necessary (Figure 3b). The openings between grippers allow easy escape of particles from inside the frock.

Sleeves. Sleeves on coveralls and frocks should be extra long to ensure coverage of the wrists when employees extend their arms. This feature is necessary to prevent skin exposure between gloves and the end of the garment sleeves.

The preferred method for adjusting sleeve hems is grippers (Figure 4a). Elastic in sleeve hems is not recommended; it may lose the ability to expand and retract after repeated processing cycles. Stockinette (Figure 4b) or knitted cuffs are not recommended; they may become lint traps. Even the best processing facilities cannot remove all of the particles which may become embedded in the open pores of stockinette and knitted fabrics.

Design features of body covers such as elastic or belted waists, elastic, stockinette or knitted cuffs, and "V-necks" compromise the cleanliness of the clean room.

Foot Covers

Shoe and boot coverings should utilize the same design characteristics as other clean room apparel. Soles should be made of a slip-resistant material that is durable, and will not contribute to particle contamination.

Shoe covers worn over employee's street shoes must be high enough to cover the entire shoe (Figure 5a).

Booties are recommended for use with coveralls. Booties must be high enough to enclose the coverall pant legs, and they should be secured to the coverall pant legs with snaps. Clean room booties may extend above the ankle (Figure 5b), or may be over-the-calf length (Figure 5c). Booties which extend over the calf generally utilize a zipper closure. Grippers are recommended for the top adjustment. Hook-and-loop fasteners should never be used in clean room environments.

Booties may utilize a polyester ribbon tie to secure the bootie across the instep. Many clean room apparel manufacturers now offer booties with a buckle/strap across the instep. This method of securing the bootie to the foot provides a more snug fit.

In order to provide support for the foot and better slip resistance, clean room shoes with knee-length tops are offered by some clean room garment manufacturers. These clean room shoes are generally available in a wide range of standard shoe sizes (Figure 5d).

The recommended foot cover for use with frocks is a shoe cover. Over-the-calf length booties sometimes are used with frocks in an effort to provide a cleaner environment. This is mostly cosmetic, since the flow of particulates from inside the frock is not controlled, and these particles fall directly into the clean work space.

Dedicated clean room shoes, instead of shoe covers or booties, are used by some clean room operators in vertical laminar flow rooms. These shoes must adhere to the same contamination control considerations as all other clean room apparel items. These may be "tennis" shoes or slip-on styles with limited linting uppers. Dedicated clean room shoes are never worn outside the clean area.

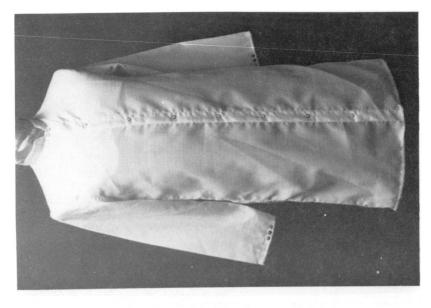

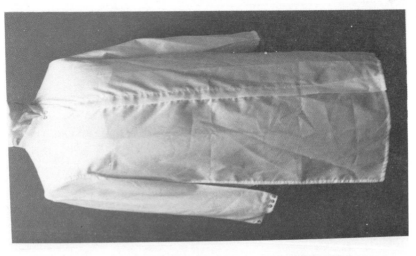

Figure 3: Frocks. (a) Zipper-fly closure. (b) Gripper (snap) closure.

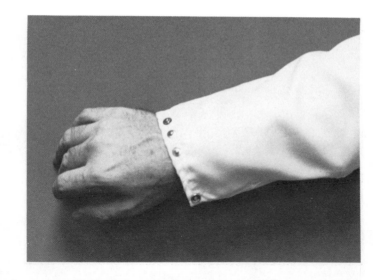

a

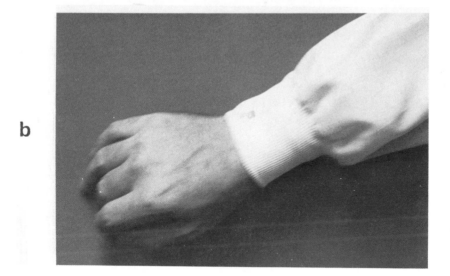

b

Figure 4: Coverall and frock sleeves. (a) Gripper adjustment. (b) Stockinette cuff.

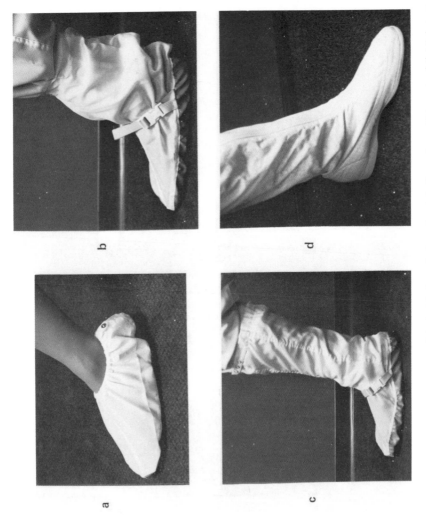

Figure 5: Foot covers. (a) Shoe cover. (b) Ankle-length bootie. (c) Over-the-calf bootie. (d) High-top clean room shoe.

Hand Covers

Gloves are selected based on the specific requirements of each clean room operation. Gloves should be long enough to enclose the coverall or frock sleeve opening. Gloves may be of reuseable or disposable materials. As with all clean room accessory items, these materials must not compromise the integrity of the clean room.

Accessory Items

All accessory items used for personnel or product protection must be compatible for use in the clean environment. Vinyl aprons, rubber gloves, face shields and other apparel items should be disposable, or withstand repeated processings for use in the clean room. These items should adhere to the same design and construction standards that are recommended for regular clean room apparel.

CLEAN ROOM GARMENT CONSTRUCTION

All garments and accessory items for use in controlled clean room environments should be constructed in such a way that will contribute minimum contamination to the work place. Fabrics and findings (thread, zippers, labels, tapes, ties, etc.) must be limited-linting. These garment components are made from materials which will not contribute to the contamination levels of clean rooms.

Seams

Sewing thread used for clean room garments should be 100% synthetic continuous multifilament yarn. Thread should be of the same synthetic material as the base fabric in the garment or apparel item. The multiple yarns in this thread will spread out to fill needle holes in the fabric. Monofilament thread should not be used unless seams are sealed on the inside of the garment.

Woven continuous multifilament synthetic fabrics fray easily when cut (Figure 6a). These frayed edges can break off and become a source of particulates. To reduce the likelihood of fraying, cut edges of continuous multifilament synthetic fabrics should be overedged (preserged) to encapsulate these loose fibers (Figure 6b).

Major joining seams on garments for use in Class 10,000 or cleaner environments are double needle barrier seams (Figure 6c). This encapsulates the cut edges of the fabric, provides a strong seam and reduces the pass-through of particles.

Hems on sleeves and pant legs should be turned under and double needle stitched. The bottom hem of frocks may be single needle stitched.

Nonwoven and coated clean room fabrics require the same garment construction considerations as woven fabrics. Because of the barrier characteristics of these fabrics, major joining seams should be sealed on the inside to cover needle holes. Bonded seaming methods may be used with these fabrics. If so, the bonded seams should be no more permeable than the base fabric.

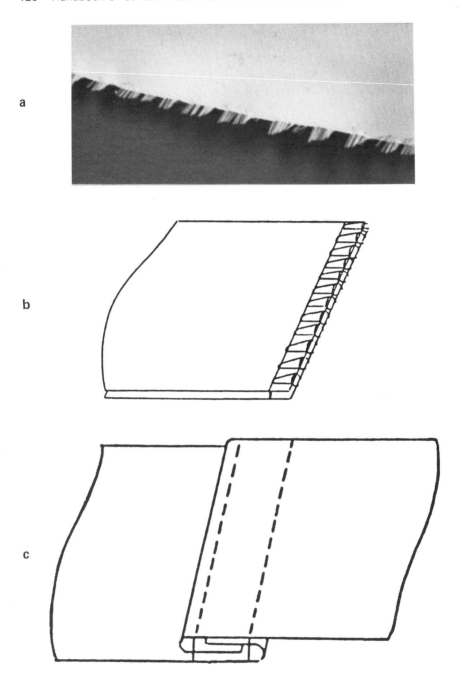

Figure 6: Seams used in clean room garment construction. (a) Frayed edges. (b) Overedged seam. (c) Double needle seam.

CLEAN ROOM FABRICS

Spun Fabrics

Regular work apparel typically is made with cotton or polyester/cotton blended fabrics. These fabrics are made with "spun" yarns. These yarns consist of short fibers which are twisted together, forming the weaving yarns. These short fibers can abrade, break off, and create particulate matter. Because these yarns tend to lint and fuzz in use, they must not be used in clean room garments (Figure 7).

Figure 7: Typical "spun yarn" polyester/cotton fabric.

Synthetic Fabrics

Clean room garments must be manufactured from fabrics that contribute the minimum possible particulate contamination. There are a number of synthetic fabrics used for clean room garments. However, synthetic fabrics for use in clean rooms must be constructed with continuous multifilament yarns, because of their limited-linting characteristics (Figure 8).

Figure 8: Typical continuous multifilament fabric.

Woven Fabrics

Continuous multifilament synthetic yarn woven fabrics are preferred for clean room use as opposed to continuous monofilament synthetic fabrics. The weaving process creates pores in the fabric. These are the openings formed by the vertical (warp) and horizontal (fill) weaving yarns as they are passed over and under each other. The multiple filaments in multifilament yarns tend to spread out and fill these openings. This provides a good filter medium, and will create a better barrier to particle pass-through from inside the garment (refer to Figure 8).

Single filament yarns can create larger pore openings, and will allow more particulates to pass through the fabric. Additionally, continuous monofilament yarn fabrics are more rigid, have a harsher hand (feel), and will be uncomfortable for the wearer.

Clean room fabrics generally are woven from nontextured multifilament synthetic yarns in either a taffeta, twill or herringbone twill weave (Figure 9). Texturized filament yarns usually provide a softer hand and tend to feel more comfortable to the wearer. However, these yarns can be abraded more easily than nontexturized yarns, and may become a potential source of lint contamination. Because of this possibility, texturized filament yarns are not recommended for use in clean room garments.

Synthetic fabrics in the twill or herringbone twill weave generally provide a tighter barrier than taffeta weaves. For this reason, coveralls made from twill weaves are recommended over taffeta fabrics. Taffeta fabrics often are lighter weight, and are the primary fabrics for use in head covers, foot covers and frocks.

Fabrics made with continuous filament synthetic yarns offer limited linting characteristics. This condition lasts until one of the filaments in the yarn is broken through abrasion, cutting, snagging, wear, flexing or chemical attack. When filaments break, they may become a source of particulate contamination. *These are not lint-free fabrics.* The terms "nonlinting" or "lint-free" should not be used with clean room apparel. Continuous monitoring of clean room apparel is necessary to prevent the use of garments which begin linting.

Polyester is the preferred synthetic woven fabric for clean room apparel. Polyester fibers are strong and nonabsorbent. These fabrics resist wrinkles to a greater degree than nylon. They drape well and recover from deformation easily. Polyester is resistant to degradation from most chemicals, and is more opaque than nylon. Polyesters wash well, dry quickly and offer durability in wear.

Nonwoven Fabrics

Nonwoven fabrics, such as spun bonded polyolefins, generally provide a better barrier to particle pass-through than woven synthetic fabrics. However, these fabrics tend to abrade more easily than woven fabrics. Flexing the fabric may cause flaking of the surface material. This abrasion and flaking results in lint generation. Because of this propensity for linting, nonwoven fabrics are not recommended for use in electronics clean room facilities where particulate control is of major importance.

Figure 9: Woven synthetic clean room fabrics. (a) Taffeta 8X. (b) Taffeta 140X. (c) Herringbone twill 8X. (d) Herringbone twill 140X.

d

c

Figure 9: (continued)

Coated (Laminated) Fabrics

Coated fabrics provide the best barrier to particle pass-through. These fabrics may utilize a film which is laminated to the base material. This may be in the form of a "sandwich" with the film laminated between two base fabrics. Some coated fabrics have the film laminated to the top of a single base material. This leaves the film exposed, and it may rupture in normal use. When this occurs, the integrity of the garment is compromised. These coated (laminated) fabrics are still in the developmental stage.

Chemical Resistance

Most woven, nonwoven and coated fabrics used in clean room garments offer good resistance to chemical degradation. If these garments are likely to be subjected to chemical exposure, they should be evaluated carefully to determine the effects of those chemicals on the fabric. Apparel selected for chemical compatibility may not provide protection for employees. Personal protection from chemicals may require the use of impermeable materials. It is important that these materials are evaluated to determine compatibility for use in the clean room so the integrity of the clean environment is not compromised.

Electrostatic Charges

Synthetic fabrics are excellent electrostatic generators. Static electricity generated by clean room garments has long been recognized as an additional "contamination source." This is not only because of the electrostatic discharge (ESD) problem. Static charges on clean room garments act as a magnet, attract particulate contamination, and carry it into the clean work place.

Employees wearing normal polyester/cotton blended work clothes may generate electrostatic charges of 5,000 volts or higher on their apparel. Cover these polyester/cotton garments with a 100% polyester clean room frock or coverall, and employees become walking "electrostatic generators." It is not uncommon to measure static charges on regular clean room apparel in excess of 30,000 volts.

ESD-sensitive electronic components brought into contact with charged clothing may be destroyed or severely damaged. Some highly sensitive electronic devices do not have to contact a charged garment. The ever-present electrostatic field may destroy or damage some of these ESD-sensitive parts.

The proper use of personnel grounding systems effectively controls electrostatic charges from the employee's body. However, the use of wrist or ankle straps has little or no effect on static charges emanating from employees' garments. Only by using inherently, permanently static dissipative apparel in conjunction with grounding techniques can control of electrostatic discharge be best achieved.

Prior to 1982, inherently static dissipative garments for clean room use were not available. The typical method of controlling electrostatic charges on 100% polyester apparel was to use topical antistats.

Topical antistats are chemical agents which, when applied to surfaces of insulative materials, will reduce their ability to generate and hold static charges. Most antistats are materials which, when combined with the moisture in the air, wet the surface of the material on which they are deposited. This "wetness" provides a conductive path for triboelectric charges to be distributed over the garment. These antistats are classified as hygroscopic. Their effectiveness is reduced in low relative humidity levels since they depend on moisture to conduct electrostatic charges. Topical antistats usually are removed when the garments are processed, and must be reapplied at each processing cycle. The duration of their effectiveness also is an aspect that must be considered.

Chemical antistats used on clothing to reduce electrostatic charges may present additional problems for use in clean rooms. Some older antistats built up residues which could flake off and become a source of contamination. Some of the newer chemical antistats may vaporize when the garments are exposed to the atmosphere, which may cause corrosion in small-scale microelectric devices.

Static Dissipative Apparel

Inherently static dissipative apparel for clean rooms is now available. Fabric mills and garment manufacturers working together developed a better method of controlling electrostatic charges on clean room garments.

Following considerable developmental work, continuous multifilament polyester fabrics with continuous filament carbonized conductive yarns were introduced. This provides all-filament synthetic yarn fabrics which are limited-linting and inherently static dissipative. Since no special finish or topical antistat is required to provide the static control properties, nothing can wash out during the laundry or dry cleaning process. The ESD-control characteristics are retained for the life of the garment. However, some of these static dissipative fabrics may lose their static control features following several processings of the garments. Clean room operators are cautioned to carefully evaluate the static control characteristics of the garments selected for their use.

Clean room fabrics with inherent static control are available in two conductor configurations—vertical conductor, and the conductor in a grid pattern (Figure 10).

The best static control is available with fabrics which utilize the grid pattern. This provides a conductive path for triboelectric charges to be distributed over the garment in all directions. Static control characteristics of static dissipative fabrics can be evaluated by comparing three standard tests: surface resistivity, static decay and voltage remaining after two seconds static decay.

Surface resistivity is a measurement of the fabric's ability to conduct electrostatic charges over the garment (AATCC Test Method 76, @ 50% RH).

Static decay is a measurement of how quickly an induced static charge will drain from the fabric (Method 4046 of Test Method Standard 101B, @ 50% RH).

Voltage after two seconds measures the electrostatic charge that remains on the material after a static decay time of two seconds.

The range of static control measurements for grid and vertical conductor fabrics is shown in Table 1.

a

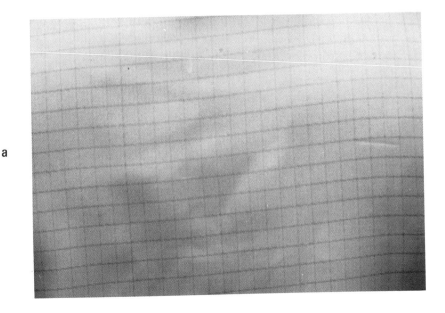

b

Figure 10: Antistatic clean room fabrics. (a) Grid conductor. (b) Vertical conductor.

Table 1: Static Control Characteristics Static Dissipative Clean Room Fabrics

Characteristic*	Grid	Vertical
Surface Resistivity:		
Fabric unlaundered	10^5-10^{10} ohms	10^8-10^{13} ohms
50 Industrial washes	10^5-10^9 ohms	10^{10}-10^{13} ohms
Static Decay:		
Fabric unlaundered	<0.1-<4 sec	<0.1->100 sec
50 Industrial washes	<0.1-<1 sec	<0.1->500 sec
Voltage @ 2 Seconds:		
Fabric unlaundered	0-<700 volts	0->400 volts
50 Industrial washes	0-<500 volts	50->800 volts

*These are the range of measurements derived from tests performed on fabrics most readily available to U.S. clean room operators.

DOD-STD-1686, and accompanying DOD-HBK-263, recommends the use of static dissipative material when available. Static dissipative materials are defined by DOD as those with a surface resistivity not less than 10^5 ohms and not greater than 10^9 ohms. Some fabrics with the conductive grid meet the DOD-STD-1686 recommendation for static dissipative materials. Some static dissipative clean room fabrics with the conductor in a grid configuration are able to control electrostatic charge retention at levels below 100 volts. None of the vertical conductor fabrics have met this recommended static control level of performance after they have been processed several cycles.

Some users of static dissipative clean room apparel have reported a 20% to 40% reduction in particles in the work place, as compared with regular polyester clean room garments. Since static dissipative garments do not generate and retain high static charges, they do not act as a magnet to attract particles and bring them into the clean work area. Static dissipative garments and accessory apparel can help maintain a "cleaner" clean room.

CLEAN ROOM GARMENT LIFE

Clean room garments, even those of the best design and manufacture, have limitations on their useful life cycle. These garments are subject to wear from abrasion in the work area, and from laundering or dry cleaning. The pores in the weave of woven fabrics may open up after numerous uses, and the garment may begin to lint and lose its barrier properties. Continuous multifilament synthetic yarns, when broken as the result of abrasion, may break into short fibers through flexing action. These fibers will result in particle contamination.

Since woven polyester fabrics tend to show little visual evidence of wear, these garments must be spot checked microscopically at regular intervals. Garments showing evidence of wear should be replaced immediately and not be used in the clean room.

Nonwoven fabrics tend to shed particles more readily than woven materials. These garments should be checked after each use to determine their lint-control characteristics. Processing of most nonwoven materials is likely to increase the level of particulation.

As stated previously, films used on coated (laminated) fabrics may rupture in use. These garments should be evaluated daily to ensure that the integrity of the material has not been compromised.

GUIDELINES FOR CLEAN ROOM GARMENTS

Different types of clean room operations require specific kinds of clean room apparel. It is recommended that all clean room facilities with controlled environments use garments made from limited-linting fabrics. This will help maintain a minimum amount of contamination in the controlled work space. Garments made from cotton or polyester/cotton spun fabrics are never recommended for use in Class 100,000 or cleaner facilities.

Recommended Clean Room Garment Usage

Clean room apparel is designed to encapsulate employees and their personal clothing. These garments utilize styling features which provide varying levels of contamination control. Garment types listed in Table 2 (see also Table 1, Chapter 4) are recommended for best contamination control.

Table 2: Clean Room Garment Classes Recommended Garment Usage

Garment Type	10	100	1,000	10,000	100,000
Head covers:					
Hood with complete facial enclosure	X				
Hood with permanent or detachable facial cover		X			
Hood, beard cover as required			X	X	X
Hat or cap with full hair cover					X
Body covers:					
Coverall	X	X	X	X	X
Zippered frock				X	X
Appropriate foot and hand covers	X	X	X	X	X

These clean room garment styles have proven effective in controlling contamination attributed to personnel. These recommendations should be considered to be minimum requirements. If the clean room is a horizontal laminar flow room, the full "bunny suit" (hood, coverall and booties) is recommended for the first two or three work stations downstream from the filter wall. This will help ensure a cleaner environment for more of the work space.

Each clean room manager must select apparel appropriate for the work done at each facility. Many clean rooms use frocks with gripper or snap front closures. The openings between the grippers become doorways for particles to escape from inside the frock. Careful consideration should be given to the clean room requirements when selecting frocks with other than zipper front closures.

Clean Room Garment Rules and Regulations

Clean rooms are restricted areas, and access to them must be limited to properly attired personnel. Protective clothing is to be used at all times for individuals entering the clean room, including repair and maintenance personnel. Visitors to the clean room must observe all the rules governing clean room employees, including *all* management and supervisory personnel.

Clean room garments always must be limited to the clean room. They are not worn outside the clean room, or in the contaminated section of the change room. Clean room garments must be stored in a manner that will not compromise their contamination control characteristics. Closed "clean cabinets" are recommended for garment storage in the change room.

Supervisor disciplines determine the quality of the clean room product. Clean room garments must be completely closed with the fasteners provided. Employees should inspect all clean room apparel items for signs of wear, abrasion or linting before dressing.

In some critical work areas, it is recommended that limited linting garments be worn under clean room apparel. In all cases, employees should wear hard-finish street clothes under clean room apparel. These garments will not shed fibrous materials as easily as loosely woven or knitted fabrics.

Employee education programs should be on-going, to reinforce the contamination control requirements for clean rooms. These programs should include how to wear clean room apparel, and proper garment donning procedures.

Recommended Clean Room Garment Changes

As important as selecting the proper clean room apparel is recognizing the need to change these garments on a regular basis. Too often clean room garments are worn beyond their capability of remaining clean. User experience has shown clean room garments accumulate particles daily. A garment with fewer than 1,000 particles of five microns or larger per square foot on Monday, may have 100,000 or more particles per square foot by the end of the week.

For clean room garments to perform as they are intended, to reduce particulate contamination in the work place, they must be changed on a regular basis. For best contamination control, recommended garment changes are shown in Table 3.

The primary reason given for not changing clean room garments on a more regular basis is cost. Because of the probability of contaminating the clean environment with "dirty" garments, this may be false economizing. The major source of contamination in clean room is *people.* If the *people filter system* does not function as designed, it will not effectively contribute to the contamination control system.

Table 3: Recommended Clean Room Garment Change Cycles

Clean Room Class	Recommended Garment Changes
10	With each entry into clean room
100	Daily
1,000	Daily
10,000	Daily
100,000	Every other day
Sterile	With each entry into clean room

Clean Room Garment Processing

Clean room environments require *clean* garments. Clean room garments cannot be laundered or dry cleaned (processed) in equipment that is used for regular work apparel. The possibility of redeposition of lint, soil and other cross contamination makes this type of processing totally unacceptable for clean room garments.

In order for clean room garments to provide the contamination control levels they are designed for, these garments and accessory items must be processed in a facility that is compatible with the clean room environment in which they are to be used. In other words, clean room garments should be processed in an environment that is as clean as the room in which they are going to be worn. This is true for on-premise laundries at the clean room facility as well as rental clean room processing facilities.

The clean room garment processor must incorporate the same environmental controls as the clean room manufacturing facility. Personnel employed in these processing facilities should be educated as to what contamination on the garments can do to the clean room facility. These employees must wear appropriate clean room apparel, and follow the same employee disciplines as manufacturing personnel. It is important that clean room garment processors recognize their limitations, and process only those garments which are within the processing facility's capabilities.

Clean room garment processors may have water wash or dry cleaning equipment. Whichever method of processing soiled clean room garments is used, all clean room garments should be laundered at least once before being processed and placed into service. This initial laundering is necessary to remove any water soluble residues that may result from the fabric finishing processes. Such residues generally are not solvent soluble, and are likely to remain on garments that are dry cleaned only.

If a rental clean room processing contractor is providing these garments, the cleanliness level of garments must be established jointly by the user and contractor. The processing contractor should be responsible for performing all necessary tests to insure that processing methods and materials are sufficient to clean garments to the required level on a consistent basis. Customers of clean room garment processing contractors are responsible for determining compliance with the requirements as agreed upon between them and the processor. The customer should perform the same tests as conducted by the processing contractor.

Each batch of garments should be sampled by random selection of two percent (2%) of the batch, with a minimum of four garments. Larger numbers of samples should be selected if experience has resulted in a low level of confidence in product acceptability. The test method described in Air Force Technical Order 00-25-203 is the most commonly used test to determine clean room garment cleanliness. This is ASTM F-51-68, "Sizing and Counting Particulates in and on Clean Room Garments." Based on this test procedure, clean room garment cleanliness levels are expressed as shown in Table 4.

Table 4: Clean Room Garment Cleanliness Levels*

CLASS	CONTAMINATION LEVEL/SQUARE FOOT OF FABRIC	
A	less than 1000 maximum	5μm and greater particle length 10 fibers
B	less than 5000 maximum	5μm and greater particle length 25 fibers
C	less than 10,000 maximum	5μm and greater particle length 50 fibers
D	less than 15,000 maximum	5μm and greater particle length 125 fibers
E	less than 25,000 maximum	5μm and greater particle length 175 fibers

Note: Obviously broken fibers and lint-bearing seams on outer surfaces of garments, wiping cloths, caps, hoods, booties, and fabrics shall be cause for rework and rejection, or both. Decontamination processed clean room fabrics shall be free of persistent objectionable odors.

* F 51-68 (Reapproved 1984) Page 5, Section X2.   Proposed Decontamination Processed Garment Classification American Society for Testing Materials, Philadelphia, PA.

All testing of garments should be performed by trained and experienced personnel wearing appropriate clean room garments. These tests should be conducted in a controlled environment equal to, or better than, that in which the garments are processed. Testing should be done after the garments are processed and packaged.

Each garment unit should be packaged individually. The garment unit may be a single item (such as a coverall, a frock, or a hood), or a complete set of garments for an individual (hood, coverall and booties). Packaging materials should not compromise the garment. All garments ready for shipment should be protected by being overbagged or placed in a suitable container to minimize the possibility of tearing, puncturing or otherwise damaging the bags. Where the bag has been torn, punctured, or otherwise damaged, garments should not be delivered to the user.

Repairs to clean room garments are to be negotiated between the user and processing contractor. All repairs are made outside the clean room. They should be made in a manner that will eliminate frayed edges, utilize materials compatible with limited-linting needs, and all repairs completed prior to final processing.

Some processes in clean rooms may result in staining of garments. Many stains will not be totally removed by processing the garments. However, stains can be "clean" and not contaminate the clean work space. Allowable stains on clean room garments should be negotiated between the user and processing contractor.

CLEAN ROOM GARMENT STERILIZATION

Some clean room facilities require sterilization of garments to remove microbial contamination. Three types of sterilization are commonly used.

Autoclaving

This is a process where garments are subjected to very high temperature steam. This is the least desirable sterilization process for clean room garments. Some fabrics used in clean room garments are susceptible to heat damage. Garment life may be reduced between one-third to one-half, and permanent creasing degrades the garment finish. Continuous multifilament woven polyester garments will shrink approximately one full size after five to seven autoclaving cycles.

Ethylene Oxide (ETO)

In this sterilization process, garments in hermetically sealed bags are placed in a sealed chamber and subjected to extended exposure to ETO. After sterilization, the garments are placed in quarantine until laboratory tests performed on samples have verified the complete destruction of all bacteria. This method has no adverse effect on clean room garments. OSHA and EPA have placed limitations on the use of ETO sterilization.

Gamma Radiation

In this sterilization process, the hermetically packaged garments are subjected to high intensity gamma radiation. This method has no known effect on 100% polyester fabrics. However, radiation causes nylon to virtually disintegrate. Garments to be sterilized by gamma radiation must not have any nylon components or accessories. Molded plastic zippers may not be compatible with gamma sterilization.

CONCLUSION

Just as important as the proper design, construction and processing of clean room garments is how the garments are viewed by employees. Properly cleaned garments, worn improperly by employees, can reduce their contamination control effectiveness. Employee discipline is perhaps one of the most important considerations for particulate contamination control in clean rooms.

Fabric mills have developed special materials for use in these clean work areas, including inherently static dissipative fabrics to help control ESD. Apparel manufacturers have developed special garment designs and construction techniques to provide clothing which is compatible with clean room contamination control requirements. Processes have been developed to effectively remove contamination from soiled garments.

Clean room garments may conflict with the basic precepts of ergonomics. However, clean room employees must be made aware of the special clothing required for this work before they accept these jobs. Then there is no excuse for not properly wearing the appropriate apparel required to maintain *contamination control in microelectronics* manufacturing.

4

Guidelines for Clean Room Management and Discipline

Anne Marie Dixon

INTRODUCTION

A manager, according to Webster's dictionary, can be defined as (1) "one who manages, or (2) one charged with the management or direction of an institution, business, or the like." This simplistic statement leaves a wide gap when applied to the demanding management of a manufacturing facility within the area of clean room. A common misconception prevails that implies only the management of "clean" air is necessary for success in today's high technology fields. In reality, the clean air is only a small fraction of the entire critical environment required to produce a quality product in adequate numbers to provide profit for a company. Today's clean room manager must be knowledgeable of, and control, a multiplicity of facets that all relate directly to the production of a quality product. In addition to the physical plant itself, the process equipment, personnel performance, repair and maintenance activities, technology evolution, quality costs, facility maintenance, and administration tasks all demand constant management. As has been written elsewhere, the entire manufacturing circle is a fragile chain that is totally interdependent upon each facet of the environment that touches the product from conception to final delivery. To further complicate this demanding management task, the critical defect causing element is not visible to the unaided human eye. This chapter is designed to provide the clean room manager with a program that allows for the old adage "divide and conquer" to be applied to quality production. In essence, by controlling those factors that can be readily managed, the clean room manager of today can systematically eliminate as many possible contamination sources as practical, and learn to circumvent those that are outside feasible human control. In addition, each corrective measure or activity must be evaluated in regard to the return on the investment concept and good old-fashioned common sense.

There is little return when we concern ourselves with sodium-free pens while we allow regular paper, cardboard boxes, and poor work disciplines to continue.

This chapter is intended as a step by step strategy for systemically controlling contamination sources on a rational basis. Combating the invisible enemy requires perseverance, diligence and infinite patience; yet the need for contamination control is unquestioned (Figure 1). This chapter will provide the manager with the important tools necessary for clean room management. Each link in the quality chain will be discussed with options available for possible improvement. Development of a "battle-plan" will be left to the individual manager since many unique factors may dictate the implementation of the program for improved contamination control. The initial section will address the area of personnel performance.

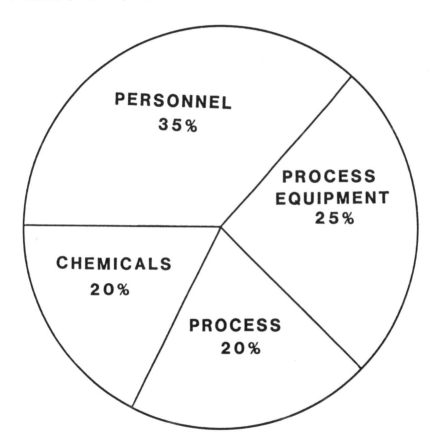

Figure 1: Causes of contamination.

CLEAN ROOM CRITERIA

Contamination is primarily the product of man and his activity; and, secondarily, the product of nature. Contamination control in a clean room manufacturing facility requires rigid discipline from the work force which includes *any person* that enters the area for whatever purpose.

Good Clean Room Work Practices

1. No eating, smoking, or gum chewing.

2. Garments specified for a given facility must be worn by *all* personnel entering the area.

3. Only approved clean room compatible paper is allowed into the area.

4. Only ball point pens are used for writing.

5. Cosmetics are not allowed. This includes rouge, lipstick, eye shadow, eyebrow pencil, mascara, eyeliner, false eyelashes, fingernail polish, hair spray, and the heavy use of aerosols, after-shave lotions, and perfumes.

6. The use of paper or fabric towels are forbidden. Wash rooms should be equipped with electrically powered filtered hand dryers. (HEPA only).

7. Gloves are not allowed to touch any item or surface that is not known to be thoroughly cleaned. Most importantly, they should never touch any part of the anatomy covered or uncovered.

8. Only approved gloves, correctly cleaned maintenance tools, or vacuum pick-up's, should touch or handle material or parts. Human skin oil is a form of contamination, and products must not come into contact with bare hands.

9. Solvent contact with bare skin should be avoided as most solvents will remove skin oils which leads to increased shedding of skin cells.

10. All personal items such as keys, cigarettes, watches, matches and lighters must be stored outside of the clean room.

11. Valuable items such as wallets and money may be taken into the clean room *provided* that they are not removed from the street clothing and are concealed by the clean room garment.

12. *No one* physically ill, especially with stomach and upper respiratory disorders, may enter the clean room.

13. Persons with severe sunburns are restricted from the clean room.

14. Workers with colds, coughing, or allergic nasal reactions should be temporarily assigned to tasks outside of the clean room.

It is important for the manager to convey to the work force the fact that a clean room is a restricted area; and, as such, is limited to authorized personnel properly gowned. The clean room must be thought of as an Intensive Care Unit, where vulnerable products are present. Just as in a hospital I.C.U., there is a list of restricted and prohibited actions. All personnel should be thoroughly familiar with the following restrictions.

Prohibited Actions in Clean Room

1. Rapid walking or any quick motions that create turbulence.

2. Wearing a torn or soiled clean room garment.

3. Wearing clean room garments outside of the clean room. Including return facility chases.

4. Storing clean room garments inside a personal locker.

5. Removal or otherwise wearing a clean room garment in a manner that allows for contamination to escape the garment. (Managers should address the safety first concept should a worker be in-inolved in an industrial accident that requires rapid removal of the garment.)

6. Improper removal or changing of gloves while in the clean room.

7. Combing, fluffing, or brushing the hair while still in the gowning area(s).

8. Use of external medications.

It is imperative for the manager to realize that the above rules are not to be presented in the context of punishment acts but inform the work force of the importance of these rules in light of the very vulnerable condition that exists for the product. A statement that places this suggestion into context is "Americans are over-managed, and underled." The enlightened manager creates a willingness on the part of the work force to be a team player with a valuable and critical contribution to make toward quality production. The need for the stringent rules and regulations should be presented in the framework of the product's vulnerability to any unseen contaminant.

PERSONNEL DISCIPLINES

Consideration should be given to the type of clothing worn under the clean room garment in relation to the particle barrier (filter) of the garment. Based upon the filtration capacity of the basic material, such articles as heavy woolen sweaters, or other shedding material may be prohibited as undergarments. Lightweight tops and pants made from dacron or cotton are acceptable. If at all possible, workers should shower prior to entering the area. The hair should be shampooed regularly with particular emphasis to dandruff control. If personal hygiene is beyond the scope of your responsibilities, a minimum protection from dandruff and hair fallout is a secondary hair covering. Hair should be washed after any haircut. Fingernails should be kept at a moderate length and care should be exercised to file any sharp edge to prohibit tearing of the glove.

Jewelry that is bulky or sharp on the edge cannot be worn beneath gloves or headcoverings.

Since the majority of today's clean manufacturing facilities contain product that is susceptible to damage from nonvisible contaminants (below 40 microns), all operational and support personnel must have a thorough understanding of how these invisible particles affect the product on a short term and a long term basis. Matter such as hair, cosmetics, dust, clothing, fibers, skin flakes, dandruff, metals, skin oils, and others directly contribute to decreased quality and low product yield. All personnel must be committed to excellence and practice defect prevention rather than defect detection at the end of the cycle. The work force must be made the key link in the entire production chain and motivated to be constantly alert to potential contamination from every source. Quality cannot and must not be the responsibility of one person. It is the total responsibility of each employee.

Training

In addition to these rules, regulations, and disciplines, we must institute modern methods of training our operators as well as all support personnel. This method can best be described as the "*adult*" method of training, whereby, one learns a concept, observes, performs, and finally management allows for feedback and additional retraining, if necessary.

In communion with modern training methods is a modern method of supervision. The elimination of the old "Carrot and Stick" philosophy. Another method that has been implemented in some organizations, is an attitude of indifference. This method is possibly more devastating to production and quality than any other. Sometimes operators and all support personnel of a clean room will fail to report any problem because of fear or because it appears that no one is interested or concerned. It is in the face of this type of continuing indifference that clean room personnel accept poor contamination control as norm and they accept rework and poor product quality.

Some New Approaches

As a modern supervisor, the establishment of a developmental review program for the clean room worker is essential. A focus or objective should be established for each worker with the goal of future improvement and performance. The method of this type of achievement should be a developmental plan for the future of each worker. All clean workers and support personnel must know what is expected of them, and equally as important, what they have achieved.

If possible, *eliminate* numerical goals for the clean room worker. If the production goal is not met, the morale of the clean room worker suffers, especially if the unreached goal was not the fault of the operator. If the goal is easily met, the good workers stagnate. The incentive to improve is removed, and product quality as well as contamination control usually suffer.

As you can see, training and supervision requires a comprehensive program that is well organized and supported by management and incorporates a follow-up system to monitor the employee job performance. Such a system will stimulate employee awareness of the overall clean room operation and eventually improve the efficiency of the entire manufacturing operation.

CLEAN ROOM GARMENTS

Even the most highly motivated and contamination control aware work force, being human beings, shed an incredible number of particles. With today's advances in the clean room design and the technology to provide ultraclean air flows well documented, the next most critical aspect of contamination control is the selection of a good clean room garment. All personnel that work in, support, maintain, or enter a clean room must be packaged to contain the contamination generated by our bodies, hair and clothing. The garment must be evaluated as a body filter to eliminate the passage of thousands of particles generated.

Classification Demands

The class of cleanliness demanded by the product should be the primary requisite for the garment. A suggested guideline for the user to select a garment filter follows and should be considered the bare minimum for good contamination control. Any additional filters such as masks, foot coverings, head coverings, etc., should offer even greater protection for the product or process. The filter effectiveness of the garment should be the initial parameter for selection. There is little justification for gowning workers in an expensive laminate style garment when the clean room is Class 100,000. Conversely, selecting a poor filter garment for Class 10 technology is counter-productive to quality. The manager must be totally objective when selecting personnel filtration garments. Comfort, style, and price should be subjugated to filtration efficiency.

Table 1: Guideline For Use of Clean Room Garments

Clean Room Class	Recommended Garment
100,000	Lab coat or frock Foot cover Head cover Gloves
10,000	Frock Shoe cover Head cover Gloves
1,000	Coverall Shoe cover Head cover Gloves
100	Coverall Hood (with or without mask) Boots Gloves
10	Coverall Boots Shoe cover or Clean Room shoes Head cover Hood - with built in face mask or full coverage hood Gloves Goggles

Once a compatible material filtration efficiency is determined, the manufacture of the garment should be considered. Special techniques are employed to ensure that the garment itself does not contribute to contamination. A double needle feld type seam should be employed to ensure that the raw or cut edge is covered within the seam. All openings such as hemmed areas should have the raw edge turned twice and single needle stitched to cover the edge. A thorough inspection should be conducted on all new garments to ensure the proper sewing technique. Periodic inspections are necessary to detect any tear or loosened seam.

A detailed and thorough discussion of clean room garments and fabrics and the clean room laundry is given in Chapter 3, *Clean Room Garments and Fabrics,* of this book. Thus they are not discussed in this chapter.

Gowning Procedures and Techniques

Once the proper garments have been selected for your facility, operational and support personnel must be educated in the correct methods for gowning. The method is predicated upon the class of cleanliness of the facility.

Class 100 to 100,000

1. Change shoes or cover street shoes with booties or clean room overshoes.

2. Put on a disposable hair cover to contain hair and aerosol hair spray products.

3. Put on selected head cover or hood.

4. Put on selected garment being careful not to allow the garment to touch or drag upon the floor.

5. If applicable, put on a face mask.

6. If applicable, put on safety goggles or glasses.

7. Put on selected clean room gloves.

For clean rooms that are designed for levels of cleanliness below Class 100, a more stringent gowning procedure is required since all contact with the outer garment must be avoided.

Class 1 to 10

1. Change street shoes or put on disposable booties upon entering change area.

2. Put on disposable hair net.

3. Next, put on a pair of inexpensive clean gowning gloves to prohibit the transfer of skin oils and flakes to the clean room garment while gowning.

4. Put on hood and face mask.

5. Put on the coverall or bunny suit, again being careful to not allow the garment to touch the floor.

6. Put on boots and insert the leg of the coverall into the boot.

7. Put on goggles or face shield.

8. Remove gowning gloves and put on the operational glove.

The above procedure ensures that the garments are not subjected to human debris while gowning. Use of the hair net is most important as recent studies have demonstrated the potential problems associated with hair sprays and styling products dislodged from the hair.

Garment Storage and Reuse

All garments received from a clean room laundry should be double bagged under Class 100 conditions. The outer layer of package should be removed in the gowning area and disposed of in the proper trash container. The inner layer or protective package should not be opened until the garment is to be worn. Disposable garments should be removed from the shipping carton outside of the gowning or change room. These garments are usually double bagged in bulk and packaged separately by the supplier. The same procedure applies to launderable garments. The outer package is removed and disposed of and the individual packages are stored unopened until used. The correct technique for opening the package requires slitting with a knife or cutting with scissors. The bag should not be ripped or torn apart as this introduces additional unnecessary contamination into the gowning area.

Secondary Usage. If garments are to be reworn, correct storage must be maintained to prohibit increased contamination from the gowning room area. A rule to follow is that garments should be hung a minimum of 3" apart under laminar flow clean air conditions. This ensures that the garments are protected from associated contamination present in the gowning area. Even under these ideal conditions, it is impossible for the garment to be returned to the initial level of cleanliness. It is, therefore, important that the surface of the garment be evaluated routinely to determine the frequency of the laundry cycle. This frequency should be dictated by the particle burden as determined by examination rather than subjective guess work or past practices.

CLEAN ROOM SUPPLIES

"Reduce the number of suppliers for the same item by eliminating those that do not qualify with statistical evidence of quality. End the practice of awarding business solely on the basis of price."[1] This is a cardinal rule of the W. Edwards Deming philosophy and is one that should well be applied to clean room supplies. All too frequently the role of supplier quality is overlooked or taken for granted. The addition of a label that states clean room supplies or clean room compatible does *not* ensure cleanliness. Clean room supplies require the most stringent quality control possible. A program for improved quality, defect prevention, and total organizational commitment to the implementation of TQC (Total Quality Control) should be the goal of all clean room suppliers. Price is not the critical factor in contamination control. Price is the final deciding factor once quality standards are met!

Wipers

Clean room wipers, sponges, and swabs are necessary items in the daily operation of a clean room. One purpose of a wiper is to soak up spills quickly and efficiently without contributing any additional contamination. Primarily, wipers, sponges, and swabs are used to clean work areas and equipment. Therefore, the characteristics of a good clean room wiper, sponge, or swab would be: (1) absorbancy both in rate and volume, and (2) cleanliness by not generating particles and metallic ionic contamination. Test data for absorbancy traits should be available from the supplier. Wet and dry particulate testing and ionic extraction testing are requirements for clean room compatibility. This data should be available from the supplier and should also be confirmed in an independent laboratory or in house. All the supplies should be properly double bagged to ensure cleanliness is maintained during shipping. The interior package should meet the clean room standard and not add contamination to the product. The packaging material should be tested for particulate contamination using a dry method.

In selecting a wiper, sponge, or swab, the initial step is the comprehensive evaluation of the job or function it is intended to perform. Rank of the characteristics required is next—particulate contamination, extractables, absorbancy and cost are considered. Once this is established, a testing program can be undertaken to ascertain which product or supplier will most effectively meet these requirements.

Gloves

Gloves are the first line of defense against skin oils, skin flakes, bacteria, and other contaminants. The sense of touch is important and thus the barrier must be thin and flexible. When chemical protection and safety considerations are important, many companies have selected the powder-free vinyl or latex glove. In addition to the powder-free glove, a "Class-100" glove was introduced for ultra-low particulate contamination. Chemical gloves are also available for clean room use. Organic and inorganic contamination derived from the gloves has become an issue of great concern. The chemical constituents of the glove are not the main concern; but, rather, the contaminants that are transferred from the gloves to the product or equipment by surface contact. Some test methods that may be considered for determining this compatibility problem are: ethyl alcohol extraction and analysis or a fingerprint sampling taken upon prewashed aluminum foil. In this manner, the user can gain the knowledge to make an objective determination of the best glove for a specific application.

Paper

Paper is a great source of contamination in today's clean room. It is critical that *all* types of paper (bond, file folders, calendars, notes, labels, etc.) be removed and replaced with nonshedding items. Equipment manuals and schematics *must* be produced on clean room paper if they are taken in or stored in the clean room. Computer form, adding machine tapes, and clip boards are available on clean room compatible material and paper. Forms for administrative activities such as payroll, vacation and memorandums should be on clean room paper or left outside the clean room. Blueprints create a real challenge to contamination control. Ideally, they should be left outside the clean area. A blueprint viewing window could be installed with the blueprint mounted on the outside pane and viewable from within the room. If this is not practical, and the blueprint must be used in the clean room, it can be sandwiched between sheets of mylar film and sealed with transparent tape, cleaned on the surface, and then brought into the clean room.

Furniture

Furniture brought into or used in the clean room must meet very strict specifications. Benches and tables must be covered with a formica type laminate on all exposed surfaces, top and bottom, sides, and back. They may be constructed of stainless steel or any other metal that is compatible with the product, process, and chemicals. Chairs are a source of contamination if improperly constructed. Particulate tests should be conducted on all clean room chairs that have cushioned seats or backs, and any rotational joints or adjustment devices. Additional testing could be undertaken for static dissipative characteristics desired. In addition the evaluation of clean room chairs should also include comfort, safety, freedom of movement, and height adjustment. Cost considerations are then appropriate.

HOUSEKEEPING MAINTENANCE[2]

This section will discuss the housekeeping maintenance procedures, frequency, and test methods for determination of surface cleanliness. It is of the utmost importance that the clean room environment be kept at the level of cleanliness for which it was designed. This provides the clean room manager with a facility that is safe for the product.

The removal of surface particles requires overcoming the adhesive forces that hold matter to the surface. The primary force by which small particles are held to a dry surface is the intermolecular London/Van der Waals force. Simply stated, to overcome this force, the cleaning force must increase as the particulate size decreases. For example, the adhesive force on a 50 micrometer particle is 2×10^2 g, a 5 micrometer particle is 2×10^4 g, and a 0.5 mictometer particle is 2×10^6 g.

Another force we must consider is electrostatic attraction. This phenomenon electrically attracts particles in the vicinity of the attracting surface. This force should be of minimal concern if the area is protected by an air ionization system. (Refer to Chapter 5 - *Electrostatics in Clean Rooms.*)

A third consideration is the phenomenon of capillary condensation—the fine liquid barrier that forms between a particle and the adjacent surface.

Cleaning the Clean Room

In order to remove surface particles, we must overcome one or all of the forces causing adhesion between the particle and the surface. Liquids assist in eliminating the surface tension adhesion. As a bonus, the liquid may be able to dissolve a portion of the soluble deposits. Clean room surface cleaning can be efficiently accomplished by using a suitable liquid solution; however, even the best cleaning procedures may prove unsatisfactory in the removal of submicrometer particles. With this fact in mind, the following methods are directed toward the removal of particles greater than 5 micrometers.

Tools and Equipment

The following list represents the tools and equipment necessary for performing housekeeping maintenance activities:

1. DI water.

2. Liquid cleaner - low particulate, low residue, nonionic detergents; Freon T.F.; isopropyl alcohol; low particulate dilute acids.

3. Clean room wipers - nonshedding and compatible with the liquid detergent or cleaner selected.

4. Mops - nonshedding material and structure.

5. Buckets - stainless steel or plastic construction.

6. Vacuum - HEPA filtered exhaust with tools constructed of nonshedding material.

Cleaning Procedures

Ceilings. The cleaning should begin at the most contamination sensitive area and proceed through the clean room toward the least sensitive area. The ceiling should be completely cleaned first. In a total HEPA filter ceiling, vacuum the grid only. If an ionization system is installed, deactivate the system prior to any cleaning. Vacuuming should follow the grid in one direction. Light fixtures should follow the grid in one direction. Light fixtures should be wiped with a clean room wiper moistened with DI water. The ceiling track, if installed, will be vacuumed and then wiped with a clean room wiper moistened with DI water and a surfactant. After the wet cleaning, the track is then revacuumed. Good maintenance procedures dictate that the ceiling in a Class 100 or 10 clean room be cleaned a minimum of once per week.

Walls. Vertical partitions including walls, windows, and doors are all cleaned in a similar manner. Beginning at the ceiling and working in a vertical line toward the floor, vacuum with overlapping strokes. This is followed with wiping or mopping utilizing DI water and a surfactant. Rinse with DI water and finally, revacuum. Good maintenance procedures in a Class 100 or 10 facility dictates that all doors and windows be cleaned daily and the walls a minimum of twice per week or as required.

Floors. Raised, grated, epoxy, and vinyl floors can be maintained in the following manner: Vacuum the floor starting in the most contamination sensitive area. Upon completion of vacuuming, wet mop the floor in the identical manner. Rinse the floor, and finally, revacuum. (Note: separate mops are necessary for the wall and the floor areas.) Good maintenance procedures in a Class 100 or 10 clean room dictate that the floors be cleaned a minimum of one per shift or as required.

Work Surfaces. The preferred method for maintaining work surface cleanliness is as follows: Using a premoistened clean room wiper that has been folded into quarters, begin at the rear of the surface and wipe in a straight line from left to right. *After each pass,* expose a fresh area of the folded wiper and with a slight overlap, wipe the next area from the left to the right. This is continued until the entire surface has been wiped. Since the work surface contacts the product and process, the selection of a cleaning solution or surfactant becomes dependent upon product compatibility factors. Good maintenance procedures in Class 100 or 10 clean rooms dictate that the work stations and surfaces be cleaned every time an operator or technician leaves and returns to the area.

Gowning Rooms. The gowning room requires special attention toward particle burden levels. In a facility, the common practice is to thoroughly clean this area a minimum of twice per shift. At all times the cleanliness level of the gowning area should be equal to or better than the area used for production. This practice greatly assists in reducing the particle burden upon clean room garments. The janitorial staff should wear clean room garments while cleaning the gowning area and any air showers and exhibit the same disciplines required in production areas.

Test Procedures. The above cleaning procedures mention a guideline for the frequency of cleaning in Class 100 or 10 facilities. However, many clean rooms may require more or less housekeeping maintenance depending upon various factors such as people, process, equipment, particle fallout, etc. In order to determine the efficiency as well as the frequency requirements in your clean room, the following test methods are suggested:

1. ASTM F-24-65 (American Society for Testing & Materials)
2. Tape-Lift method
3. Inspection witness wafer test

Numerous other test methods are also available. Perhaps the least costly and simplest is the use of an ultraviolet inspection lamp. This method is not quantitative or qualitative but does offer the manager an insight into the conditions within the room that may require additional cleaning prior to testing by one of the conventional methods.

FACILITY START-UP PROCEDURES

The correct start-up of a new facility is a critical first step toward production. Improper cleaning methods and materials increase the potential for product loss through contamination. The cleaning must be conducted in the correct order to decrease cross-contamination and eliminate wasted time and effort. The correct sequence for cleaning a new facility is a planned series of steps that lead from gross clean to precision cleaning without continuous remedial efforts. A suggested sequence for the actual start-up of a facility follows:

Gross Cleaning

1. Continuous gross cleaning of debris from the air return system.
2. Covering the floor to protect the vinyl or raised floor tiles.
3. Blow-down of the air handling system.

4. Thorough cleaning of the area after HEPA filter connection.

5. Initial precision cleaning.

6. Final precision cleaning.

In addition to the cleaning procedures mentioned in previous sections, it is important to note the procedures for cleaning the tiles in a raised floor system. During the gross cleaning of the area special care must be exercised to thoroughly vacuum the area below the tiles. Special care should be exercised to clean all stanchions and the adjacent floor. Each of the tiles must be removed from the clean room and thoroughly cleaned on all sides. The tiles are then replaced and the floor surface cleaned following the standard floor cleaning method.

Prior to beginning any precision cleaning, the following conditions must exist:

1. Visible dirt must be completely removed from all surfaces.

2. Surface testing conducted to determine that no particles greater than 10 micrometers are detected.

3. Airborne particle counts indicate compliance with the design standards in all areas.

4. Velocity, parallelism, and recovery are within specifications throughout the facility.

Precision Cleaning

The precision cleaning phase is similar yet more extensive than the initial cleaning phase. All surfaces are vacuumed, washed, rinsed, and revacuumed. Cleaning is accomplished with strict adherence to the overlapping strokes described earlier to ensure complete cleaning coverage. During the precision cleaning phase only one cleaner is allowed per 1,000 feet of space to control potential contamination. Whenever anyone exits and reenters the facility a new clean room garment will be donned. Only the highest grades of cleaning products and materials will be utilized during this precision cleaning phase. This procedure is time consuming and exacting to guarantee maximum particle removal. Once the entire facility has been precision cleaned, it is now ready for surface particle monitoring and testing. Upon the evaluation of your specific clean room, this procedure may require some modification to enable you to achieve the levels of cleanliness desired. This decision should be based upon objective evaluation of defect density criteria and an understanding that the clean room itself does not guarantee perfect conditions.

Equipment Installation

Process equipment should, if possible, be brought into the facility in stages. The equipment should be uncrated and all the packaging material removed in a receiving area. The equipment is then gross cleaned of all large visible debris. Cleaning will entail vacuuming as well as wet cleaning with Freon or other acceptable solutions. The equipment can be transported to a staging area, i.e., an equipment pass-thru, chase, R&M shop, or any other area adjacent to the clean room. The cleaning will proceed from the top to the bottom. The sequence: vacuum, clean room wiper saturated with an acceptable solution or cleaner, revacuum. While performing this secondary cleaning, the technician must wear compatible clean room gloves, head covering, shoe covering and face mask.

Upon conclusion of stage two, the equipment can be relocated into the clean room facility. Electrical, plumbing, and piping installation may be accomplished with a minimum of contamination. All tools and installation support equipment must be recleaned prior to future clean room use. At the conclusion of the installation, the surfaces of the equipment and the surrounding areas must be recleaned. A precision clean of the equipment and all areas concludes the final stage.

EQUIPMENT INSTALLATION AND REPAIR—EXISTING FACILITY

One of the most difficult tasks of a clean room manager is to control excess contamination during repairs and installation of equipment in an operational clean room. The quality of the product cannot be sacrificed as a compromise to "quick-fixes" and additional maintenance costs. As the clean rooms of today become more contamination control sensitive, procedures must be established within the facility to accommodate the event of repairs and installations. If a Utopia could exist, we would. . . . remove all product, repair or install, reclean the entire facility, perform air and surface analysis; and, if acceptable, return to production. With the requirements of the semiconductor industry, this is not a practical solution. However, in theory, this is the best contamination control practice.

All tools and service equipment must be thoroughly cleaned prior to admittance to any clean room area. The proper procedure would utilize a vapor degreaser for tools that are submersible, staged clean those that cannot be Freon degreased, and package the remainder. After each use, the tools must be recleaned. All tools and service equipment, if possible, should be maintained in the clean room areas and not utilized in other areas of the plant.

Major repairs and installations can be accomplished by the "isolation" method. By isolation, a portion of the clean room is "curtained-off" utilizing antistatic cleaned screens, or cleaned antistatic film that is mounted from the ceiling of the clean room. Inside the isolation area equipment repair and installation can be undertaken without contaminating the entire clean room. Within this area, however, technicians are not exempt from good clean room practices.

Once the required tasks are complete, the equipment as well as the immediate areas are cleaned. The procedure followed is identical to our previous discussion on equipment installation and housekeeping maintenance. The repair tools are removed from the area and recleaned prior to readmittance. Air quality within the isolation chamber is monitored until achieving acceptable particle count limits. The isolation chamber is removed from the facility and the equipment as well as the surrounding areas are precision cleaned prior to production.

CONTAMINATION CONTROL MONITORING

Once a clean room is certified, the monitoring requirements of the facility have only just begun. The quality of the air at the filter level and the work bench height must be monitored and recorded daily. The certification data becomes the "thumb-print" of a clean room to which all future monitoring is compared. In addition to particle monitoring, air velocity measurements should be taken and recorded every 3 months for each filter. Any unusual declines in air velocity or increases in particle counts should be investigated immediately and corrected. An on-line, real time particle monitoring system will provide the clean room manager with continuous air quality measurements. Threshold settings can be selected to alert the supervisor of potential contamination control violations.

In addition to the classic particle counter, the manual method ASTM-F25-68 can be utilized for microscopic examination of the air quality. This has proven useful in determining a particular airborne contaminant within the clean room standard of air quality.

A particle fallout test is performed by measuring the levels of large particles and small fall-out that are not detected effectively using airborne optical counters. This measurement is obtained by allowing a witness wafer to collect particles which settle out in the clean room environment. The particles are then counted and the fallout rate is determined. An automated silicon wafer surface particle detector is utilized as the detection instrument.

The semiconductor industry has recognized control of particulates in chemicals is a matter of primary concern. Higher device yields and ultimately higher profits are intimately related to the maintenance of a low level particulate environment. Thus, the monitoring of such becomes a vital issue to the clean room manager. Routine batch monitoring techniques have been utilized in semiconductor clean rooms for the past several years. The onset of real-time, on-line monitoring at the levels desired is now available. Low-range parts/per/billion analysis for certain parameters is now achievable. It is important to realize that on-line monitoring is important, but the critical analysis must be done before the chemicals are accepted by the clean room manager. Each shipping container must be analyzed for compliance to specifications.

DI water analysis for low level particulates and contaminants has been available in the low ranges for years. On-line, point-of-use analysis should not be limited to pH and resistivity alone. Silica, sodium, TOC, chlorine and other parameters should be monitored continuously with specification for acceptable limits.

Ultrapure gases, like chemicals and DI water can be monitored on-line and continuously. Numerous parameters are available; one of the most common is the monitoring for water in gases.

Contamination control monitoring has evolved beyond simple air detection into a complex science. With the assistance of monitoring analytical equipment, the clean room manager can control the processes as well as the air.

The Clean Room Manager

In Peter F. Drucker's book on Management,[3] there are five basic operations in the work of the manager; "a manager, in the first place, sets objectives; second, a manager organizes; next a manager motivates and communicates; the fourth basic element in the work of the manager is measurement; finally a manager develops people, including himself." All of these basic elements can be discussed in the light of clean room management.

Set Objectives. There are many objectives to be established, but often overlooked are those of the clean room. What type of production can be achieved? Have the guidelines and standards for the clean room been established? These are important objectives and can only become effective by communicating them to those capable of attaining these goals and the superiors demanding them.

Organization. Analyze activities, make decisions, classify the work and select the leadership for the clean room. How and when are repairs of equipment handled? Housekeeping maintenance - a necessity, if so, how often? What type of training is involved for the clean room workers? When and how are they certified?

Motivation and Communication. Contamination control is a state of mind and is as important as the investment in capital equipment. Motivation requires physical presence in the clean room. The offices are often too comfortable to leave and don those white restrictive garments. The total responsibility for production does not rest with the hourly employee. It requires constant communication, not only to our superiors, but of our subordinates. The supervisors must spend at least 50% of their time in the clean room, training, guiding and above all exemplifying good clean room practices.

Measurement. The appraisal and interpretation of individual performances is the requirement of all management. The interpretation of the clean room air, gas, water, chemical and all process related monitoring measurements is part of the daily scope of the clean room and, together with the engineering departments, a production guideline is established.

The clean room manager works with a specific resource: man. The human being is a unique resource requiring development. In the clean rooms of today as well as the distant future, total robotics may not be in your company's future. People will be needed in the clean room, if for no other purpose than to repair the robotic systems. New job descriptions and assignments, the psychology of man working in communion with robotics, and many others all present challenges to the development of your staff and your role as the clean room manager.

Quality and Productivity

Quality improvement is a way to increase productivity and profit potential. Changing the clean room environment from one in which management and employees rank quality third when compared to cost and schedule into one where quality is given equal billing is a major problem in today's clean rooms. Emphasis must be placed on long term improvement, not short term weekly or monthly outputs or profits.

Many clean rooms today do not consider rework or scrap a matter of importance. Corrective action should be taken to resolve the cause for rework and scrap and prevent a recurrence instead of acceptance as the norm for all clean rooms. In many cases, the clean room manager ignores the costs of rework and scrap in the overall cost of quality. Many managers prefer the method of "getting more wafers out the back door" to any preventive action. The days of production at any cost are behind us, and quality is the lifeline of the future. The primary requirement is total commitment by management to the production of quality.

The building of the clean room is only the beginning in the long road of quality production. As Thomas Alva Edison stated, "There is no substitute for hard work." A long road for the clean room manager, but ever increasing quality and production goals can and are being met in today's clean rooms.

THE FUTURE

Today a hierarchical approach to automation is emphasized by many semiconductor experts, that of top-down planning and bottom-up implementation. Automation in many aspects is reviewed and questioned by clean room managers today. The definition of process automation should reflect a realistic, but reachable level over a defined period of time. It is important to realize that the automation of a clean room represents a change in the philosophy of manufacturing within the company. It therefore requires the support and commitment of the management team. The development of automation for the semiconductor industry is a requirement for the future. The how, when and where are the questions of the near future.

The goal of the clean room manager is in reality the challenge of the industry - quality production and the commitment to excellence!

REFERENCES

1. Mann, N.R., *The Keys to Excellence,* Prestwick Books (1985).
2. Dixon, R.C. and Dixon, A.M., *Clean Room Management Manual,* Cleanroom Management Associates, Inc. Tempe, Arizona (1985 Edition).
3. Drucker, P.R., *Management,* Harper and Row (1973) (1974).

BIBLIOGRAPHY

Bartasis, J., *Microelectronics Manufacturing and Testing,* Vol. 8, No. 13, pp. 12-14 (December 1985).

Beeson, R., *Microcontamination,* pp. 41-44 (December 1983/January 1984).

Hayes, G.E., *Quality Progress,* pp. 42-46 (October 1985).

Swick, H. and Brinton, S.J., *Proceedings - Institute of Environmental Sciences,* pp. 163-165 (1984).

Mann, N.R., *The Keys to Excellence,* Prestwick Books (1985).

Bowser, J.J., *Proceedings - Institute of Environmental Sciences* (1985).

"The Renaissance of American Quality," Fortune Magazine (1985).

Institute of Environmental Sciences Recommended Practices, *RP-CC-006-84T,* Institute of Environmental Sciences (1984).

5

Electrostatics in Clean Rooms

Rollin McCraty

INTRODUCTION

In today's clean room the effects of static charge cannot be overlooked. Static charge can cause surface contamination of the parts the clean room was built to protect, and in many cases the devices are also sensitive to damage from ESD (electrostatic discharge). Elimination of these static charges is often more complicated in the clean room due to the large amounts of nonconductive materials used in the clean room. Teflon and polypropylene are commonly used throughout the clean room and are both high static generators that can not simply be grounded. Additionally the materials used to eliminate the static and protect ESD sensitive devices must not add to the contamination problem.

The majority of this chapter is dedicated to the instruments and equipment needed for the elimination of static charges from the clean room.

Static Electricity

Most of the observed electrostatic phenomena are explicable in terms of electric charge either stationary or moving. Static charges give rise to an electric field of force and particle charging. This chapter will explore the role static electricity plays in contamination in the clean room, rather than ESD.

Helpful to the understanding of electrostatics is a review of the background of the fundamental concepts, that is, our current knowledge of atomic structure and the existence of tiny indivisible particles called electrons, protons, and neutrons. The existence of these positive and negative electrical particles is inherent in the structure of matter, and it is accepted that they possess a mass and quantity of electrical charge.

The smallest positive or negative charge is the elementary charge unit of 0.16×10^{-18} coulombs. The coulomb is a large unit. It takes 6.28×10^{18} electrons to make up a negative charge equal to 1 coulomb. A surface can either be charged or lose its charge in quantities which are integral multiples of this elementary charge. A surface is electrically neutral when the number of positive and negative charges is equal and their random distribution over the body is in a state of equilibrium. When a surface is charged it possesses either more or less electrons than protons. On nonconductive surfaces these charges can be localized or uniform (Figure 1).

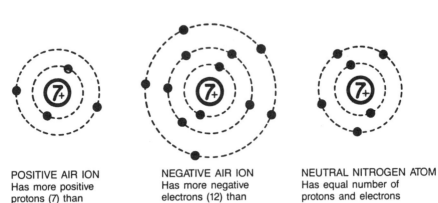

POSITIVE AIR ION
Has more positive
protons (7) than
negative electrons (4)

NEGATIVE AIR ION
Has more negative
electrons (12) than
positive protons (7)

NEUTRAL NITROGEN ATOM
Has equal number of
protons and electrons

NITROGEN
Atomic No. 7
Atomic Wt. 14.0067

An "Ionized" atom is an ion, which has acquired an electric charge by gain or loss of electrons surrounding its nucleus.

Figure 1: Three electrical states of the nitrogen atom.

Atoms and molecules which are not neutral are termed ions. Here a distinction between ions and large charged particles is made. An ion is defined to be a charged particle, either positive or negative, of atomic or molecular size. Ions exist either in a liquid or gas and occasionally in a solid. The formation of ions of either polarity is always accomplished from the removal or addition of electrons. It should be noted that with many materials (e.g., oxygen) it is easier to produce negative ions, and with others (e.g., hydrogen) positive ions are more readily produced.

Forces Between Charges

Numerous experiments can be constructed to demonstrate that charged bodies experience a force of attraction or repulsion. If a negative charge (excess of electrons) is placed on two bodies, these bodies will repel each other. If electrons are removed creating positive charges on the two bodies they will again repel each other. Experimentation with oppositely charged bodies shows that in this case the bodies will attract each other. The electrostatic force between two charged bodies is proportional to the product of their charges and inversely proportional to the square of their separation distances.

The force varies as a function of the medium in which they exist. These forces are greatest in a vacuum. The reduction in force is a measure of the dielectric constant of the medium.

Electric Field

The interaction between charged bodies gives rise to the concept of electrical field. The direction and intensity of an electric field created by charges depends on their polarity, quantity, and distribution as well as on the geometrical conditions and properties of the material in the field. All points in a field possess a potential. Generally, the ground potential is equal to zero. Accordingly, the potential of a point is its voltage relative to ground (Figure 2).

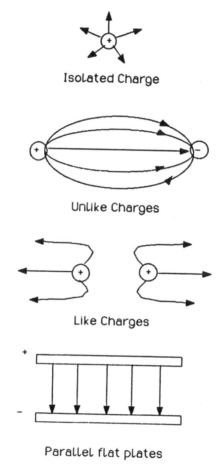

Isolated Charge

Unlike Charges

Like Charges

Parallel flat plates

Figure 2: Charge diagram.

Capacitance

The geometry of a group of conductors determines its energy-storing capability. Energy may be stored with some conductors charged and others discharged or it may be stored with some conductors grounded and others charged to some potential. All electrical energy is stored in fields. All conductor geometries have capacitance and therefore these geometries allow energy to be stored in their surrounding space. The energy stored thus rests both in the physical space of the conductors and in the space between interconnecting conductors.

Charge (Q) and potential difference (V) are related linearly to one another. Capacitance (C) is the constant of proportionality between Q and V. This may be expressed by the equation $Q = CV$. Capacitance is measured in a specific unit called farad. $1F = 1C/V$ and is a very large and impractical quantity. We therefore use $1 \mu F = 10^{-6} F$ and $1 PF = 10^{-12} F$.

Induction Charging

When an electric field is created in an area containing an isolated conductor or an isolated conductor is moved into an electric field, charges are moved to new locations within the conductor. This movement of charge initiates a current in the conductor which continues until its field is reduced to zero. Some of the free electrons drift in the direction of the positive pole of the field where they are fixed in position by the charges of opposite polarity from which the field emanates. On the opposite side of the conductor, excessive positive charges are held by the negative field (Figure 3).

The charge is said to be induced. Upon removal of the external field the induced charge disappears as the electrons return to their neutral positions. If the conductor is temporarily connected to ground while it is in the field, charges of one sign will be eliminated and when the field is removed, a charge of the opposite sign will remain.

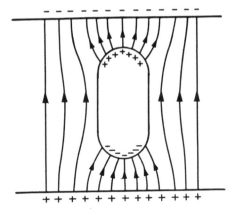

Induction Charging

Separation of induced charge on an isolated conductor in an electric field.

Figure 3: Induction charging.

Triboelectric Charging

Electrostatic charging following the mechanical separation of solids or the separation of solids from liquids is termed triboelectric charging. Current ideas on solids and liquids consider their surfaces as consisting of electric double layers. When two surfaces separate, electrons transfer from one surface to the other. This results in the creation of a negative charge being created on the one which has the excess electrons and a positive charge on the other. The quantity of charge depends on a number of factors: conductivity, number and size of contact points, speed of separation, and surface temperatures. If the conductivity of the material separated is sufficiently large, the charges will be totally neutralized as the electrons recombine with the positive charges. When the conductivity is such that more charges are separated than recombine, the material will become charged. For this to occur one of the two separating surfaces has to be a poor conductor. If both surfaces are poor conductors, then the two surfaces are always oppositely charged after separation. Figure 4 illustrates the polarities that different materials will charge when separated. The material closest to the positive end of the table will become positively charged. The farther apart the materials are in the series, the higher the voltage that will be generated when the materials separate as shown in Figure 4.

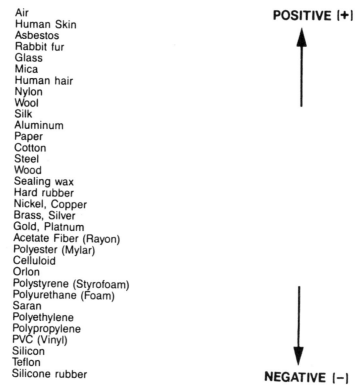

Air
Human Skin
Asbestos
Rabbit fur
Glass
Mica
Human hair
Nylon
Wool
Silk
Aluminum
Paper
Cotton
Steel
Wood
Sealing wax
Hard rubber
Nickel, Copper
Brass, Silver
Gold, Platnum
Acetate Fiber (Rayon)
Polyester (Mylar)
Celluloid
Orlon
Polystyrene (Styrofoam)
Polyurethane (Foam)
Saran
Polyethylene
Polypropylene
PVC (Vinyl)
Silicon
Teflon
Silicone rubber

POSITIVE (+)

NEGATIVE (−)

Charge-generating capability of different materials. The material closest to the positive end of the table will become positively charged while the other material will become negatively charged. The farther apart the materials are in the series, the higher the voltage that will be generated when the materials separate.

Figure 4: Triboelectric series.

Surface Contamination From Static Charges

Once a surface has obtained a charge, its resulting electric field can attract oppositely charged particles or polarize and attract neutral particles to its surface. Once a particle is bonded to a surface it becomes extremely difficult to remove. Electrostatic forces can overcome the laminar flow air streams, or gravity and particles will end up on charged surfaces that would otherwise not have been attracted to the surface. Isolated conductors such as silicon wafers can be charged from friction, heat, evaporation or by induction charging from a charged wafer carrier or other charged object (Figure 5).

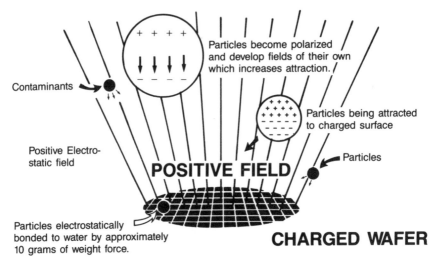

Figure 5: How particles are attracted to charged objects.

ELIMINATING STATIC CHARGE

The removal of the charges from both conductive and nonconductive surfaces is the most obvious and practical approach to reducing contamination from electrostatic forces. The following conditions should be considered in order to reduce or eliminate static charges from the clean room environment:

- Air Ionization - neutralizes nonconductive materials
- Grounding of conducting surfaces
- Increasing conductivity of materials
- Humidity control
- Surface treatment with topical antistat solutions.

AIR IONIZATION

Ionization is primarily employed to remove static charges from the surfaces of nonconductive materials. It can, however, also be used to neutralize isolated conductive or semiconductive materials.

Ionization equipment is available in many configurations for use in the clean room. Room ionization systems can ionize an entire clean room, ion grids and bars are available for all types of laminar flow hoods, desiccator box ionizers, blow off guns, and work bench ionizers are all commonly used throughout the clean room.

The word ion is derived from the Greek word meaning "traveler" and was first used in connection with electrochemistry. "Ion" is also used to describe the electrical balance of atoms and molecules of all kinds. When we speak of air ions we are referring to the electrical balance of the air molecules. Nitrogen makes up 78% of the air and is the primary air molecule that is ionized. The electrical balance of the atoms and molecules is changed by the addition or subtraction of electrons. When a molecule has more protons or positive charges than electrons or negative charges, it is said to be a positive ion. If electrons are added to the molecule, it then becomes a negative ion. It is always the movement of electrons which forms a positive or negative ion.

Charged surfaces are neutralized when there are sufficient numbers of air ions in the area near the surface. The electrostatic field radiated from the charges will attract oppositely charged air ions and repel like polarity ions. When a positively charged surface draws to itself a negatively charged air ion, the extra electrons surrounding the air ion are transferred to the positively charged surface (Figure 6).

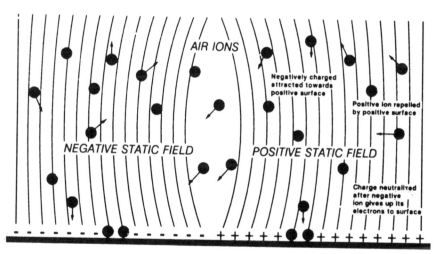

Figure 6: How air ions are attracted and repelled by static field.

Both the air ion and the surface charges are neutral after this transfer takes place. A more intense electrostatic field will result in a greater attraction of oppositely charged ions. As the charges are neutralized by the ions, the field is also reduced and the ions are attracted at a slower rate. This can be seen in the resulting static neutralization curve shown in Figure 7. Unfortunately, the natural air ion levels of 2000 to 7000 ions/cm^3 found in unpolluted air are reduced to zero as air passes through filtering. Even in nonclean room environments it is rare to find more than 50 ions/cm^3. In addition to static, human activities, industrial pollutants, smog, and stray electrical fields all remove air ions from the environment. Adding to the problem of ion depletion, a building forms a faraday cage which insulates us from the natural ion generating phenomena of nature.

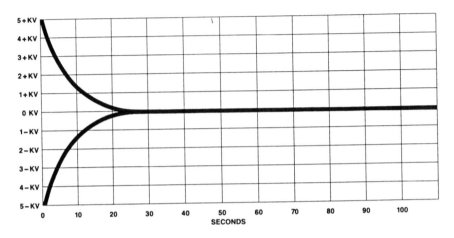

Figure 7: AC system neutralization curve.

ION GENERATION

There are two basic methods now employed to produce air ions in electronic production and clean room environments: nuclear isotopes that produce alpha particles and electrical ionization which intensifies a high voltage electrical field. Both methods impart energy into the surrounding air molecules dislodging electrons from neutral molecules. The molecule from which the electron is dislodged becomes a positive ion. The dislodged electron finds itself surrounded by 30 quintillion (3 x 10^{19}) neutral molecules per cubic centimeter of air. These neutral molecules are colliding at a rate of one billion contacts per second. The departed electron is quickly captured by one of the neutral molecules and a negative ion is formed. The coexistence of both positive and negative ions in the same relative space is possible because of the vast number of neutral molecules which separate and hold them apart. In normal air, there are 10 quadrillion (1 x 10^{16}) neutral molecules for every charged ion. When opposite charged ions meet they do neutralize one another as the negative ion transfers its extra electron to the positive ion. It is this same recombination process which neutralizes the static charges on the surface of nonconductors (Figure 8).

Nuclear Ionizers

Nuclear ionizers can only be leased and it is essential to keep track of the required yearly radioactive element replacement. The spent radioactive cartridges must be shipped back to the manufacturer for disposal. The radioactive cartridges are reported to be safe, but employees are often alarmed by their presence. EMI

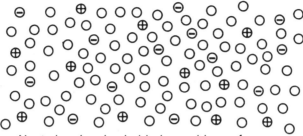

Neutral molecules hold charged ions of opposite polarities apart allowing both polarities to coexist. In normal air, $3x10^{19}$ neutral molecules exist for each charged one.

Figure 8: Ion coexistence.

and ozone are not generated and nuclear ionizers do provide a balanced ratio of positive and negative ions which must be distributed by moving air currents. Nuclear ionizers can be used in blow off nozzles and bench blowers but are not practical for laminar flow hoods or total room ionization.

Electrical Ionization

Electrical ionization is the most common method used to generate air ions. There are three categories of electrical ionization: AC, Bi-Polar Constant DC, and Pulse DC. Each method has its advantages and disadvantages. The electrical methods can produce a great number of ions and is always used in laminar flow hoods and room ionization systems. Ionizing blowers and blow off guns are also available in all three types. The main advantages of electrical ionizers is their efficiency and ease of use. The disadvantage is the need to maintain proper ion balance and cleaning of the electrodes.

All electrical methods of generating air ions impart energy to the air by intensifying an electric field at the sharp point or edge until it overcomes the dielectric strength of the surrounding air causing an ion current to flow. A field strength of approximately 20,000 volts per centimeter is required to accomplish this process which is termed corona discharge. Negative corona occurs when electrons are flowing from the electrode into the surrounding air and positive corona is the flow of electrons from the air molecules into the electrode. The quantity of ion current flow is determined by the applied voltage, electrode sharpness, humidity, atmospheric pressure, and proximity of a ground plane. An electrode tip of 0.010 inch with a negative 12 kV DC potential applied to it, in 40% relative humidity in open air (no ground plane closer than one meter) will draw approximately 0.000002 amp.

Space Charge. If the net number of positive and negative air ions in the same relative space is equal, the overall net charge of the air is zero. If, however, there is an imbalance in the ion ratio, the atmosphere will become charged relative to ground. This charge is termed "space charge." When a space charge exists the air will seek to neutralize itself by either giving up or acquiring electrons from the surrounding environment. If there are isolated conductors in the environment they will become charged from the addition or subtraction of electrons from their bodies. The charging energy of the space charge depends on the total number of air ions of imbalance. The safe level of space charge depends on the devices being protected, if surface contamination is the only concern, higher space charges can be allowed.

AC Ionization. AC (Alternating Current) generates both positive and negative ions by applying a high voltage AC waveform to a series of electrodes. During the positive half cycle of the waveform, positive ions are generated, and likewise, negative ions are generated during the negative half cycle. AC, like that of nuclear isotopes, requires a moving air current to deliver the ions to the area where they are needed. This type of ion generation is usually found in blowers and laminar flow grids. AC has the advantage of simplicity, but the emitter systems are usually large and bulky. The electrodes can be capacitively coupled to eliminate shock hazard which is important to insure safety, as the AC power supplies will produce a hefty shock. The ion balance is difficult to adjust, but can be balanced. Many of the AC systems do not have a way to balance them if they drift out of balance (Figure 9).

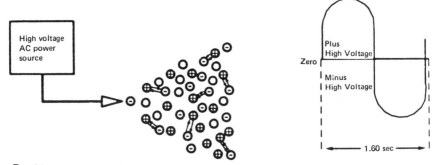

Positive and negative ions attract each other. Both polarities together create zero space charge and the ions never leave close proximity of the discharge point unless moved by air currents.

Figure 9: Alternating current (AC).

Constant DC Ionization. Constant DC incorporates two independent power supplies, one negative and one positive to generate both polarities of ions. Each power supply is connected to a series of electrodes. This type of ion generation is used in blowers and some room ionization. It works best in conjunction with air flow, but can also be used in environments without air flow, if the electrodes are spaced far enough apart. The ion balance can be easily controlled by adjusting one of the power supply voltages up or down.

The ionized air blowers normally use a low current power supply and do not require current limiting resistors in series with each electrode, however this is not the case in room ionization systems. Each electrode should be resistively coupled to the main feed line.

Pulse DC Principles of Operation. A pulse DC ionizer uses two independent high voltage DC power supplies and a power supply control circuit to generate both positive and negative ions. Pulse DC is used in most total room ionization systems and the majority of the laminar flow systems. Pulse DC also works well in non-air-flow areas such as a class 10,000 clean room. Positive and negative ions are alternately generated by turning the positive power supply off while the negative is on. The power supplies are switched by a multi-vibrator control circuit. The plus and minus power supplies use separate electrodes to generate the positive and negative ions (Figure 10).

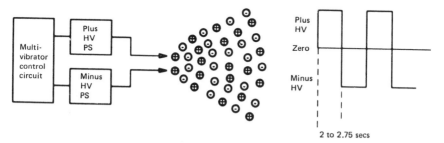

Alternate waves of ions are generated by allowing enough time for the space charge to spread the ions. The positive and negative ions intermix as they move away from the electrodes by their attractive forces.

Figure 10: Pulse DC technology principles of operation.

The electrodes are normally arranged in electrode pairs, one plus and one minus. By switching between polarities, ions of one polarity are generated at a time, which allows the ions to be distributed without the need of moving air. The ions are dispersed throughout the area by repulsion from one another. In still air applications the distance the ions travel and the space charge is a function of the pulse frequency and the voltage applied to the electrodes. A slower pulse frequency will increase the distance the ions travel by allowing a higher space charge to form. To prevent the space charge from reaching too high a level, electrode pairs are often placed at spacings which lower the distance the ions must travel.

For example, in room ionization systems the electrode pairs are placed typically on 5' centers to keep the space charge directly under the electrode pairs at a safe level. If the electrodes were spaced at 8' centers the ions would need to travel 4' instead of 2½' to provide ions in between the electrodes. The pulse period used to provide proper ionization and maintain safe space charge levels in still air varies from 0.5 to 6 seconds depending on electrode pair spacing and the distance between the electrode pairs and the area being ionized. In laminar flow applications, the air streams distribute the ions in the same way that AC and nuclear ionizers do. Faster neutralization times are often achieved due to the higher concentration of ions which can be generated by the pulse DC ionizers. This happens because there is less ion recombination and a high number of ions can be generated with the higher potentials possible with pulse DC. The ion ratios and levels are adjusted by varying the output potentials of one or both power supplies and the pulse frequency and pulse duty cycle. The pulse frequency in laminar flow applications must be much higher than that used in still air, or high space charges will result. Typically, a 5 Hz to 1 second period is used depending upon air flow volume and the application.

Electrodes. Electrical ionization normally uses a point electrode to intensify the field strength in order to generate the ions. The selection of electrode material is important in order to minimize point breakdown, which decreases the field intensity and results in lower and unequal ion output (Figure 11).

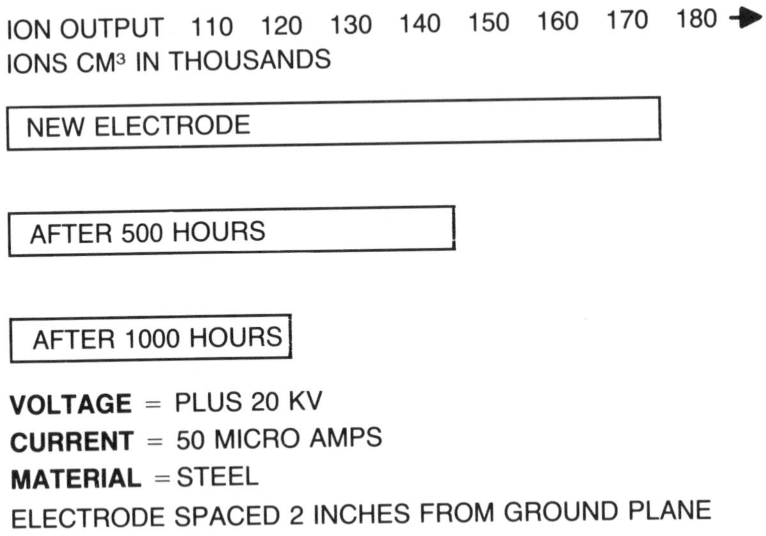

ION OUTPUT   110   120   130   140   150   160   170   180 ➤
IONS CM³ IN THOUSANDS

NEW ELECTRODE

AFTER 500 HOURS

AFTER 1000 HOURS

VOLTAGE = PLUS 20 KV
CURRENT = 50 MICRO AMPS
MATERIAL = STEEL
ELECTRODE SPACED 2 INCHES FROM GROUND PLANE

Figure 11: Effects of steel electrode erosion on ion output.

When negative ions are generated, electrons flow out and away from the electrode (Figure 12). Positive ion generation behaves in an opposite manner; the electrons flow into the electrodes. Point erosion of the positive electrode is hundreds of times faster than that of the negative electrode when separate electrodes are used. In single electrode systems the electrode will erode even faster. To prevent electrode breakdown, it must be made of a material which resists breakdown. Presently the best known material is tungsten (Figure 13). Electrode cleaning is necessary to maintain ionization efficiency. The time interval between cleaning varies from quarterly to yearly depending upon the cleanliness of the air. Most electrodes can be easily cleaned with a mild solvent.

TOTAL ROOM IONIZATION

Total room ionization systems are designed to completely fill a room or

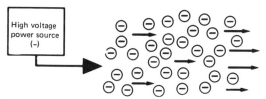

Negative-electrons flow from point into air.

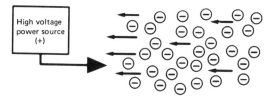

Positive-electrons flow from air into point.

Figure 12: Electron flow.

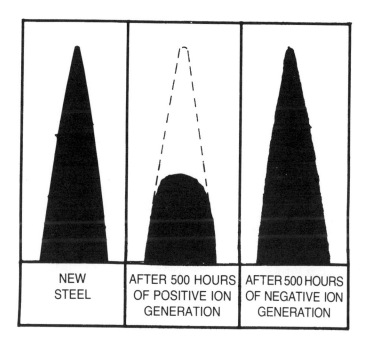

Figure 13: Steel electrode wear.

part of a room with air ions. In rooms without total laminar flow such as a class 10,000 clean room pulse DC or in some cases dual DC type systems are used in order for the ions to be properly distributed (Figure 14). If laminar flow air is present throughout the area being ionized AC grids or pulse DC bars can be used to achieve total room ionization.

The systems designed for use in partial laminar flow rooms normally use emitters mounted from the drop ceiling panels and are spaced every five to six feet. If the emitters are spaced too far apart there will be areas between the emitters where the ions will not reach without creating too high a space charge directly below the emitters (Figure 15).

The total room systems designed for total laminar flow rooms require closer emitter spacing because the laminar flow air carries the ions straight down and does not allow them to spread out as far. The spacing depends on the clean room air flow patterns. The emitters used are also different than ones used in still air room in that they are normally thin bars which hook to a power source that drives many bars or bars with self-contained power supplies. If AC grids are used, every other two x four HEPA module typically has a grid which covers the entire area of that module.

Auto Balance

Total room ionization systems, whether designed for laminar flow or still air are available with ion sensors which monitor the ion ratios and automatically maintain the ion balance. If the ion ratio should change due to a change in humidity or electrode contamination, the resulting space charge will be immediately detected and a signal fed back to the power supplies where the necessary adjustments are made.

Two different principles are currently being employed as ion imbalance detectors. The first monitors an electrically isolated metal plate which is monitored by an electrostatic field meter. When the ion ratio is in balance, there will not be any charge transfer from the air molecules to the plate. When a space charge does exist, however, a charge will be transferred to the plate. The resulting voltage is measured by the meter and a signal is sent to the power source (Figure 16).

The second method directly detects the ion balance. The charged molecules which create a space charge when out of balance either give or acquire electrons from the ion sensors "detector" creating an ion current proportional to the ion imbalance. This current is measured by a high impedance pico amp meter where it is translated into a plus or minus DC voltage that is sent to the power source. The ion imbalance can then be displayed in ions of imbalance in millions of ions per square centimeter per second (Figure 17).

DESICCATOR CABINETS

Parts are often stored in desiccator cabinets to keep them free from moisture or contamination. If the devices are charged before they enter the cabinet any particulate in the cabinet or nitrogen supply can be attracted to the device. Many of the desiccator cabinets are made of plastic and high static charges can be

Figure 14: Pulse DC technology total atmosphere room ionization system installed at Litton Aero Products, Moorpark, California.

COURTESY OF STATIC CONTROL SERVICES, INC.

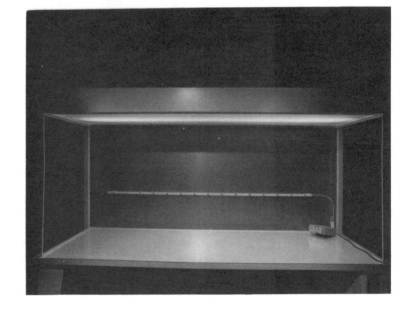

Figure 15: (a) AC laminar flow hood ionization system horizontal bench. (b) pulse DC laminar flow ionization system horizontal bench.

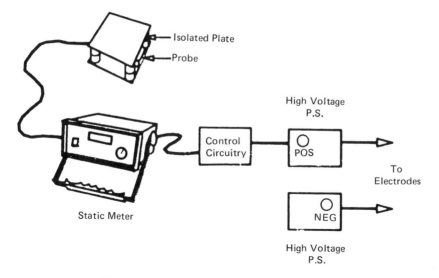

Figure 16: Automatic balance (method I–isolated plate).

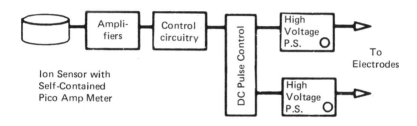

Figure 17: Automatic balance (method II).

generated on their surfaces. The outsides of the cabinets can be treated with a topical antistatic solution, but often this is not a viable solution for the inside because of the possibility of contamination. Additionally, the devices would not be neutralized if they were charged. Ionization is often used to keep the inside of the box and the parts neutral. Both AC and pulse DC ionizers are available. If a large amount of nitrogen flow is not used the AC ionizers may not be very effective as AC requires air flow to distribute the ions (Figures 18 and 19).

BLOW OFF GUNS

Ionized air guns are often used to remove both static charges and particulates from objects in the clean room. These guns are available with plastic coatings and air filters. Typically, either pulse DC or nuclear isotopes are used to ionize the air stream. The nuclear ionizers do not require a power supply and associated wiring, but should not be used in areas where solvents could enter the nozzle.

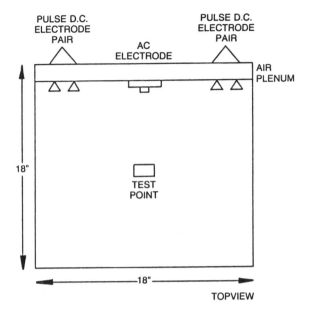

Figure 18: Desiccator box test location.

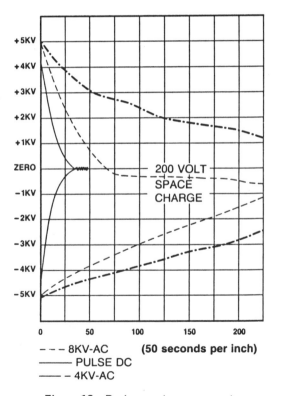

Figure 19: Desiccator box test results.

They also have a life of only one year before they need to be returned to the manufacturer for replacement (Figure 20).

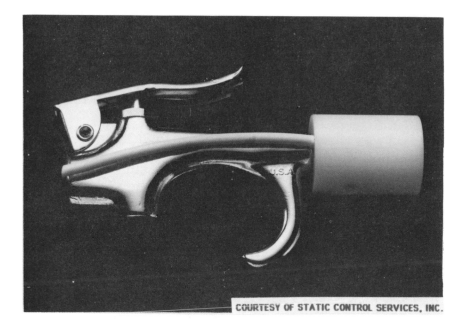

COURTESY OF STATIC CONTROL SERVICES, INC.

Figure 20: Pulse DC technology ionizing blow off gun.

GROUNDING

Historically, grounding requirements arose from the need to provide protection against lightning strikes and industrially generated static electricity. Structures as well as electrical equipment were connected to earth, i.e., grounded to provide a path for lightning and static charges. As utility power systems developed, grounding to earth was found necessary for safety. All major components of a system, such as generating stations, substations, and distribution systems, are earth grounded to provide a path back to the generator for the fault in case of line trouble. With a properly installed third wire ground network, faults internal to a building are rapidly cleared regardless of the resistance of the earth connection.

The required low resistance return path inside of a building is provided by the grounding (green) conductor and the interconnected facility ground system. Two different methods of grounding work stations are currently employed. First, independent ground stakes are driven for each work station. The table top, floor mat, and wrist strap are then tied to this ground. This is done to prevent small voltage and current fluctuations that can occur in the power system ground during a fault. The main drawback with this method is the possibility of having considerable different resistance and also voltage potential between earth grounds. Soil resistivity varies with moisture, soil chemistry, and temperature. A difference in ground potential can under certain conditions be a hazard. If equipment which is grounded to power system ground is used on a conductive work surface which is tied to a separate earth ground, potentials can develop between the two grounds.

The second method ties all the work stations to the building electrical ground. This ensures that all grounds will be at the same potential. The national electrical code states that, "Electrical systems are grounded to limit voltage due to lightning, line surfaces, or unintentional contact with higher voltage lines and to stabilize the voltage to ground during normal operation." As long as the facility ground system is properly installed this method is generally accepted as the most practical and reliable (Figure 21).

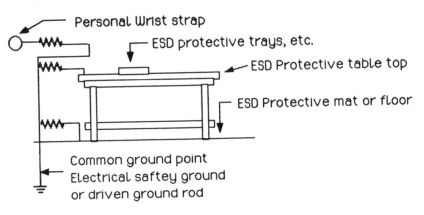

Figure 21: Wrist strap diagram.

Wrist Straps

Wrist straps are used to ground personnel handling ESD sensitive devices. They are of prime importance and considered the first line of static defense and should be used whenever possible. If personnel mobility requirements restrict the use of wrist straps, alternate grounding such as conductive flooring and heel straps or conductive footwear should be incorporated instead. Companies doing work for the military are not only required to use wrist straps, but also to check their operation once or twice daily depending on the sensitivity of the devices being handled (Figure 22).

There are a number of wrist straps on the market. The more popular ones can be divided into two categories, the twist watch band and the elastic or velcro bands. In clean room applications, particle entrapment and sloughing must be considered as well as the electrical specifications. Most of the elastic type bands cannot be used in clean rooms because of particle sloughing and their inability to be cleaned. The twist bands can be cleaned and are available in stainless steel to prevent particle sloughing. If the personnel are handling equipment under power where they may come in contact with a high potential, the strap should have an outside coating of a nonconductive material to help prevent shock hazard. Paint should be avoided, as it can chip. Preferred coatings are of an electrostatically bonded teflon material. All wrist straps should have a 1 mega ohm resistor in series between the strap and ground to protect the person wearing the strap from hazardous current level should they come in contact with a high voltage potential.

The lead wire should tie to a common ground connection, preferably the same point where the work surface is tied to ground. The strap should not be grounded directly to the work surface unless it is to the same point that the ground wire is connected to the work surface.

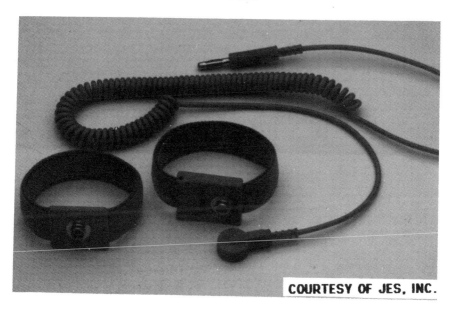

COURTESY OF JES, INC.

Figure 22: Photo of wrist strap.

SURFACE RESISTIVITY

As previously mentioned the ability of a material to allow electrons to flow is determined by its conductivity. The surface conductivity is a major factor which determines the ability or speed that a material will discharge a static charge. Thus, the greater the material resistivity, the longer it takes for the static charge to decay. The choice of resistivity depends on the environment and the types of electronic devices handled. Materials are divided into three ranges; conductive, static dissipative, and antistatic. The category a material falls into is determined by its surface resistivity which is the inverse of conductivity. If the surface resistance is below 10^5 ohms per square, it is said to be conductive. The range 10^6 to 10^9 ohms per square is considered static dissipative while 10^{10} to 10^{14} is called antistatic. Measured resistance may not always correspond to the nominal values of those attributed to a material because of the dominant effect of surface contamination. For example, the presence of a surfactant residue on the surface might cause an antistatic material to register in the static dissipative range.

Bench Top Materials

The primary purpose of a static control work surface or bench top is to drain static electricity. In all cases the static charge must be transferred from a charged object through the work surface to ground. If an ESD sensitive device has obtained a charge and comes in contact with a grounded surface with a high conductivity, the charge will be rapidly discharged and the current levels may reach sufficiently high levels as to damage the device. Static dissipative materials are employed to slow down the discharge rates so as to reduce damage from discharging too rapidly. Antistatic materials are low in static generating ability and also drain off a charge relatively slowly. It is possible to make a surface antistatic by treating it with an organic material specifically selected for its ability to absorb water. The absorbed water combined with free ions in the additive creates a sweat layer that acts as a path to conduct static electricity to ground. Antistatic materials will also discharge without being grounded by absorbing air ions which neutralize the charge (Figure 23).

Conductive	Static Dissipative	Antistatic
1. Dissipates charges rapidly throughout the material and to ground, and will not maintain a high static voltage.	1. Charge dissipation rate generally adequate for most ESDS parts.	1. Provides slow bleedoff of static charges. If ground straps are used by personnel working at the work bench high ESD voltages should be rapidly dissipated through the ground strap.
2. Could discharge an ESD in form of a spark causing EMP.	2. Provides greater resistance for personnel protection from high voltages or hard grounding if the table top is contacted with test equipment ground.	2. Eliminates sparks from ESD.
3. Could cause a high current discharge through an ESDS part.	3. Reduces discharge currents through ESDS parts.	3. Limits discharge currents through ESDS parts to low levels.
4. Could present a safety hazard or short if a high voltage source contacted the bench top. Could hard ground the table top if test equipment with grounded chassis contacted the bench top surface.	4. Safety could require that series resistances be provided in connection to ground where high voltages can be contacted by personnel.	4. Generally provides adequate resistance for personnel safety.
5. Safety could require that series resistances be provided in connection to ground where high voltages can be contacted by personnel.		

Figure 23: Types of ESD protective bench tops.

Antistatic materials work well in conjunction with air ionizers because the ion concentration of the surrounding air is increased. Triboelectric charge generation is also greatly reduced by the decreased friction caused by the thin sweat layer; however, since the moisture is drawn from the air, a drop in relative humidity can cause a loss of antistatic properties. Cleaning of antistatic tops with freon or a similar solvent can remove the sweat layer and antistatic agents. It can take several days for the antistatic agents within the material to migrate to the surface and form a new moisture layer.

Conductive Flooring

Most floors, whether painted or sealed concrete or vinyl tile can contribute to ESD problems. If ESD is a concern in the clean room the floors should be covered with conductive flooring or mats, or be treated with a topical antistat. For them to be effective the workers must be connected to the floor by heel straps. Like the work surfaces conductive flooring is available in conductive, static dissipative, or antistatic. These floors are available in the form of mats, vinyl tiles, and terrazzo.

The disadvantages of floor mats are that they need to be removed to clean the floor and they can be a tripping hazard. If conductive floor tiles are used care must be taken to insure that it is not waxed as wax will make the floor nonconductive.

Humidity

By maintaining a high relative humidity in the environment a thin layer of moisture will form on the surface of objects in the area. This thin film provides a conductive path for a charge which may be generated to flow across its surface thereby reducing the electrostatic charge. Triboelectric charging is also reduced by the added lubricity to the surface. In general, the lower the humidity the higher the charge that can be generated. Unfortunately even high humidities will allow significant static voltages to be generated (Figure 24).

Means of Static Generation	Electrostatic Voltages	
	10 to 20 Percent Relative Humidity	65 to 90 Percent Relative Humidity
Walking across carpet	35,000	1,500
Walking over vinyl floor	12,000	250
Worker at bench	6,000	100
Vinyl envelopes for work instructions	7,000	600
Common poly bag picked up from bench	20,000	1,200
Work chair padded with polyurethane foam	18,000	1,500

Figure 24: Electrostatic voltages.

Workers can also be quite uncomfortable, parts can rust, and solderability problems can arise in high humidity environments. Humidity control is expensive, normally costing one dollar per square foot per year in energy costs and in dry areas the energy costs can double that amount. Room ionization systems are often used instead of humidity control and offer much faster static neutralization.

Topical Antistats

Topical antistats are chemicals which when applied to a surface reduce the coefficient of friction and increase the surface conductivity. Topicals are generally in liquid form and consist of a carrier such as water, alcohol, or other solvent along with an antistatic material which, when deposited on a surface, draws moisture from the air. The antistatic agents work in conjunction with free air ions to neutralize static charges and in many applications do not require grounding. For this reason a topical's performance is enhanced in an ionized environment. In the clean room environment, care must be taken to choose a topical that will not add contaminants to the area. Some contain chloride and nitrate salts and should be avoided in many applications. Topicals can be used on the sides of a laminar flow hood and other nonconductive surfaces in the clean room and in many applications. It is, however, a surfactant and although inert to most electronics processes can be a contaminant to processes such as optics. Normally, topicals should be avoided in areas which can come in contact with materials being processed and limited to floors, clothing, walls, etc.

PROTECTIVE PACKAGING

Protective packaging is used to keep susceptible items out of the path of any ESD current. Proper packaging should be used whenever susceptible items are removed from a static protected environment. Adequate packaging should totally enclose the device in a sufficiently conductive sheath so that a charge on the sheath will exert no electrical influence on the device inside. The path of lowest resistance should be around the outside of the package. The inner layer of the package should be a nonstatic generating material so that internal movement will not generate a charge.

Packaging which enters a clean room must also have the same specifications as the class of the clean room. There are two basic types of packaging and handling materials available which offer static protection. Conductive materials are those whose surface resistivity are less than 10^4 ohms/square. These materials are normally constructed of metal or a plastic base material that has been coated or filled with carbon or metal. Conductive materials must be used to provide shielding. The most complete shielding is afforded by thick layers of metal or a container made completely of metal or foil.

Antistatic materials have a surface resistivity of between 10^{10} and 10^{14} ohms per square. Antistatic materials are very effective at inhibiting charge generation and retention. Antistatic materials are often used on the inside layers of a conductive package. They are also very useful in packaging of nonsusceptible hardware and components when these packages must enter areas where sensitive items are being processed.

They can also be used in the manufacturing of paperwork holders and other plastic items such as cushioning material (popcorn) which enter static sensitive areas.

Antistatic plastics are relatively inexpensive, but their effectiveness depends on additives that draw water from the air. A decrease in relative humidity can reduce their effectiveness and these additives can come into contact with other materials, causing contamination. This loss of additives can eventually lead to loss of antistatic properties.

MEASURING AND TESTING METHODS

Static Fieldmeters

Fieldmeters are available in three basic variations. They are the electrometer type, the radioactive sensor type and the modulated capacitor or AC carrier type. Stability, accuracy, size, cost, and usage shuld be considered when choosing a fieldmeter. The electrometer or pocket size electrostatic locator type is basically a capacitively coupled DC amplifier with a shunt capacitor for calibration. All amplifiers draw a finite amount of current, therefore this type of instrument's reading will drift over time. To counteract this problem the shunt capacitor must be periodically discharged in a zero field condition. This type of meter cannot be used in ionized environments, because even a minute current transferred to the input of the DC amplifier from the air ions will cause extreme drift. The main advantage of this type of fieldmeter is the low cost, simplicity, small size, and the ability to make fast measurements (Figure 25).

COURTESY OF STATIC CONTROL SERVICES, INC

Figure 25: Photo of hand held static locator.

The second type of fieldmeter utilizes a radioactive sensor. This sensor ionizes the air in its immediate vicinity. When this is exposed to an electric field, a current will flow that is proportional to the electric field reading. This is a very simple system and it is DC stable. (Its drawbacks are the fears associated with radioactive material and neutralization of the charge you are measuring by the ionized air.) The half life of the radioactive materials also requires that the meter be returned to the manufacturer for sensor replacement every year.

Fieldmeters utilizing an AC carrier type system are the most common in the high quality electric field monitoring systems. The AC signal is produced by modulating the capacitance pickup probe in an electric field. This is normally accomplished by one or two means. The first and most commonly used for long term monitoring is the modulated capacitor fieldmeter. This utilizes an electrode which is vibrated perpendicularly to an electric field. To maintain good stability the modulation amplitude must remain stable. This can be accomplished by utilizing a null seeking feedback technique that minimizes any calibration variations due to modulation amplitude changes. These instruments can be made rather small in size with the probe being one or two cubic inches. They are extremely reliable and require little power. The speed of response is typically around 0.5 second (Figure 26).

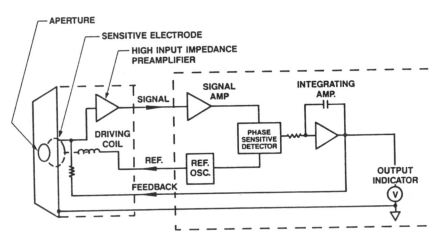

Figure 26: Simplified functional block diagram of electrostatic fieldmeter.

The second type of AC carrier fieldmeter utilizes a rotor that chops the electric field and a stator which acts as the electrode in this system. These meters are simple in design and principle. The drawbacks are that they are usually large in size and may have limited motor life. The speed of response is limited by the chopping speed of the rotor (Figure 27).

Charge Plate Monitor

To insure and validate any ion producing device or system, test and measuring methods were developed. A charge plate monitor is an instrument designed to

COURTESY OF MONROE ELECTRONICS, INC.

test both the ion balance and static neutralization times of ionization equipment. It consists of a metal plate of a given size and spacing to ground which is connected to plus and minus high voltage power supplies through a series of high voltage relays. The plate is then monitored by noncontact fieldmeter, usually an AC carrier type. Ion balance can be checked by monitoring the voltage transferred to the isolated metal plate from air ions but the voltage seen on the plate is only the net difference between the positive and negative ions and does not give us the actual quantity of ions. The plate can also be charged by the internal power supplies and the positive and negative neutralization times determined by timing the time it takes the plate to neutralize. The capacitance of the plate with respect to ground must be considered, a higher value of capacitance will store more energy and thus require more ions to be neutralized resulting in a longer neutralization time (Figure 28).

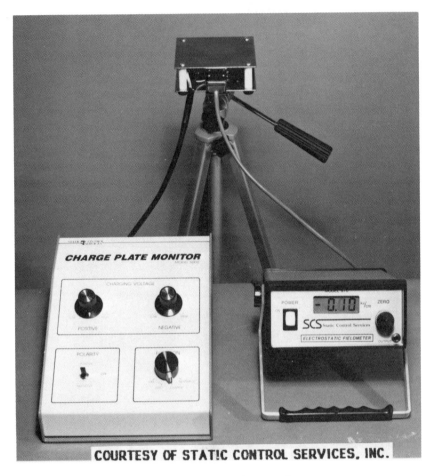

COURTESY OF STATIC CONTROL SERVICES, INC.

Figure 28: Photo of charge plate monitor system.

If the distance between the charge plate and ground is kept constant and no external capacitance is added the area of the plate can change with little effect on the neutralization times. This is usually not the case however as the open contacts of the switching relays add external capacitance to the plate. When comparing neutralization times obtained with different charge plate monitors, the capacitance, plate area, distance to ground, and the radius of the plate edges must be considered as they all affect the results.

The IES (Institute of Environmental Sciences) has established an industry standard for the Charge Plate Monitor and standard test locations and methods for evaluating ionizers. The standard specifies the charge plate size to be 6 inches by 6 inches with a separation from a ground plate of 1 inch with a total capacitance of 20 pf. The plate is charged to 5 kV and the time taken for the charge to be neutralized to 10% of the charging value (500 volts) is termed the neutralization time.

Ion Counter

The number of air ions in a given volume of air can be determined with an air ion counter. This device measures the concentration of ions per cubic centimeter of air. The volumetric method separates ions by their polarity and mobility (size). An internal blower draws the ambient air into the instrument at a given rate across a series of parallel collector plates arranged in a rectangular assembly. Alternating between the collector plates is a series of additional plates upon which a DC is imposed. The potential difference between the polarized plates and the collection plates establishes an electrical field which exerts a deflecting force of ions of the same polarity as they enter the field. As the ions are deflected into the nearby collector plates they give up their charge to it. The collection of charges given up from numerous ions forms a minute current which is measured with an extremely sensitive electrometer. The current is converted to ions per cubic centimeter which can be read on a built-in meter. There are several variations of this same principle available but they all measure the number of electrons per given rate of air and translate it into ions per cubic centimeter of air (Figures 29 and 30).

COURTESY OF STATIC CONTROL SERVICES, INC.

Figure 29: Photo of air ion counter.

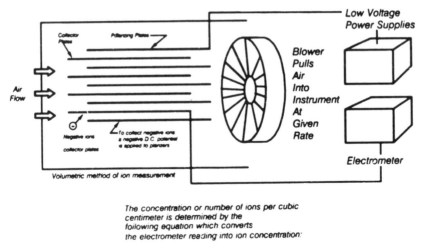

The concentration or number of ions per cubic centimeter is determined by the following equation which converts the electrometer reading into ion concentration:

$$N = \frac{I}{qvA}$$

Where N = number of ions cm³
I = ion current (electrometer reading), ampheres
q = charge on one ion, coulombs (1.6 x 10⁻¹⁹)
v = velocity of air flow through the collector (cm/sec)
A = effective entrance area of the collection plates (cm²)

Figure 30: Air ion counter – volumetric method of ion measurement.

Surface Resistivity Meter

Surface resistivity measurements of materials such as work surfaces, floors and packaging materials are obtained using a surface resistivity meter. Surface resistivity is measured in ohms per square. Surface resistivity is divided into three categories based on their resistivity, conductive (less than 10 to the 5 ohms per square), static dissipative (10 to the 5 to 10 to the 9 ohms per square), and antistatic (10 to the 9 to 10 to the 14 ohms per square).

Several types of instruments are available, but most apply a voltage to one of two parallel rails and the current flowing between the rails is detected by the other one, thus defining the resistivity of the surface. ASTM Standard D257 defines the weight to be applied to the rails and applicable formulas for the rails or concentric circles (Figure 31).

Wrist Strap Testers

Wrist straps must be checked periodically to be sure they are doing their job. As mentioned earlier, companies doing military work under military standards are required to test wrist straps daily and keep an ongoing log of each strap. There are two types of wrist strap checkers in use: resistance and capacitance. The resistance checkers test for continuity and the operator must touch the tester to complete the path. These testers are normally set to insure that at least 200 kilohms is present between the operator and ground, but no more than 2.5 megaohms. The capacitance tester, monitoring the individuals capacitance through the wrist strap, generates an alarm if the strap loses continuity or is being worn incorrectly. Constant monitoring systems continuously sense the resistance or capacitance and generate an alarm if 10 megaohms or more is measured.

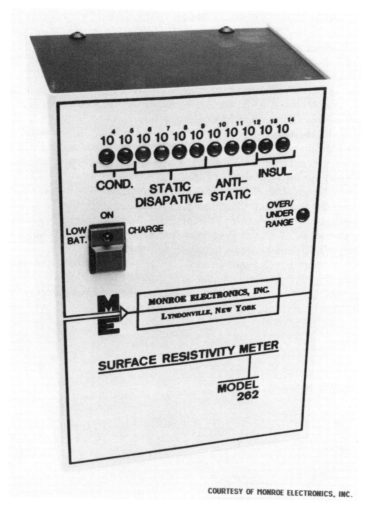

Figure 31: Photo of surface resistivity meter.

Ion Balance Meter

Ion balance meters are used to survey ion balances throughout the environment. They are much more sensitive than charge plate monitors when determining ion ratios. These instruments consist of an electrode which is tied to a picoamp meter. Electrons are either given to or removed from the sensor creating an ion current which is proportional to the difference in the ion ratio.

BIBLIOGRAPHY

Anderson, Dan C., *Booming Electronics Industry Spur Antistatic Packaging,* Packaging Engineering (Feb. 1982).

Denny, Hugh W., *Grounding in the Design of Buildings and Facilities for Safety Protection,* EMC Technology (March 1983).

Dillenbeck, Keith, *Characterization of Air Ionization in the Cleanroom,* Microcontamination (June 1985).

Gary, O., *Drastic Loss of Conductivity in Antistatic Plastics,* EOS/ESD Symposium Proceedings (1982).

Kabaker, Brian, *Pulse DC Ionization for Work Stations,* EOE Proceedings (1986).

Kolyer, John M. and Watson, Donald E., *Cost Effective Methods of Testing/Monitoring Wrist Straps,* EOE Proceedings (1986).

McCraty, Rollin, *Methods for Controlling Ion Levels,* EOE Proceedings (1985).

Military Handbook DOD-HDBK 263, *Electrostatic Discharge Control Handbook for Protection of Electrical and Electronic Parts,* Assemblies and Equipment (May 2, 1980).

Mykkanen, Fred, *The Room Ionization System,* An Alternative to 40% RH, EOE/ESD Symposium Proceedings (1983).

Ott, Henry, *Ground-A Path for Current Flow,* Bell Laboratories.

Unger, Burton A., *A Room Ionization System for Electrostatic Charge and Dust Control,* EOS/ESD Symposium Proceedings (1984).

Vosteen, Robert E., *Capabilities and Limitations of Electrostatic Voltage Followers and Electrostatic Fieldmeters,* Monroe Electronics.

Vosteen, William E., *A Review of Current Electrostatic Measurement Techniques and Their Limitations,* EOD Proceedings (1984).

6

Ultra High Purity Water—New Frontiers

Terry L. Faylor and Jeffrey J. Gorski

PURE WATER AND THE SEMICONDUCTOR INDUSTRY

If you manufacture semiconductors you must manufacture pure water. If you manufacture pure water, 99% of you use the same technology and, therefore, you each experience the same problems with both the technology and the quality.

Like never before chip makers are focusing on the purity of materials they use and the cleanliness of the environment in which they use them. This march toward ever-more-compact circuitry makes contaminants, that weren't a concern previously, a new problem to be solved today. A wayward particle from a chemical bath, a fleck of dust, an impurity in a gas used during manufacture or organics leached from a pure water distribution system can render todays one micron technology inoperable.

As a rule of thumb, chip makers can tolerate contaminants about one-tenth to one-third the size of circuit features. So with a one micron feature contaminants larger than 0.10 micron to 0.33 micron become a problem.

All the important process steps associated with the manufacturing of semiconductors are affected primarily by the same contaminants in water, air, gases, and chemicals. This "motley crew of equation spoilers" is the basis for impurity. These "spoilers" consist of inorganic ions, organisms, organic ions, colloids, nontoxic organics, and particulates.

The Pure Water Process

The purity of water is directly proportional to its initial clarity and the number of filtration steps it has passed through before it arrives at the process sink. It also depends on the number of chemical treatments or filtration and settling steps it takes to dispose of your industrial waste, and the number of steps taken by the local municipality before it is returned to your company. You can begin to see that a water purification system is created on a step-by-step component basis. No single component can accomplish the purification process by itself. Each piece of equipment in a system plays a specific and vital part in achieving the overall system goal of pure water.

However, there hasn't been a new look in water treatment since 1973 when Reverse Osmosis/Deionization (RO/DI) became the standard for a conventional system. If you have an ultrapure water system, you can bet it's conventional and that it's design was based on source water quality, flow rate volume requirements, pressure losses and automatic or manual mode of operation. Typically, these conventional systems vary only in the selection of the pretreatment filtration methods and the degree of redundancy provided as a protection against the risk of down time. Look at what makes up a conventional system:

1. Granular activated carbon filtration (redundant and optional).
2. Multimedia filtration (redundant).
3. Cartridge filtration (redundant).
4. Injection of acid, chlorine and sequestering agents.
5. A degasifier with redundant transfer pumps.
6. Reverse osmosis with redundant pumps.
7. A deionizer system (redundant).
8. Ultraviolet sterilization.
9. Cartridge filtration (redundant).
10. Chemical storage and containment (acid and caustic).
11. Waste water neutralization.

Since pure water is supplied on a demand basis, it must be stored and distributed via:

1. Fiberglass tanks and redundant distribution pumps.
2. Polishing portable exchange deionizers.
3. Ultraviolet sterilization.
4. Cartridge type microfiltration.

Each component consists of pressure vessels or tanks, internal distributors, external piping, media, membrane and cartridge filters, manual or automatic valves, electrical control panels and pumps. There are standard formulas for the design of the tanks, valves, internals and control functions, and there are specific capacity formulas for the filtration media, ion exchange, and membrane applications.

The Pure Water Industry Today

A summary of interviews with a cross section of notable names in the high-tech industry in 1985 regarding their experience with conventional pure water systems, told pretty much the same story:

1. It's no easy task and very time consuming (two to three months) for facilities, operations and manufacturing personnel to establish projections and guidelines for volume, flow rate and expansion requirements for a high tech water treatment system.

2. Buying, installing and maintaining a conventional ultrapure water system is a hassle! Once you've decided what your needs are, you either have to hire an outside consultant or utilize internal experts to prepare a specific RFQ. This takes another month or two.

3. Vendor selection is complicated because it is difficult to compare not only the material and design of the specific equipment but the installation, start-up and operational costs. Vendor selection takes another month, and delivery anywhere from four to six months.

4. Conventional systems require substantial space and are delivered in hundreds of pieces to be skillfully and expensively installed, which takes another one or two months.

5. Start up problems are the rule, not the exception.

6. Skilled, attentive, operative, and maintenance labor is critical to success and trained personnel are scarce.

7. Conventional systems depend heavily on the complex, multistep ion exchange process which generates large quantities of acid and caustic waste from on-site regeneration.

8. Chemical waste treatment problems are common. Transfer of "bulk" quantities of hazardous chemicals are required to pretreat not only the raw water, but also post treat the waste stream generated by the chip fabricated process and the cleaning operation of the system itself.

9. Continuous monitoring is required not only to assure compliance with your own pure water quality specifications, but also state and local waste treatment regulations.

10. The frequency of replacement and disposable filters and membranes and the regeneration of ion exchangers cause many interruptions in the pure water production process. Contamination control experts know that each interruption creates a contamination alert in their process loop.

11. Changes in the feed water quality often render custom designed conventional systems ineffective.

It is this continuous demand for labor, chemicals, replacement parts and dependency on ion exchange and monitoring that cause high technology users to label ultrapure water as a "critical process."

The High Purity Water Improvement Industry

With the exception of improvements in reverse osmosis, and ultrafiltration membrane technology and "reversal" technology for electrodialysis, there just have not been many breakthroughs in water treatment in well over a decade.

Look at how the pure water improvement industry is structured. The majority of this industry is made up of fabricators and assemblers. There are few real manufacturers. Those that do exist are limited to manufacturing membranes and ion exchange media. All major components supplied are basically the same. A prospective user can purchase one or many purification components from several companies (all providing similar or identical equipment), who in turn purchase and assemble components all from common vendors. In the semiconductor industry, changes during the last 15 years have been staggering, while there has been virtually no change in the water treatment industry. While the biggest manufacturers have continued to be profitable, the fabricators and assemblers, the key suppliers to the high tech industry, have continuously been plagued with both financial and quality problems associated with a typical "me too" industry. The majority in the pure water improvement industry are trying to serve a "third wave" high tech industry, while operating at a "second wave" level.

Some New Prospects in Pure Water Technology

Let's take a look at what's new and improved in the industry. Two new concepts in packaging technology have been introduced beginning in 1982. The first is the introduction of a double pass reverse osmosis system. In the double pass system, the product water from the first stage is sent to a second stage RO. In the conventional reverse osmosis systems, the first stage concentrate is sent to the second stage reverse osmosis. When utilizing double pass reverse osmosis, the efficiency of rejecting dissolved silica and ionic species is dramatically improved over the conventional method. Application of this technology reduces the requirement for the complex chemical and labor intensive regeneration procedures of ion exchange resin. Application of this technology can replace the need for make-up ion exchange and reduces the load on the polishing loop by delivering cleaner water. This "two-membrane-is-better-than-one" concept is a significant improvement in the manufacture of ultrapure water.

Another great breakthrough in high purity water technology is the development of an automated, off-the-shelf, modularly expandable, portable, ultrapure water pretreatment system that incorporates all three major membrane separation processes. These separation processes are ultrafiltration, electrodialysis reversal and reverse osmosis. This technology produces pure water with no buttons and no chemicals and no labor and it's housed in an environmentally controlled mobile trailer. It can literally be rolled onto a customer site, hooked up to the water source and plugged in. These units take half the space of similar conventional systems of the same capacity, are more energy efficient, cost 80% less to install and can be universally applied on any water supply. These systems do not require the addition of chemicals and their nontoxic waste discharge can be reused for normal plant purposes without further treatment.

PURE WATER TECHNOLOGY

The rest of this chapter will describe high purity water and its contaminants and the principle water purification processes. We will also take a look at a historical perspective of the application of water purification technology and describe conventional water treatment systems and the new hybrid multimembrane high purity water systems. A description of the pure water industry and its structure will be provided. Finally, we will focus on the alternatives that are immediately available to the high-tech user to modify and upgrade their conventional high purity water systems to meet the demands of the 1980's and 1990's.

Contamination Problems

Let's take a closer look at the meaning of high purity water. Perhaps we should take a look instead at the meaning of impure water since in a literal sense, absolutely pure water is a laboratory curosity. The instant pure water is created, it becomes contaminated. What are those contaminants? First, they are the same contaminants that spoil the purity of cryogenic gases, air, and chemicals for semiconductor processing. They differ only in size and quantity and in fact are the same contaminants that create the term "water pollution." Size and quantity is the only difference. If you wish to improve the quality of the water that you are receiving in your plant, you are faced with the removal of inorganic ions, organisms, organic ions, colloids, nontoxic organics, and particulates.

There are two types of organic contaminants in water, nonionic and ionic. These vary in size and are referred by molecular weight and range from 20 to over 100,000 MW. In Table 1 below, the projection for the approximate size of various organics as a function of the molecular weight is provided. Since two molecules with the same molecular weight can vary in size as one can be linear and one can be branched, these values are given as approximate.

Table 1: Approximate Effective Radius of Organic Molecules

Ionic, Å*	Molecular Weight of the Organics*	
	Ionic Species	Nonionic Species
3.5	40	
4.0	50	100
4.5	75	
6.0	100	400
7.5	150	
8	200	1,000
10	300	2,000
18		10,000
22		20,000
50		100,000
65		200,000

*Approximately

Organics. The molecular weights of humic acid found in water is about 1,000 to over 100,000. Fulvic acid, also found in water, has a molecular weight of about 300 to 2,000. At a low pH they are weak acids; at high pH they are salts, therefore ionized. If you study the table closely, you can see that ionized organic species tend to have a greater effective radius for a given molecular weight than nonionic organics. Ionic species have water of hydration and, due to the like charges they have in their molecule, they repel internally causing the molecule to stretch. The organics in water can also have high molecular weights.

Colloids. Colloids are aggregates of molecules too small to be seen with an optical microscope. A large percentage are dissolved in solution. The surface of colloidal particles carries either a positive or negative charge. This results in the charged particles repelling each other. The charged surfaces prevent the particles from becoming large enough to settle. Colloids have a range size from 0.005 to 0.2 micron or 50 A to 2,000 A. These colloids are constantly in motion.

Organisms. There is a wide range of organisms that can be found in water i.e., protozoa, bacteria, viruses, fungi, cysts and their metabolic products. Their sizes vary also, and most of these can be seen or photographed under an optical microscope if proper techniques are used. The approximate size ranges are given in Table 2.

Table 2: Size of Various Organisms

Organism	Size Range, μm	Å
Bacteria	0.3–30	3000–300,000
Fungi	6	60,000
Viruses	0.003–0.05	30–500
Smallest visible particle	40	400,000

Particles. Particles can vary in size too. They can be classified in both micron and submicron sizes. These can be found in water in the form of silt which is typically greater than two micron, and clays, less than two micron. They can come from the water treatment processes themselves like activated carbon, broken ion exchangers, and sloughings from membranes, filters and material of construction as well as precipitates formed such as oxides.

Removal Technology

All of the above mentioned contaminants; inorganic ions, organisms, organic ions, colloids, nonionic organics and particulants can be altered or removed by the following materials or media. Ion exchangers remove ion species and particulates greater than 5 micron. For lower sized particulates, powdered ion exchangers are used. Some organics and colloids are absorbed by activated carbons and polymeric adsorbants. Some are removed by filter media, others by submicron filters and ultrafilters. Membranes like reverse osmosis remove ionic species and organic species. The molecular weight of the organics removed depends on membrane type. Electrodialysis removes ionic species and low molecular weight organic ionic species less than 200 MW. Table 3 provides information relative to the pore size of various media. In studying this table, remember that: 1 micrometer = 10,000 Angstroms = 1/1000 of a millimeter = 1/10,000 of a centimeter.

Table 3: Pore Sizes of the Media

Ion Exchangers	Size in Å
Inorganic ion exchangers molecular sieves	4-13
Organic ion exchangers microporous-gel type	<2
Macroporous resins	
extraction type (mesoporous)	10-25
standard resins (phase sep.)	50-800
organic scavengers	
extraction type	25
phase separation	250-1,000
colloid removal	25,000-1,000,000
Adsorbents	
Activated carbons	
type (1)	10-40
type (2)	150-400
type (3)	500-2,000
Polymeric	
XAD-2	90*
XAD-4	50*
XAD-7	80*
XAD-8	250*

*Approximate Average

Details of the Purification Process

Due to the different natural water sources (wells, reservoirs, rivers, etc.) and geographical locations, the range of contaminants which municipal water contacts and absorbs varies tremendously. The fact that municipalities frequently have a number of different water sources almost guarantees that influent water will fluctuate over time in terms of quality.

Consequently, water purification is one of the few industries that has no control over the raw material that is to be processed. Other industries typically specify the quality of raw materials and design their manufacturing processes accordingly. In the past a system capable of removing all types of contaminants in any possible quantities would be prohibitively expensive. Therefore, the design of an ultrapure water purification system was based upon a best *estimate* of what type of water would be provided.

Each water purification system must be created on a step-by-step, component-by-component basis. Each piece of equipment in a system plays a specific and vital part in achieving the overall system goal—high purity water. No one component accomplishes the purification process in itself. Rather, each alters the water to some degree as it passes, so that the next component, or the one after, can effectively do its job.

From a technological viewpoint, there have been few breakthroughs in water treatment process methods during the past 50 years. The principal water purification processes are as follows:

1. Initial Filtration
   a. Media
      1) Diatomaceous earth
      2) Activated carbon
      3) Sand
      4) Multimedia

   b. Membrane
      1) Cartridge filtration
      2) Ultrafiltration

2. Deionization (DI or demineralization)
   a. Ion exchange
   b. Electrodialysis
   c. Reverse osmosis
   d. Electrodeionization

3. Softening
   a. Ion exchange
   b. Lime/soda

4. Chemical Treatment
   a. Acid
   b. Caustic
   c. Sequestering agents
   d. Polymers

5. Bacteria Treatment
   a. Chlorination
   b. Ozonation
   c. Iodination

Deionization. From a historical viewpoint producing pure water started with distillation then deionization. Let's take a chronological look at these technologies and their applications.

Deionization became commercially available during the early 1930's replacing distillation as the primary method of pure water production. Today's deionizers, with improvements in resin, materials of construction and automation, utilize the same basic design of the 1930's. Deionizers, however, still needed to be protected from particles, colloids, organics, silica and microorganisms. For the production of ultrapure water in the semiconductor industry, the only method of achieving the purity required is deionization (DI). There is no other technology available to reach this asymptotic plateau, plain or simple. However, DI is not "an island" unto itself and in practice, it was found that it required pretreatment to extend its life and reliability. The DI resin was susceptible to rapid fouling if used directly on well or municipal water. It needed to be protected from the likes of particulates, colloids, organics, silica and microorganisms.

Flocculation. On the large industrial systems of the 1960's, flocculation was the state-of-the-art. This process was very good at removing particulate matter in the form of colloidal and micronic particles and helped to remove some of the dissolved organic matter. However, flocculation was not without its inherent problems. Typically, it required constant surveillance of the incoming water supply along with sand filtration and activated carbon as mandatory post-treatment to act as buffers prior to DI. Due to its cost, its capricious temperment, its high manpower requirements and its overall unreliability, another pretreatment process to DI was sorely needed.

Reverse Osmosis. On the large industrial systems of the 1970's, RO became the state-of-the-art. Not only did RO do everything a flocculator could do, it could also remove microorganisms, pyrogens, silica and organics. It also had the extra added benefit of removing the active ions. It was found to remove 98% of hardness, total dissolved solids and silica. Old DI systems now found that their time between regenerations was increased as much as tenfold; that the DI resin itself had a much longer useful life; that maintenance was cut drastically; that DI water costs were less; that the manpower requirements were reduced; that the reliability of the product water was improved, and, that the chemical requirements were lowered by 90 to 95% reducing costs and minimizing waste disposal problems. RO seemed to be the ultimate pretreatment to DI for a continuously reliable source of ultrapure water. The balloon was soon to burst.

It was discovered that RO was not a "black box" means of pretreatment. Any RO unit itself requires more than the originally quoted "minimal" level of maintenance. The level of chemical pretreatment can be costly and the membranes themselves are quite fragile requiring their own level of pretreatment. Typically, this was a sand filter post-treated by a cartridge type filter and finally chemically treated with acid, chlorine and a sequestering agent. Even at that, membrane life had never exceeded 2 to 3 years of reliable operation.

Ultrafiltration. During the last decade, an economical process for the treatment of raw water to remove suspended or colloidal matter was developed. This process, ultrafiltration, is the most effective replacement for media and cartridge filters. Its ability to perform in the 0.05-0.2 mm or 15-50 Å range is a vast improvement over the 3 micron to 200 micron range of the aforementioned filtration techniques.

Ultrafiltration can also be utilized to replace microfiltration which is the conventional method of removing suspended solids and particles in the final polishing loop of an ultrapure water system. A UF system is essentially maintenance free requiring only periodic back washing and cleaning with small quantities of chlorine.

Electrodialysis. The oldest and most reliable membrane technology, electrodialysis, has been in use since the early 1950's. This ion transfer process, using ion selective membranes and small quantities of electrical energy to remove or concentrate dissolved ions/salts in water, underwent a major product redevelopment in the early 1970's with the addition of a reversal technique which provides self-cleaning of the vital membrane surfaces as an integral part of the demineralizing process. Because this proprietary process is so effective in removing high concentrations of dissolved solids (1,000 to 10,000 ppm) it has become an important, reliable and economical method of demineralization to produce fresh potable water and industrial pure water. The EDR's ability to produce 1 to 10 ppm water from municipal sources containing 10 to 1,000 ppm with outstanding process ability, stability and reliability has led to the increased use in pretreatment to not only ion exchange but reverse osmosis.

EDR membrane technology offers many advantages over osmosis. It requires virtually no chemicals for normal operation; operates at low pressures (50 to 60 psi), eliminates the need for degasifiers, operates at high water recovery rates, is not subject to silica scaling, is easily cleaned and has actual membrane life expectancies of 7 to 10 years.

The typical conventional industrial high purity water system utilized for semiconductor use since the early 1970's, consists of five subsystems, as shown in Figure 1.

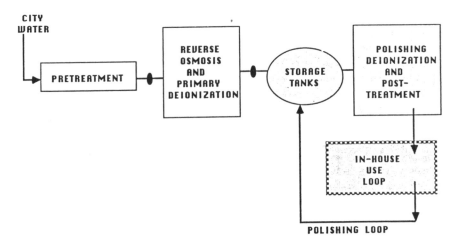

Figure 1: Typical conventional ultrahigh purity water treatment system.

The Pretreatment Dilemma

Each of the subsystems shown in Figure 1 is made up of a variety of components. Since every system design depends upon the particular requirements of the application, i.e. initial water quality, volume, flow rate, and desired quality specifications, not all systems will contain all of these components. Certainly some systems will contain equipment not mentioned here.

A major function of the pretreatment system as shown in Figure 2 is to protect the reverse osmosis membranes. Obviously the weakest link in the RO/DI system is the pretreatment system that must be individually designed in accordance with the ever changing water analysis of the incoming municipal water supply. The problem with designing pretreatment to a particular water analysis is the fact that water is ever changing and the sample water analysis used for its design is never typical for the water supply year round. Until recently, you could not economically design a pretreatment system ahead of an RO unit to cover all possible levels of contaminants.

Pretreatment, as shown in Figure 2, is the process of preparing the raw municipal feedwater to meet the tolerances of the purification equipment which follows.

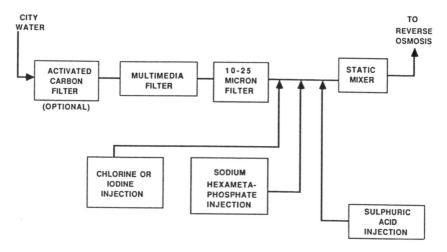

Figure 2: Typical reverse osmosis pretreatment procedure.

The Steps for the Pretreatment Process

Most pretreatment begins with some combination of filters to remove most of the suspended solids, organics and microorganisms, as follows:

1. Activated carbon filter to remove nonionic organics with a high molecular weight (the reverse osmosis membrane downstream removes organics over 200 MW).

2. Multimedia filter, comprised of 6 to 8 levels of aggregates from gravel through sand and quartz, to remove silt and suspended solids (typically in the 20 to 50 micron range).

3. 5 to 10 micron cartridge prefilter to remove more particles in the 5 to 10 micron range.

4. Chemical pretreatment. To control pH, inhibit bacterial growth and prevent formation of scale on the RO membrane, chemicals are injected into the stream before the water enters the reverse osmosis unit:

    a. Chlorine, or iodine is injected in front of the media filter to prevent the possibility of slime, algae or bacteria growth on the media filter and in the reverse osmosis unit.

    b. Sodium hexametaphosphate is injected upstream of the RO unit to control scale formation.

    c. Acid injection, either hydrochloric or sulfuric, controls pH and minimizes RO membrane hydrolysis.

    d. To assure pH control efficiency and reliability, a static in-line mixer allows the chemical feed pump to operate against a constant backpressure.

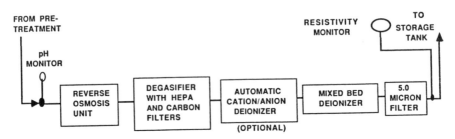

Figure 3: Typical reverse osmosis/deionization/make-up technique.

Degasification

Carbon dioxide, produced by the acid pretreatment, must be removed from the product stream by a degasifier. The basic principle is to pass water through a tower packed with plastic saddles which break the water into fine droplets, lowering the partial pressure of the CO_2 gas in the water. Air is then forced through which has been filtered as follows:

1. HEPA filtration to remove airborne particulates down to 0.2 micron absolute.

2. Activated carbon filtration to remove noxious gases generated by the typical parking lot environment of the system. Activated carbon filters are provided not only on the bottom of the degasifier, but also on the top of the tower.

After the reverse osmosis unit reduces impurities by an approximate order of magnitude, the primary deionization tank(s) (this can be cation followed by anion and/or mixed beds) removes most of the remaining dissolved solids, raising purity to the 10 to 18 megohm range (0.5 ppm to 0.027 ppm).

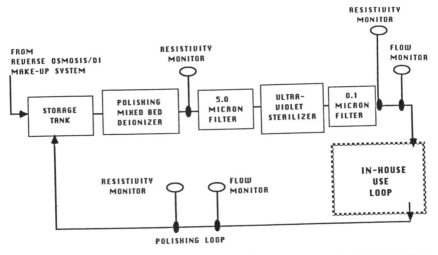

Figure 4: Typical storage and polishing deionization and post-treatment system.

Storage and Final Polishing

A storage tank, which varies from a few hundred to many thousands of gallons depending on need, acts as a buffer, allowing water to be used at a faster rate than it can be treated on a continuous basis.

From the storage tank, the high-purity water goes through a final deionization recirculation loop. Again, the pure water is recirculated at a flow rate that is 25% to 50% greater than actual maximum usage. Unused water is returned to the storage tank and kept in a state of constant motion to maintain maximum purity. Velocities of 5 to 10 feet per second are recommended in the pure water piping loop. The polishing mixed-bed deionizers produce the desired 18 megohm quality.

The polishing system includes:

1. Ultraviolet sterilization which destroys over 99% of bacteria, virus and algae present after polishing DI.

2. 0.45 to five micron roughing filters to protect the final filters by removing any matter, such as resin fines from the deionizers, that could clog or foul the finer matrix of the final filters.

3. Membrane final filter cartridge to remove particulates (0.1 to 0.2 micron rating absolute).

4. Resistivity monitors which provide instant visual reading of water purity in three locations: entering the plant, returning to storage, and in the filter rinse lines.

REQUIREMENTS FOR IMPROVED WATER

Ultrahigh purity water quality specifications have changed considerably since 16K bit drams were being manufactured. Although resistivity has not changed much, very severe standards have been applied to the numbers of particulates, microorganisms, and bacteria. This is true since a VLSI chip is greatly affected by impurities such as particulates with the decrease in the dimensions of the VLSI pattern due to the continuance of integration in VLSI designs. The requirements for the particulate size to be counted have been changed from 0.2 μm to 0.1 μm. The specification for the bacteria count per ml has also been changed since the 256K bit dram production commenced. These are common specifications and as such the standard has been changing to incorporate these more severe requirements.

Table 4 was produced to illustrate the relationship between the quality of ultrapure water and the growth of VLSI's. This data, which varies from manufacturer to manufacturer, serves as a good example of the past, present, and future water purity requirements for the semiconductor industry. No attempt is made to explain the various analytical methods used to confirm test results conforming to the recommended water quality specifications shown in this Table.

Table 4: Ultrapure Deionized Water Quality

Relationship Between Quality of Ultrapure Deionized Water and Intensity of Integration of LSIs

ITEMS	64K	256K	1M	4M	16M	
Specific resistance (MΩ-cm at 25°C)	>17*	>18*	>18*	>18*	>18*	A
Particulates (number/ml)	<50	<20 <50*	<10 <30*	<1 <10*	——	
Object particulates (m)	(0.2)	(0.2) (0.1)	(0.2) (0.1)	(0.2) (0.1)	(0.05)	A
Bacteria (number/100ml)	<25*	<10*	<1*	<0.5*	<0.1*	A
TOC (C/l)	<200*	<100*	<50*	<30*	<30	B
Silica (SiO$_2$/l)	<20*	<10*	<5*	<3*	<3	B
DO (mg O/l)	<0.2*	<0.1*	<0.1*	<0.05*	<0.05	
Sodium (Na/l)	<1*	<1*	<1*	<0.1*	<0.05	
Potassium (K/l)	<1*	<1*	<1*	<0.1*	<0.1	
Chlorine ion (Cl/l)	<5*	<5*	<1*	<1*	<0.1	
Copper (Cu/l)	<2*	<2*	<1*	<1*	<0.1	
Iron (Fe/l)	——	——	<1*	<1*	<0.1	
Zinc (Cr/l)	<5*	<2*	<1*	<1*	<0.1	
Chromium (Cr/l)	——	——	<1*	<0.1*	<0.02	
Manganese (Mn/l)	——	——	<1*	<0.5*	<0.05	
Deionized water, Ultrapure deionized water facilities, and range of record of performance	Deionized water	Ultrapure deionized water				
		—— Existing performance record (including planned) →				

NOTE: 1.   Asterisk mark (*) indicates the items required by users.
          2.   A or B indicates the priority.

Early in this chapter we established that we all have common problems with our conventional water treatment systems and point-of-use water quality. About five years ago the authors concluded that a new approach was needed in the application of water purification technology if the "Pure Water Industry" was going to keep pace with the demands of the semiconductor industry in the 1980's and 1990's.

Borrowing the concept of "packaging" from high tech clients, we began working on the multimembrane concept in 1981.

The principle design criteria was to package a water treatment system utilizing the best available proven technologies. This system would preferably have the following characteristics:

● Virtual elimination of chemical feeds and regeneration.

● Virtual elimination of operating and maintenance labor.

● Drastic reduction in operating costs for module replacement, resin replacement, final submicron filters replacement, etc.

● Substantially improved reliability to handle unforeseen changes in feed water.

● Modular sizing for rapid expansion.

● Maintenance of the highest quality ultrapure water under varying conditions.

- Reduction of space requirements.

- Reduction of energy costs related to electrical power.

- Off the shelf (standardized) and pretested.

- Portable and containerized.

- Ease of installation.

- Focus on reuse of waste water without further treatment for other important plant processes, i.e. scrubbers, cooling towers, vacuum pumps, sanitary use, irrigation use, etc.

- Streamlining the selection process for procurement of ultrahigh purity water treatment systems for the high tech user.

THE DOUBLE PASS REVERSE OSMOSIS SYSTEM

It has already been established that efficient water purification is created on a step-by-step component basis and that reverse osmosis and deionization is the focal point and accepted standard of every conventional water treatment system in the semiconductor industry. Our initial focus was to add additional membrane technology, recognizing that two-membranes-are-better-than-one would provide a significant improvement in the manufacture of ultrapure water. It was discovered that as early as 1976 consultants and manufacturers had taken a serious look at the concept of double pass reverse osmosis. In 1975, the *U.S. Pharmacopoeia* published a new set of standards allowing the use of reverse osmosis in the production of both "purified water" and "water for injection" for pharmaceutical use. The U.S.P. set standards mainly for the method of producing various types of water as well as water quality requirements. Other industries outside the pharmaceutical field were upgrading some of their water requirements to meet the U.S.P. standards. Several of the major pharmaceutical manufacturers expressed interest in the use of reverse osmosis to produce a low total dissolved solids water containing few organics and bacterial impurities. This would eliminate deionizers which were well known for the growth of biological contaminants on the resin surface. Field tests and experiments were carried out which developed into commercial double pass reverse osmosis systems. In a typical double pass reverse osmosis system, product water from the first stage is sent to a second stage reverse osmosis. In conventional reverse osmosis, the first stage concentrate is sent to the second stage reverse osmosis. No water is lost from the double pass system, since the reject, usually being of a higher quality than the feed water, is recycled back to the suction side of the first system's high pressure pump.

Installation of double pass reverse osmosis systems in the late 1970's in the pharmaceutical industry were particularly effective in their ability to filter small organics and particles. They certainly reduced the ion exchange requirements by as much as 90% compared to the conventional reverse osmosis systems. The double pass system also allowed for a much higher feedwater concentration than the single pass system. But, since reverse osmosis efficiency is specifically related to the effectiveness of the pretreatment, these early double pass systems required constant surveillance and frequent replacement of the reverse osmosis membranes. They were certainly not space saving, energy efficient, chemical-less, labor-less, or simple to maintain. They did require substantial replacement parts and did keep ultrahigh purity water on the "critical" list in the production of ultrahigh purity water for the microelectronic industry.

Double Pass RO for Microelectronics

Since 1984, the double pass reverse osmosis concept reentered the micro-electronics marketplace. With improved reverse osmosis membranes, namely spiral-wound polysulfone, a double pass system was introduced primarily to focus on the rejection of dissolved silica and ionic species. The objective was to eliminate the need for deionization on the make-up section of existing purified water systems. The parameters considered were the reduction of specific ions, dissolved silica, particles, and organics (both purgeable and nonpurgeable). These units, not unlike the earlier aforementioned models of the late 1970's consisted of RO units in series. The first stage was a parallel series array similar to those used in conventional systems. The product from the first stage was directed into the second pass unit, while the concentrate from the second pass is fed to the first pass feed as shown in the concept Figure 5.

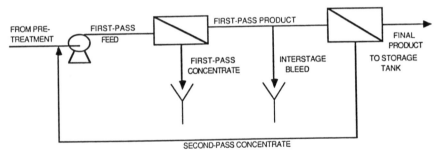

Figure 5: Simplified process diagram of double pass reverse osmosis system.

The first pass reverse osmosis operates at a pressure of 500 to 600 psig and is fed to the second pass at a pressure of 200 to 250 psig. The results of applying this double pass concept have been quite favorable compared to conventional single pass systems. Technical data has shown that the reduction of dissolved silica is dramatic. The rejection of ionic species is quite high on a continuous basis greater than 99%. To date there are no firm conclusions as to the im-provement of rejection of purgeable organics. All RO units are expected to be efficient in removing particles when properly maintained. There is no evidence at this writing that the double pass system is more effective in particle removal as it has been found that the particle concentrations are roughly the same as the concentrations found in final filter water for the semiconductor industry.

An important consideration here is that a RO unit can by itself supply polished quality water in so far as the level of particles is concerned. Use of a double pass RO system, reduces contamination and can provide three major benefits:

- It replaces the work done by conventional make-up system components, thus eliminating or reducing dramatically the need for make-up ion exchange.

- Utilization of multi-membranes reduces the need for chemicals and thus the need for a degasifier, a well documented source of contamination.

- It reduces the load on the polishing deionizers by delivering cleaner water and does not add contaminants of its own.

This last point is particularly important since every system component that is used in an ultrahigh purity water treatment system has the potential to contribute contaminants, i.e. ion exchange resins slough organics, organic vapors and particles can enter a system from the environment thru degasifiers; filters can shed particles and piping products can leach organics.

However, when comparing this new application of double pass RO with our set of desirable criteria, the results are similar to those referred to in the early attempts of the late 1970's with the application of double pass RO technology. Ineffective pretreatment continues to be the most significant problem associated with any type of RO application.

THE DEMAND FOR CONTINUED IMPROVEMENTS

In 1982, we refocused on the fact that pretreatment still remained the single greatest problem associated with conventional ultrahigh purity water treatment systems including double pass reverse osmosis. The major objective of pretreatment is to protect the RO unit downstream. Reviewing early pretreatment techniques found users reducing the size of particles from 25 microns to as low as 5 microns with mechanical cartridge filters but at great cost for replacement cartridges. Sand filters with an effective range of particle reduction of only 20 to 50 microns were not satisfactory. Finally diatomaceous earth was found to be most effective. Although diatomaceous earth had been around a long time in virtually every municipal water treatment plant, its successful utilization in achieving RO pretreatment for semiconductor production was somewhat new in the middle 1970's. Proper implementation of DE filtering was found to extend the life of a RO unit downstream and eliminate silt buildup which caused high pressure differentials across RO membranes. Diatomaceous earth, it was found, could stop particles down to one micron size. However, after several years of application, it was found that DE filtration was a source of contamination in itself. Tiny holes in diatomaceous filter covers occasionally would allow diatomaceous earth to escape and coat the expensive RO membranes rendering them inoperable by raising the pressure and causing expensive replacement.

USING RO UP FRONT

In the early 1980's, one of the major manufacturers of water purification equipment applied ultrafiltration technology to raw water. Little data was available about the application of ultrafiltration on raw water applications and most consultants considered the concept a poor application since ultrafiltration is capable of removing filter particles in the 0.05-0.2 mm or 15-50 Å range. When comparing filtration of ultrahigh purity water in the semiconductor industry in the 0.1 to 0.2 micron range, it seemed unlikely that ultrafiltration on raw water could work at all, cost effectively. However, for years engineers and designers of filtration equipment clearly stated that a filter designed to operate at a given flow rate will operate much more efficiently if the design flow is reduced dramatically. So, in the early 1980's one U.S. major supplier of demineralization equipment tested a series of ultrafilters by dramatically reducing the input feedwater to a rate 3 times less than that specified for the particular ultrafilter. After 6 to 8 months of testing, it was documented that an ultrafilter, operated at as much as 3 times less than specified rates and cleaned on a weekly basis with small quantities of chlorine solution, could be operated effectively at the beforestated-particle-removal levels. After cleaning they could be returned to service without reduction in their ability to remove particles in the 0.05-0.2 mm or 15-50 Å range. Silting density index tests clearly identified this process to be effective in the removal of particulates and organics as pretreatment for RO/Deionization.

THE UPSTREAM REQUIREMENTS FOR DEIONIZATION

While these tests were being conducted on ultrafiltration, the authors reviewed the most common methods of demineralizing water for industrial purposes. The most common method had been the utilization of cation, anion and mixed bed ion exchange deionizers (DI). The relative ease with which DI units reduce the ionic and silica concentrations in water to the part per billion range has made this process indispensable in meeting the more and more stringent purity specifications for deionized water in the power, process, pharmaceutical, and electronics industries. Demineralization of water by properly regenerated ion exchange columns is a simple and reliable process. However, the process of regeneration itself is a complex multistep process requiring error-free control of flow rates, concentrations and sequencing of operations with acid and caustic soda. The reliability and efficiency of ion exchange in the demineralizing process is inversely proportional to the frequency of ion exchange regeneration. We've stated earlier there's a great interest in methods to reduce this frequency.

Over the past decade and a half, membrane demineralization processes, electrodialysis (ED), reverse osmosis (RO) and more recently electrodialysis reversal (EDR) have begun to be used as roughing demineralizers ahead of DI plants. These processes have reduced the load on downstream DI plants making possible reduced regeneration frequency, chemical consumption and waste regenerate disposal while simultaneously improving capacity, product quality and reliability. Since the early 1970's the standard demineralizating method in the semiconductor industry has been the RO/DI plant in which RO not only reduces the ionic load on the DI plant, but also achieves quantitative reductions in particulates, bacteria, and organics above the 300 molecular weight level. However, for optimum operation, RO units require a high level in consistency of pretreatment to remove substances which foul membranes. The next step in developing multimembrane technology then was to apply the concept of electrodialysis reversal.

A TRIPLE MEMBRANE FIVE STEP CONCEPT

Obviously, the application of ultrafiltration to remove particulates, bacteria, nonreactive silica, and some organics and deliver water with an SDI of less than 1.0 as well as an electrodialysis removal unit to remove approximately 90% of the TDS in the feed water would maximize the efficiency of a standard RO/DI system. Thus, a triple membrane process was developed by Ionics, Inc. of Watertown, Massachusetts, one of the world's largest water treatment companies in 1982. The results from this development have been spectacular. The-two membrane is better than one-concept became a-three membrane is better than two-concept. The triple membrane process allowed use of a widely varying feed water quality in terms of both TDS and particulates. The following is a description of the triple membrane concept as shown in Figure 6.

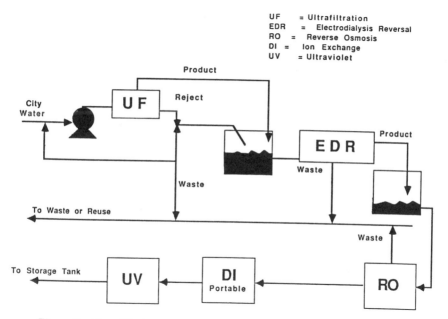

Figure 6: Simplified process diagram of triple membrane demineralizer.

Ultrafiltration

As pretreatment, a relatively coarse (50,000 molecular weight cut-off) spiral-wound polysulfone ultrafiltration unit of very conservative design was chosen. The conservative design made possible low temperature operation, and use on high silt density index (SDI) waters. Untreated surface water supplies (such as those of the Boston and San Francisco Bay areas) often have five minute SDI values as high as 17 to 18. The spiral-wound UF system eliminates concern about broken fibers (which can cause by-passing) and allows high velocity recirculation to minimize flux decline caused by build-up of removed particles. The polysulfone allows sterilization with sodium hypochlorite to control bacteria and slimes. At the same time, the ultrafilter provides very low SDI water to the units which follows and takes the first big cut on entering bacteria, large organics, or particulates greater than about 0.05-0.2 mm or 15-50 Å. Unlike the media filters or 5 to 10 micron cartridge filters which it replaces, the ultra-filter does not require manipulation by backwashing or correct placement of cartridge filters. The use of an ultrafilter eliminates both media and micron cartridge filters which can cause greatly increased bacterial, organic, and particulate matter in conventional systems, and are difficult to sterilize.

Electrodialysis Reversal

As a primary demineralizer, the EDR process was chosen. EDR demineralizes water by passing small amounts of direct current electricity through the water, moving ions out of the water through alternating cation and anion exchange membranes. EDR was chosen to do the primary demineralization (approximately 90%) because it is capable of accepting a wide range of entering feed waters, demineralizing them without chemical feeds, and reducing the pH of its product water to the 5.5 to 7.0 pH range which is desirable for CTA hollow fiber RO feed. EDR also removes TOC, with removal values in the 30% to 60% range being typical. While the organic compounds removed have not been identified, they are almost certainly materials of low molecular weight which might not be removed by UF or RO. Thus it appears that the organic removal achieved by EDR is probably complimentary to that achieved by the other membrane processes. EDR is also exceptionally durable and reliable, capable of accepting continuous free chlorine residuals of 0.5 ppm or less without damage, and of periodic sterilization with considerably higher chlorine or sodium hypochlorite levels. Demineralization by EDR leads to preferential removal of hardness - so that the changes of calcium carbonate, calcium, strontium or barium sulfate scaling in the RO unit which follows are very remote. If the EDR unit should scale up, it is capable of being easily descaled with acid flushing without damage to the membranes. Finally, if it should be necessary to by-pass the UF pre-treatment unit, the EDR unit will accept water with a five-minute SDI as high as 17 to 18 without damage and still provide the specified demineralization.

Reverse Osmosis

The third element in the process design for this system is hollow fiber RO with cellulose triacetate (CTA) membranes. This type of RO was chosen for its compactness, its cost effectiveness, its ability to operate at low pressures, and its ability to operate in the conductivity range 1 to 190 microsiemens/cm without damage to the membranes. The tendency to plugging which is one disadvantage of hollow fiber RO is not a problem in this design because of the UF and EDR pretreatment. The RO unit removes another approximately 90% of the re-- maining minerals, plus silica, bacteria, particulates and organics with molecular weights over 300. The availability of highly pretreated water of low SDI and low scaling potential as RO feed eliminates the necessity of utilizing 5 to 10 micron cartridge filters and of feeding acid or inhibiting agents for scale control. Thus the reliability of the RO operation is greatly enhanced.

Deionization

The fourth element in the new design is ion exchange (DI) polishing to reduce ionic constituents to the lowest possible level and to achieve the 18.3 meg-ohm resistivity which is one indicator of the maximum requirement for ultrapure DI water. After the triple-membrane treatment as described above, the water will typically have a conductivity of 0.5 to 10 or 15 microsiemens/cm—depending on the feed water conductivity—roughly 0.5% to 1.5% of the entering figure. The water after the three membranes and prior to DI will be very low in organics, bacteria, and particulates, typical of the levels from an RO unit with very clean feed. The only other significant contaminants will be CO_2 and SiO_2. The CO_2 levels will probably be in the order to 5 to 10 ppm and SiO_2 0.5 to 10 ppm depending on feed water levels. Because acid is not added to the feed stream, a degasifier is not required, thus eliminating a potential source or recontamination. Because of the low levels of ionic constituents to be removed, it would be cost effective to use off-site regenerated portable ion exchange rather than conventional on-site regenerated ion exchange for this step. Portable exchange is particularly advantageous in areas such as Silicon Valley where high quality off-site regeneration service is available at relatively low cost. The use of off-site regenerated portable ion exchange also radically lowers capital cost through elimination of chemical storage, chemical feed, mixing, monitoring waste disposal, and regeneration piping, valves, and controls. Reliability is increased by the use of portable exchange units which have been subject to positive quality control testing of product water quality prior to delivery to site. Elimination of the necessity for on-site regeneration eliminates the possibility of human or mechanical errors in regeneration, e.g. over backwashing of resin to waste, under or over regeneration, under or over rinsing, failure to achieve good separation and remixing of mixed beds, etc.

Bacteriological Control

Since bacterial control of water is essential for the semiconductor industry, an untraviolet sterilizer was incorporated as the fifth step in the process. When there is no call for water from the triple-membrane unit, the water from the RO unit is constantly recirculated through the ion exchange units and UV unit to control bacteria and ion build-up. We have also found that bacterial control is enhanced in this system because of all the principal membrane elements used are compatible with continuous free chlorine residuals. It is not necessary or desirable to dechlorinate a municipal supply prior to treatment in this system.

THE NEW CONCEPT IN PURE WATER SYSTEMS

The incorporation of the above five steps was based on the result of the survey previously mentioned in this chapter citing the common problems with conventional ultrahigh purity water treatment systems. These criteria were selected to eliminate those common problems. The most common failures encountered with conventional systems are:

- Unanticipated changes in in-coming water quality including the unanticipated and intermittent appearance of random trouble elements.

- Errors of operators, or mechanical manipulations. For example, in backwashing of media filters, positioning of filter cartridges, regeneration of DI plants, chemical feeds to RO plants, etc.
- Problems with bacteria in semiconductor DI water supplies.

The five step system as described does not use chemicals on a routine basis. Neither the UF, EDR or RO units have chemical feeds, this means that the combined waste from the system can be used for normal plant purposes such as sanitary, cooling, irrigation, etc. The waste water differs from the feed only in higher concentration of TDS and other impurities. There is no need for on-site chemical regeneration of ion exchange resins or subsequent neutralization of the chemical waste. All ion exchange resin used is placed in portable units which are removed when exhausted by local portable exchange DI service companies. This system consists of a unique combination of membrane and ion exchange processes which have been preinstalled and pretested in an over-the-road trailer. This portable mobile system can supply make-up ultrapure water to an in-house polishing loop within a week of arrival on job site. The system operates unattended and starts and stops via a signal from the polishing loop storage tank. Resistivity control is included to prevent poor quality water from entering the polishing loop, should excursions of any process parameters or malfunctions occur they are logged on an event time recorder. Malfunction alarms are transmitted to security personnel and an automatic telephone dialing system advises the service vendor. These systems can treat almost any municipal or industrial water supply and produce ultrahigh purity specification grade water effectively and continuously because no chemicals are added. The waste water from these units has high environmental acceptability and value for reuse. Reuse of all of the waste water from these units can save money and help achieve the water conservation goals which are so important today. These ultrahigh purity water systems offer the semiconductor industry convenient, cost effective systems which virtually eliminate or drastically reduce custom engineering, purchasing, project management, installation, commissioning, start-up and space cost, and the substantial time delays involved in these activities. It is literally possible to "plug-in" this packaged concept in water treatment in less than a week after arrival on a job site and produce specification grade pure water from a standard, factory-built, pretested, modularily expanded unit with all the redundant back-up and reliability factors built in. It takes less energy, less space, less labor, and all of its waste water can be used for practical purposes within the plant itself. Most significantly it improves drastically the overall quality of the water making it possible to meet the demands of the semiconductor industry for the 1980's and the 1990's.

SUPPLIERS FOR HIGH PURITY WATER

The Water Quality Association lists over 175 manufacturers and suppliers of water purification equipment. With few exceptions , most of these companies build specific components for water purification systems.

A manufacturer/supplier is any individual or firm engaged in the manufacture or assembly and sale, through established channels of distribution, of a complete unit of equipment; and any individual or firm engaged in the manufacture, production or sale of components, parts, equipment, commodities or other merchandise or services for other members of the industry or for users of industry equipment, products and services.

A retailer/distributor is any individual or firm engaged in the retail selling of industry equipment, products and/or services to the user, and any individual or firm that buy industry equipment and products on their own account for resale to retailers, in the course of which they warehouse and promote the sale of such equipment and products.

In reality, there are few manufacturers. The majority are fabricators and assemblers. Manufacturing is limited to products such as ion exchange resin, membranes, chemicals, tanks, valves, and instruments. No one company is completely vertically integrated (i.e., manufactures all basic components such as resin, membranes, tanks, valves, etc.).

The water purification industry serves several major markets; household, commercial, and industrial/municipal by supplying components, services and systems.

Like most industries, there are companies which specialize in each marketplace, and there are some which serve all markets with one or more product lines or services.

There are a few companies which concentrate their marketing on custom components and systems for the municipal/large industrial (over 200 GPM) water treatment marketplace. Pure water systems evolved naturally from the industrial systems business.

Typically, these companies provide design, engineering, and fabrication/assembly services for each major system component. Equipment and components are purchased for each job upon receipt of a purchase order. Company personnel then fabricate and assemble major subsystem components. All subsystems are then shipped to the customer's site where the installation is typically completed by a mechanical contractor selected by the customer. These companies normally provide start-up supervision and personnel training for their customers.

As reported earlier, technological breakthroughs in the Water Purification Industry have been few. Improvements in membrane technology for ultrafiltration, reverse osmosis, and electrodialysis represent the state-of-the-art in water purification technology.

A "ME TOO" INDUSTRY

Water treatment is a "me too" industry with little uniqueness except perhaps the scope of the product line for each assembler/fabricator and the marketplace served.

From an engineering viewpoint, all major components supplied in the industry are basically the same. A mixed-bed deionizer, regardless of the designer or assembler, is built the same today as it was 30 years ago.

It is important to focus on this fact; that a system is compiled of many components, each of which are designed based on influent water quality, flow rate, volume requirements, pressure loss, and automatic or manual mode of operation.

Thus, each component consists of pressure vessel or tanks, internal distributors, external piping, media, manual or automatic valves, and electric control panels if applicable.

There are standard formulas for the design of tanks, pipes, valves, internals and control functions. There are specific capacity formulas available for filtration media, ion exchange, and membrane application. In short, a prospective customer can purchase one or many water purification components from several companies, all of which provide similar or identical equipment.

Proprietary products or process patents are held primarily by the resin and membrane manufacturers which serve most suppliers in the industry.

When reviewing the list of 175 plus manufactuers/suppliers, it becomes obvious that most companies in this industry were formed to take advantage of a creative uniqueness in a particular market or product line. A classic example is Aqua-Fine Corporation, the major supplier and one of only three or four fabricators/assemblers of ultraviolet sterilizers in the industry. Aqua-Fine has standardized its product line by paying the overhead burden of standardization, while simultaneously serving specialized applications in the marketplace for bacteria control. Aqua-Fine sells to other suppliers in the industry as well as to end users.

Sophisticated engineering departments with graduate engineers have not been widely utilized in the Pure Water Industry, since most components can be copied or designed by application of simple formulas.

The major suppliers of components and systems in the Pure Water Industry differ only in product lines, standardization, and primary markets served.

A company's success is largely dependent on utilizing a standard line of simple components which can be fabricated on a repetitive basis. Maintenance of current pricing and delivery information from two to three other vendors for each component is mandatory to establish competitive pricing and terms.

Vertical integration is the other key! The greater the ability to manufacture internally, the greater the profits.

If a given supplier can manufacture the tanks/vessels, linings, resin, membranes, valves, panels, etc., it has a distinct advantage over its competitors who use distributors for each of these components.

SUMMARY

We have now reviewed the step by step processes utilized over the past fifty years to purify water. In the very beginning of the chapter, it was suggested that 99% of all of the persons who read this section of the book utilize typical conventional high purity water treatment systems and that each of you face the day to day common problems that keep water on your "critical" list as it relates to the production of your microelectronic devices. It might be obvious at this point that most of the problems related to water that you are experiencing can be put to rest by simply the addition of an ultrafilter and an electrodialysis reversal unit ahead of your existing reverse osmosis systems and by adding a second pass reverse osmosis to your already existing RO system you can meet your demands for ultrahigh purity water for the 1980's and the 1990's.

In the process, you can expect to reduce energy, chemicals, labor, ion exchange, replacement parts and waste water problems all at a dramatic level while improving your overall water quality and reliability while reducing the risk of down time and management headaches.

If you want to remove the label "critical" from your ultrahigh purity water system process you may want to consider other concepts and application of technology that are catching on in the search for the ultimate in pure water. Take a close look at the utilization of new piping products covered in another chapter of this book. The greatest pretreatment system in the world combined with standard reverse osmosis deionization would be to no avail if your storage/distribution systems fails to provide contamination free water. Check the velocities of the water as it flows to your process areas. Consider the use of hot pure water—it will not only improve cleaning operations dramatically but also will reduce your DI volume requirements at the same levels—consider ultrafilters and reverse osmosis systems in your process distribution loops—consider what ozone can do for oxidation and sterilization purposes—consider more energy efficient pumps—and most of all consider greater reuse of reclaimed water.

Water is no longer in unlimited supply!! Your action to upgrade your existing systems utilizing these breakthroughs and application of technology will dramatically impact your community by reducing contaminant levels from your existing RO/DI plants by 80% to 90% while at the same time reducing chemical consumption and subsequent toxic chemical pollution.

Not only will your water be more pure and more reliable than ever before, but you will have made a personal contribution in reducing water pollution in our rivers, streams, lakes and oceans. You'll save your company money, reduce management stress and feel good about your action.

7

Deionized (DI) Water Filtration Technology

Mauro A. Accomazzo, Gary Ganzi, and Robert Kaiser

1. INTRODUCTION

1.1 Semiconductor Product Characteristics and Trends

Advances in the semiconductor industry over the past two decades have increased the purity requirements of process fluids that are used in the manufacture of semiconductor devices of ever increasing complexity. Deionized (DI) water is one of the most critical fluids used in the manufacture of semiconductor devices. Because a wafer may be rinsed with DI water many times, it may come into contact with from 10 to 500 liters of DI water during the manufacturing process. Because of this repetitive exposure, any impurities in the DI water at points-of-use will have numerous occasions to be transferred to the wafer surface, and thereby significantly influence device quality and yield.

As the capabilities and complexity of semiconductor devices have increased, their geometry has shrunk. One of the most significant changes in the semiconductor industry of the past decade has been the change in design rule requirements. According to one industry expert,[1] as shown in Figure 1, industry design rules which were in the 5 to 8 micron size range in the mid-seventies, have dropped to the 1 to 3 micron range currently, and are expected to drop by an additional factor of three by the year 2,000.

1.2 Effects of Contaminants on Product Performance and Yield

As design rule geometries decrease, so do the levels of contamination that can potentially damage a device based on that geometry.

Suspended particles are the contaminants of greatest concern to the electronics industry. Particulate contamination has been blamed for up to 80% of process yield losses, and it is this yield that is the principal driving factor in the final cost per device.[2] Once particles have adhered to the surface, it is difficult

210

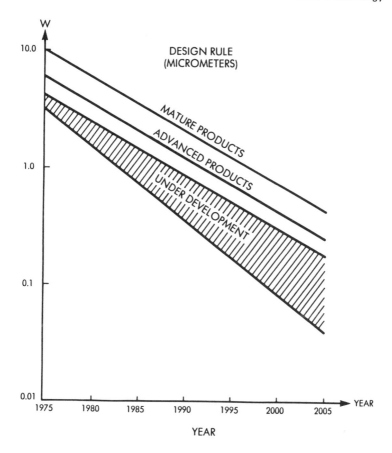

Figure 1: Evolution of microelectronic device geometry.[1]

if not impossible to remove them. This is particularly true of sub-micron particles, as the ratio of adhesive forces to drag forces which can be applied to a particle increases as size decreases.[3] An unfortunate consequence of this is that the yield loss from contamination at various steps in the process is cumulative,[4] and the yield loss is exponentially dependent on the cumulative defect level.[5]

Particles can create device failures by causing a mechanical defect in a device structure or by chemically contaminating the device. The former is more generally recognized. A defect of lateral dimension of one-tenth to one-fifth of a feature's width can cause a device to behave in an unacceptable manner. This results in a minimum critical particle size, assuming 1 μm device features, of 0.1 to 0.2 μm.[3,6] In most devices, however, the smallest features are not the widths of the lines but the thicknesses of critical films, most commonly gate insulators in metal-insulator-silicon (MIS) devices. These are typically 0.01 to 0.04 μm, reducing the critical dimension for a particle by an order of magnitude.

The latter mechanism, chemical contamination, is less well understood and somewhat more difficult to identify, as the particles may result in aggregated but separate defects and be completely absorbed by a film[7] or active region.[8] Because the impurity level in a film varies with the size and composition of the particle, not all defects result in failures which occur in testing; they may produce failures years after manufacture, presenting both a yield and a reliability problem.[3] The effect on critical device parameters depends on the nature of the contaminant and the inherent sensitivity of the device design.[8]

The breakdown voltage of a film is extremely sensitive to its composition. Change in average breakdown voltage of 50%[7] and of breakdown voltage dependent yield of 40%[8] have been reported. Ionic contamination of junctions causes increased leakage current and results in decreased yield.[8]

1.3 Process Water in Semiconductor Manufacturing

1.3.1 Purity Requirements: The application of new and improved methods of chemical analysis has confirmed that minute levels of water impurities are detrimental to both the function and production yields of microelectronic devices. As device geometries have shrunk, so have the requirements for water purity become more stringent. This historical trend is reflected by evolution of proposed ASTM standard specifications for electronic grade water (Type E-1) since 1979, as indicated in Table 1.[9]

Table 1: Proposed ASTM Standard Specification for Electronic Grade
Water Type E-1

. . . . Draft Date . . . .
	6–79	1–80	1–81	1–82	2–83
Resistivity (megohn-cm	15–18	17*	18** 17*	18** 17*	18** 17*
TOC (ppb)	500	75	200	200	50
Particles >1 μ (max/ml)	2	2	2	2	2
Dissolved SiO$_2$ (ppb)	50	–	–	–	–
Total SiO$_2$ (ppb)	–	5	75	75	5
Living organism (max/100 ml)	1,000	0	1	1	100
Total solids (ppb)	800	50	–	10	10

*Minimum.
**90% of the time.

Source: Reference 9.

Suggested guidelines for pure water for semiconductor processing were recently approved by the SEMI Subcommittee of Standards for Chemical Reagents.[10] These suggested guidelines are based on the quality of final "filter" water found at various U.S. companies involved in manufacturing integrated circuits and on data presented in the technical literature, particularly papers presented at the Annual Semiconductor Pure Water Conferences. As shown in Table 2 four categories of criticality were established. Ranked in decreasing order of stringency, the following categories were established:

Attainable Acceptable Alert Critical

The attainable column presents the highest quality of water that is currently being produced by semiconductor manufacturing plants. It represents the current practical limits of the state of the art. The acceptable column was so labeled since the values represented water quality most often found in manufacturing houses where yields were acceptably high. The quality of water during a "yield bust" provided the limits for the critical condition. The alert condition is representative of situations where there was a growing level of malfunction but where the consequences had not yet become evident in the processing area.[9]

Table 2: SEMI Suggested Guidelines for Pure Water for Semiconductor Processing

Test Parameter	Attainable	Acceptable	Alert	Critical
Residue, ppm	0.1	0.3	0.3	0.5
TOC, ppm	0.020	0.050	0.100	0.400
Particulates, counts/ℓ	500	1,000	2,500	5,000
Bacteria, counts/100 ml	0	6	10	50
Dissolved silica (SiO_2), ppb	3	5	10	40
Resistivity, megohm-cm	18.3	17.9	17.5	17
Cations, parts per billion				
Aluminum (Al)	0.2	2.0	5.0	*
Ammonium (NH_4)	0.3	0.3	0.5	*
Chromium (Cr)	0.02	0.1	0.5	*
Copper (Cu)	0.002	0.1	0.5	*
Iron (Fe)	0.02	0.1	0.2	*
Manganese (Mn)	0.05	0.5	1.0	*
Potassium (K)	0.1	0.3	1.0	4.0
Sodium (Na)	0.05	0.2	1.0	5.0
Zinc (Zn)	0.02	0.1	0.5	*
Anions, parts per billion				
Bromide (Br	0.1	0.1	0.3	*
Chloride (Cl)	0.05	0.2	0.8	*
Nitrate (NO_2)	0.05	0.1	0.3	*
Nitrate (NO_3)	0.1	0.1	0.5	*
Phosphate (PO_4)	0.2	0.2	0.3	*
Sulfate (SO_4)	0.05	0.3	1.0	*

*Values not assignable at this time.

Source: Reference 10.

These categories have also been correlated with the dimensions of the devices being manufactured. For geometries with line widths of 1 to 3 μm, water quality should be in the acceptable to attainable categories. For line widths greater than 3 μm, quality in the alert category or better should be adequate.

1.3.2 Production of High Purity Water: Most wafer fabricators obtain their water from local utilities that are concerned with providing potable water from wells or surfaces. Potable water at point-of-delivery is a complex mixture of impurities, including living organisms, organic and inorganic compounds, ranging from simple salts and low molecular weight hydrocarbons to macromolecules of limited solubility. While fit for human consumption, as received water is not suitable for the production of microelectronic devices, and extensive water purification systems are integral parts of semiconductor fabrication facilities.

It is useful to consider Balazs and Poirier's interpretation of different classes of electronic purity proposed by the American Society for Testing Materials (ASTM) as a means of describing the major elements of a water purification system.[9] A water purification system may be considered to consist of primary treatment section, a secondary treatment section, and a tertiary treatment section, as indicated in Figure 2. Table 3 provides the relationship of the different segments, or levels of treatment, with proposed water quality standards presented in Table 4. Primary treatment consists of those processes required to upgrade incoming water to an Type E-IV purity level. This section may include a variety of pretreatment processes in addition to reverse osmosis (RO) and/or two bed deionization. The secondary treatment section is defined as those processes, mainly mixed bed deionization, needed to upgrade the effluent of the primary section to the resistivity level designated for the desired higher grade (Types E-III, E-II, or E-I). Lastly, tertiary treatment consists of those purification means required to comply with the remaining specifications that do not affect resistivity.

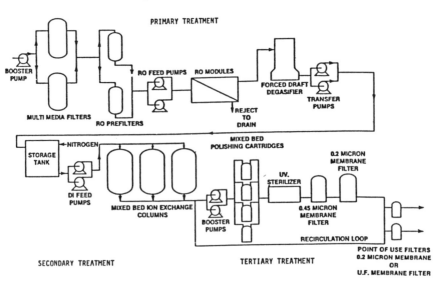

Figure 2: Typical treatment process to produce high purity water for electronics manufacturing.

Table 3: Water Quality vs Water System Components

Water System Section	Purpose	Typical Components
Primary	Upgrade incoming water to Type IV specifications	Sand, filtration, GAC, RO, deionization
Secondary	Upgrade primary effluent to required resistivity (Type E-III through E-I)	Mixed-bed deionization
Tertiary	Upgrade secondary effluent to specifications of desired level that do not affect resistivity	UV, submicron, filtration, etc.

Note: This table represents an interpretation of a proposed ASTM specification for electronic grade water.

Source: Reference 9.

Table 4: Proposed ASTM Requirements for Electronic Grade Water

	Type E-1	Type E-II	Type E-III	Type E-IV
Resistivity, minimum, megohm-cm	18*	15**	2	0.5
SiO_2 (total, maximum, micrograms per liter)	5	50	100	1,000
Particle count (particles larger than 1 micrometer), maximum per milliliter	2	5	100	500
Microorganisms per milliliter	<1	10	50	100
Total organic carbon, maximum micrograms per liter	50	200	1,000	1,000
Copper, maximum, micrograms per liter***	<1	5	50	500
Chloride, maximum, micrograms per liter***	2	10	100	1,000
Potassium, maximum, micrograms per liter***	<2	10	100	500
Sodium, maximum, micrograms per liter***	<1	10	200	1,000
Total solids, maximum, micrograms per liter***	10	50	500	2,000
Zinc, maximum, micrograms per liter	5	<20	200	500

*90% of time with 17 minimum.
**90% of time with 12 minimum.
***Concentrations to be measured periodically at the option of the water system used to aid in the diagnosis of system problems.

Source: Reference 9.

Primary and secondary treatment operations are already discussed in detail in the preceding chapter of this monograph, and will not be further considered here. The rest of this discourse will focus on membrane filtration systems which are the primary elements of tertiary systems.

2. DEIONIZED WATER FILTRATION EQUIPMENT

2.1 Definitions and Requirements

Deionized water filtration equipment includes, for the purpose of this chapter, any filtration equipment installed in a pure deionized water distribution system. This equipment may include the relatively large, multitube filtration housings that may be installed in a major DI water recirculation loop, as well as the "point-of-use" filtration modules that may also be installed at the egresses of the distribution system, immediately prior to the introduction of the DI water in a wafer processing operation. Both types of units serve the same function, which is to eliminate the last traces of foreign contaminants that may be present in what would normally already be considered to be very pure water. More bluntly put, deionized water filtration equipment is a semiconductor manufacturer's ultimate line of defense against waterborne particulate contamination.

The nominal requirements placed on deionized water filtration equipment, which are easy to state but difficult to achieve, are that it totally remove any suspended particulates that may be present in the incoming water stream without itself introducing any suspended or dissolved contaminants into the filtered product water stream.

A conceptual representation of the major elements of a deionized water filtration system are presented in Figure 3. The major element of the system is a thin polymeric membrane of controlled porosity through which the DI water stream is forced to flow, and thus removing suspended particles. Depending on the size and characteristics of the pores, this membrane is considered to be a microfilter, and ultrafilter or a reverse osmosis membrane. The properties and characteristics of commercial point-of-use filtration systems based on these three types of filtration membranes form the basis of this chapter.

The other essential features of a deionized water filtration system are the structural elements used to support the filtration membrane, the housing used to contain the water stream so that it passes through the supported filtration membrane without by-passing and leakage, and the process piping used to convey the filtered water from the point-of-use filter to the wafer processing unit.

A number of contamination control processing units, such as mixed bed ion exchange resin polishing units, other adsorption units, and U.V. sterilizers, are usually used in conjunction with deionized water filtration equipment in a main DI water recirculation loop. Depending on the purity of the incoming DI stream, a number of the process units that are often used at entry of a water distribution system, may also be used in association with a point-of-use filtration system. An example would be small mixed-bed ion exchange resin cartridges to remove trace quantities of dissolved ions from the aqueous stream prior to filtration.

The other important elements of a DI water filtration system that will be considered are the equipment and methods used to measure and monitor the efficacy of the system, particularly in terms of the process and system limitation that these impose.

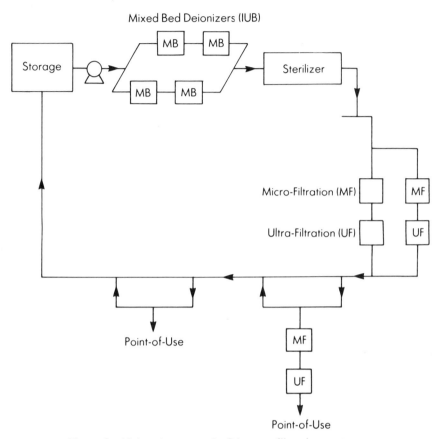

Figure 3: Major elements of a DI water filtration system.

2.2 Microporous Filters

2.2.1 Definition and General Description: Microporous filters (MF) are thin, 50 to 150 μm thick, membranes of polymeric materials that have a substantially continuous matrix structure that contains pores or channels of small size. Filtration occurs because of this open pore structure which allows a fluid to pass through the membrane while capturing suspended solids. The size range for pores of microporous filters is not precisely defined, but is usually understood to extend from about 0.05 to about 10 μm. Two recent papers[11,12] provide an excellent review of membrane technology.

Microporous membrane filters used in DI water distribution systems typically have a pore size rating of 0.22 μm or less, as indicated in Table 5. This table summarizes the characteristics listed in the sales literature of various suppliers for the microporous membrane filtration products they recommend for filtration of DI water in the electronics industry.

Table 5: Characteristics of Commercial Microporous Membrane Filter Cartridges Recommended for the Filtration of Deionized Water in the Electronics Industry*

DESCRIPTION			
MANUFACTURER	Brunswick	Brunswick	Gelman
FILTER NAME	BTSM	BTS1	Acroflow
			Super E
RATED PORE SIZE, µm	0.1	0.2	0.2
CARTRIDGE CONFIGURATION	Pleated	Pleated	Pleated
HEIGHT, IN	10	10	9.8
DIAMETER, IN	2.7	2.7	2.6
FILTRATION AREA, FT²	5.8	5.8	5.2
MATERIALS OF CONSTRUCTION			
MEMBRANE	PS	PS	ACR+/PES
VERTICAL SUPPORTS	PES	PES	PES/PP
CORE	PS	PS	ACE+
SLEEVES	PS	PS	
END CAPS	PS	PS	PS
SUPPORT DISCS			
O-RINGS	S	S	EP
POTTING ADHESIVE	PU	PU	PU
PERFORMANCE PROPERTIES			
INTEGRITY TESTABLE	YES	YES	YES
BUBBLE POINT TEST VALUE			
PSI			
MAX. DIFFERENTIAL PRESSURE			
PSIG	80	80	80
@ TEMPERATURE, 0 C	20	20	88
MAX. BACK PRESSURE			
PSIG	100	100	60
@ TEMPERATURE, 0 C	20	20	88
MAX. OPERATING TEMPERATURE, OC			
CONTINUOUS EXPOSURE	126	126	88
INTERMITENT			
SANITIZABLE	YES	YES	YES
WATER FLOW, ΔP(PSI)/GALLON	0.7	0.3	0.9
PERMEABILITY, GPM/PSIΔP-FT²	0.259	0.552	0.226
RESISTANCE RECOVERY TO 17 MEGOHM, GAL	5	5	20
EXTRACTABLES, MG/FT²	2.5	2.5	

*Key to Table 5 is on page 223.

(continued)

Table 5: (continued)

DESCRIPTION			
MANUFACTURER	Millipore	Millipore	Millipore
FILTER NAME	Durapore	Durapore	Waferguard-40
	CVDI	CWI	WGGL
RATED PORE SIZE, μm	0.2	0.1	0.2
CARTRIDGE CONFIGURATION	Pleated	Pleated	Stacked Disk
HEIGHT, IN	10	10	4.0
DIAMETER, IN	2.9	2.9	2.5
FILTRATION AREA, FT^2	7.4	7.4	2.2
MATERIALS OF CONSTRUCTION			
MEMBRANE	MPVDF	MPVDF	MPVDF
VERTICAL SUPPORTS	PET	PET	
CORE	PP	PP	
SLEEVES	PP	PP	
END CAPS	PP	PP	PS
SUPPORT DISCS			PS
O-RINGS	V	V	V
POTTING ADHESIVE	TP	TP	TP
PERFORMANCE PROPERTIES			
INTEGRITY TESTABLE	YES	YES	YES
BUBBLE POINT TEST VALUE			
PSI	40	70	45
MAX. DIFFERENTIAL PRESSURE			
PSIG	50	50	60
@ TEMPERATURE, 0 C	23	23	20
MAX. BACK PRESSURE			
PSIG	50	50	10
@ TEMPERATURE, 0 C	23	23	20
MAX. OPERATING TEMPERATURE, 0C			
CONTINUOUS EXPOSURE	80	80	100
INTERMITENT		145	
SANITIZABLE	YES	YES	YES
WATER FLOW,ΔP(PSI)/GALLON	0.9	2.0	2.9
PERMEABILITY, GPM/PSIΔP-FT^2	0.150	0.068	0.159
RESISTANCE RECOVERY TO 17 MEGOHM, GAL			
EXTRACTABLES, MG/FT^2			

(continued)

Table 5: (continued)

DESCRIPTION MANUFACTURER FILTER NAME	Millipore Waferguard-40	Nucleopore Polycarbonate	Nucleopore Polycarbonate
	WGVL	QR	QR
RATED PORE SIZE, μm	0.1	0.2	0.1
CARTRIDGE CONFIGURATION	Stacked Disk	Pleated	Pleated
HEIGHT, IN	4.0	10	10
DIAMETER, IN	2.5	2.8	2.8
FILTRATION AREA, FT²	2.2	18	18
MATERIALS OF CONSTRUCTION			
MEMBRANE	MPVDF	PC	PC
VERTICAL SUPPORTS		PP	PP
CORE		PP	PP
SLEEVES		PP	PP
END CAPS	PS	PP	PP
SUPPORT DISCS	PS		
O-RINGS	V	BN	BN
POTTING ADHESIVE	TP		
PERFORMANCE PROPERTIES			
INTEGRITY TESTABLE	YES	YES	YES
BUBBLE POINT TEST VALUE			
PSI	60		
MAX. DIFFERENTIAL PRESSURE			
PSIG	60	100	100
@ TEMPERATURE, O C	20	25	25
MAX. BACK PRESSURE			
PSIG	10	10	10
@ TEMPERATURE, O C	20	25	25
MAX. OPERATING TEMPERATURE, OC			
CONTINUOUS EXPOSURE	100	79	79
INTERMITENT			
SANITIZABLE	YES	YES	YES
WATER FLOW, ΔP(PSI)/GALLON	2.3	0.2	
PERMEABILITY, GPM/PSIΔP-FT²	0.200	0.278	
RESISTANCE RECOVERY TO 17 MEGOHM, GAL		2	
EXTRACTABLES, MG/FT²			

(continued)

Table 5: (continued)

DESCRIPTION			
MANUFACTURER	Pall	Pall	Pall
FILTER NAME	N-66	N-66	N-66
	Posidyne	Posidyne	Ultipore
	NAZE	NIZE	NAE
RATED PORE SIZE, μm	0.2	0.1	0.2
CARTRIDGE CONFIGURATION	Pleated	Pleated	Pleated
HEIGHT, IN	10	10	10
DIAMETER, IN			
FILTRATION AREA, FT^2	9	9	5
MATERIALS OF CONSTRUCTION			
MEMBRANE	NYL/PET	NYL/PET	NYL/PET
VERTICAL SUPPORTS	PES	PES	PES
CORE	PP	PP	PP
SLEEVES	PP	PP	PP
END CAPS	PES	PES	PES
SUPPORT DISCS			
O-RINGS	S	S	S
POTTING ADHESIVE	TP	TP	TP
PERFORMANCE PROPERTIES			
INTEGRITY TESTABLE	YES	YES	YES
BUBBLE POINT TEST VALUE			
PSI			
MAX. DIFFERENTIAL PRESSURE			
PSIG	80	80	80
@ TEMPERATURE, 0 C	50	50	50
MAX. BACK PRESSURE			
PSIG	50	50	
@ TEMPERATURE, 0 C	20	20	
MAX. OPERATING TEMPERATURE, OC			
CONTINUOUS EXPOSURE	50	50	50
INTERMITENT	125	125	125
SANITIZABLE	YES	YES	YES
WATER FLOW, ΔP(PSI)/GALLON	0.6	1.5	1.7
PERMEABILITY, GPM/PSIΔP-FT^2	0.185	0.074	0.120
RESISTANCE RECOVERY TO 17 MEGOHM, GAL			48
EXTRACTABLES, MG/FT^2	4	4	

(continued)

Table 5: (continued)

DESCRIPTION MANUFACTURER FILTER NAME	Pall N-66 Ultipore NIE	Sartorius Sartobran II	Sartorius Sartobran II
RATED PORE SIZE, μm	0.1	0.2	0.1
CARTRIDGE CONFIGURATION	Pleated	Pleated	Pleated
HEIGHT, IN	10	10	10
DIAMETER, IN		2.8	2.8
FILTRATION AREA, FT²	7.5	6.4	6.4
MATERIALS OF CONSTRUCTION			
MEMBRANE	NYL/PET	CA	CA
VERTICAL SUPPORTS	PES	PP	PP
CORE	PP	PP	PP
SLEEVES	PP	PP	PP
END CAPS	PES	PP	PP
SUPPORT DISCS			
O-RINGS	S	S	S
POTTING ADHESIVE	TP		
PERFORMANCE PROPERTIES			
INTEGRITY TESTABLE	YES	YES	YES
BUBBLE POINT TEST VALUE			
PSI			
MAX. DIFFERENTIAL PRESSURE			
PSIG	80	74	74
@ TEMPERATURE, 0 C	50	20	20
MAX. BACK PRESSURE			
PSIG			
@ TEMPERATURE, 0 C			
MAX. OPERATING TEMPERATURE, 0C			
CONTINUOUS EXPOSURE	50		
INTERMITENT	125		
SANITIZABLE	YES	YES	YES
WATER FLOW, ΔP(PSI)/GALLON	0.6	0.48	1.25
PERMEABILITY, GPM/PSIΔP-FT²	0.240	0.328	0.125
RESISTANCE RECOVERY TO 17 MEGOHM, GAL	30		
EXTRACTABLES, MG/FT²			

(continued)

Table 5: (continued)

Key to Table 5

ACE+: ACETAL COPOLYMER
ACR+: ACRYLIC COPOLYMER
BN : BUNA-N
EP : ETHYLENE-PROPYLENE COPOLYMER
CA : CELLULOSE ACETATE
MPVDF: MODIFIED POLYVINYLIDIENE FLUORIDE
NYL: NYLON
PBT: POLYBUTYL TEREPHTHALATE
PC: POLYCARBONATE
PES: POLYESTER

PET : POLYETHYLENE TEREPHTHALATE
PFA : POLYFLUOROALKOXY
PP : POLYPROPYLENE
PS : POLYSULFONE
PTFE: POLYTETRAFLUOROETHYLENE
PU : POLYURETHANE
S   : SILICONE
TP : THERMOPLASTIC
V   : VITON

2.2.2 Material Properties: *Materials of Construction* — While originally made from cellulose esters such as cellulose acetate, cellulose nitrate and mixed esters of cellulose, microporous filters are now made from a wide variety of polymeric, mainly thermoplastic, materials. Examples of these polymers include polyolefins such as low density polyethylene, high density polyethylene and polypropylene; vinyl polymers, acrylic polymers such as polymethylmethacrylate and polymethylacrylate; fluoropolymers, such as polytetrafluoroethylene (PTFE) and polyvinylidene fluoride (PVDF); and condensation polymers such as polyethylene terephthalate, polyamides (nylons), polycarbonates and polysulfones.

Microstructure — Most commercially available microporous membranes have a fairly uniform, cellular, sponge like microstructure, as shown in Figure 4. The porosity of such microporous membranes is usually very high (>70 volume percent). Asymmetric membranes with a graded pore size distribution are now also commercially available. In these membranes, the pore size increases across the thickness of the membrane, as shown in Figure 5. Microporous filters with very uniform cylindrical pores (or channels) also are available (Figure 6). These microfilters are made by a unique track-etch process, in which thin films of polycarbonate or polyester are bombarded with high energy electrons to form latent pores which are then developed by etching. These microporous filters differ from other microporous filters in that the membrane is traversed by parallel cylindrical pores rather than consisting of a three-dimension web. These track etched microporous filters are further characterized by having a narrow pore size distribution and a low pore volume concentration (porosity <10%).

2.2.3 Microporous Filter Configurations of Industrial Interest: For laboratory scale work, microporous membranes are usually used in disc form that range from 13 to 293 mm in diameter. These discs are supported in massive hardware to provide mechanical stability. This configuration is not a practical one, however, for industrial process applications which entail significant flow rates. With a typical 0.2 μm rated filter and a nominal pressure drop of 10 psig (69 kPa), the filtered water output with even the largest discs, which have

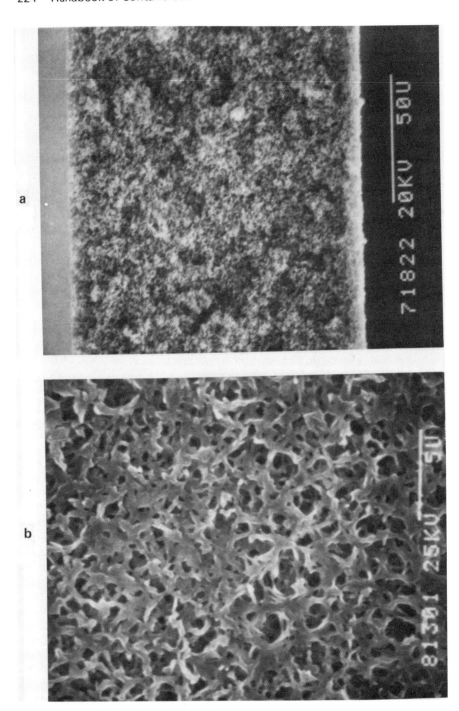

Figure 4: Photomicograph of Millipore PVDF microporous membrane (0.2 μm) (a) cross-section (b) surface.

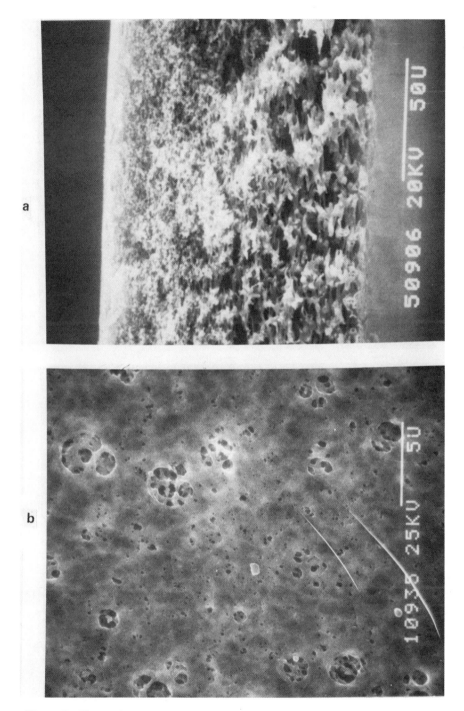

Figure 5: Photomicrograph of asymmetric Brunswick polysulfone microporous membrane (0.1 μm) (a) cross-section (b) surface.

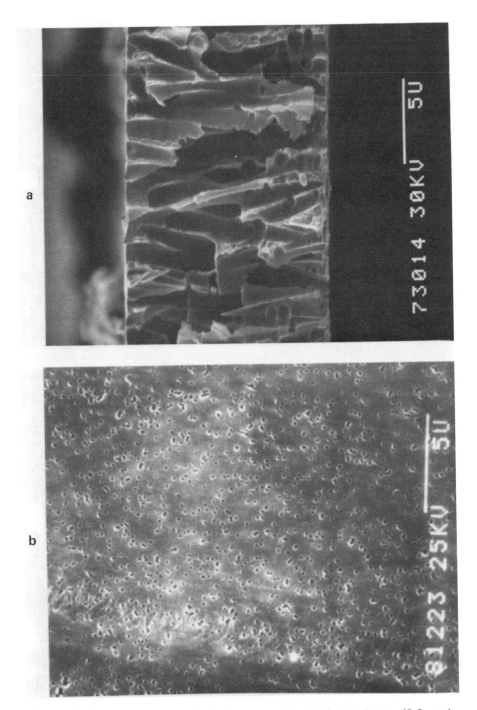

Figure 6: Photomicrograph of Nuclepore track-etched membrane (0.2 µm) (a) cross-section (b) surface.

an effective filtration area of about 500 cm^2, would be less than 6 ℓ/min. This is insufficient capacity for most semiconductor DI water process streams.

For process applications, microporous membrane filters are normally used in the form of cylindrical cartridges, which is an efficient way of providing increased surface area in a small package. Extended filtration area is provided by using either a pleated membrane configuration, or by using parallel membranes in conjunction with a plate and frame arrangement.

Pleated membrane cartridges will contain, in addition to at least one layer of a microporous filtration membrane, whose ends are bonded to each other to form a continuous enclosing surface, an inner support core around which the membrane is wrapped, inner and outer porous support layers, a protective outer sleeve, and o-ring fitted end caps to seal the whole assembly, and to enable its positioning in a filter holder. These supports and caps are made from a variety of thermoplastics, typically polypropylene, polysulfone, or polyfluoropolymer. Thermoplastic or thermosetting compounds are used to bond these various components to each other. While different cartridges can vary significantly in details of construction, most cartridges of this type have a similar external appearance and physical dimensions, and with minor modifications, different suppliers' products are designed to be used in standard sized filter housings, and thus are physically interchangeable with one another. Standard cartridges are hollow cylinders, approximately 10 inches (25.4 cm), or a multiple of 10 inches [up to 40 inches (101 cm)] long, and approximately 2.7 ± 0.2 inch (6 to 7 cm) in diameter (see Figure 7). The available filtration area for a 10 inch long cartridge is typically of the order of 7.5 ft^2 (0.7 m^2).

Figure 7: Photograph of pleated MF membrane filter cartridge.

Cartridges can also be configured with these membrane materials by arranging a plurality of parallel membranes discs in parallel in a plate and frame configuration, as shown in Figure 8. This configuration is believed to provide more rigid and uniform support to the microporous membrane filter than can be achieved in a pleated configuration. This configuration also eliminates any cartridge assembly problems that may be due to inherent membrane material properties such as brittleness.

Figure 8: Photograph of stacked disc cartridge.

2.2.4 Particulate Removal by Microporous Filters: *Capture Mechanisms* — Microporous filters can remove suspended solid particles from a liquid stream by either mechanical entrapment or a variety of adsorption mechanisms. Mechanical entrapment or sieving is the dominant capture mechanism for particles equal to or larger than the surface pore openings. These particles are retained on the filter surface as the liquid stream flows through the filter. Some particles smaller than the pore size can also be mechanically entrapped in the channels of the pores as they attempt to travel the tortuous path through the filter.

Microporous filters can also remove particles that are significantly smaller than the pore size because of the tendency of small particles to spontaneously adhere to any solid surfaces with which they come into contact. As the particles attempt to travel through the filter pores, they will periodically approach or come into contact with the pore walls. If the approach distance is sufficiently small, adhesion forces can develop between the particle and the pore wall that are greater than the thermal, electrostatic and shear (and possibly inertial) forces that tend to keep the particle in suspension. At least two types of adhesion forces can be postulated: electrostatic forces, which requires the particles and the filter matrix to exhibit opposite charges,[13] and secondary valence, or Van der Waals forces, which are nonspecific in nature.[14] To enhance the potential of electrostatic capture, microporous filters can be chemically modified so that they exhibit a residual surface charge. This is usually a positive charge since waterborne suspensoids often exhibit a negative surface charge (e.g., Zeta potential). Retention of particles retained by electrostatic or secondary valence forces can be significantly affected by changes in process conditions that influence the balance between the forces that tend to retain the particles to the matrix surface and those that tend to cause their resuspension in the liquid being filtered. Significant parameters include sudden increases in flow rate or pressure differential across the filter, which result in increasing fluid shear, as well as changes in liquid composition which influence the surface properties of the particles in suspension and of the filter matrix. The latter include, in particular, pH variations and the presence of surface active agents.[15,16]

Pore Size Ratings — The pore size rating of a microporous filter is a measure of its ability to remove particles of known sizes from a liquid stream. The industry practice is to assign a single pore size rating to various products as an indication of the size of the particles that are retained by a certain microporous membrane. This practice has its antecedents in the use of membranes by the pharmaceutical industry to effect the removal of certain species of bacteria in cold sterilized solutions. Thus, a 0.22 μm microporous membrane "sterilizing filter" is defined by HIMA as a filter capable of retaining 1 x 10^7 *Pseudomonas diminuta* per cm^2 of membrane surface at differential pressures of up to 30 psig (2 bar).[17] Microporous membrane filters specified in this manner, however, may not be absolute filters for 0.22 μm particles. *P. diminuta* bacteria are significantly larger than 0.22 μm in size, being approximately 0.3 to 0.4 μm in diameter and 0.6 to 1 μm long,[18] and studies have shown that so-called absolute 0.22 μm filters will allow particles as large as 0.5 μm to pass.[19]

Characterizing the pore size rating of a microporous filter by a single number is, in of itself, a misleading concept because variations in pore size distribution exist in any filter, and because microfilters can effectively capture particles that are smaller than the size of the pores, as discussed above. The situation is rendered more difficult because different manufacturers of microporous filters may not all use the same tests to establish the particle capture efficiency of their products.

Qualitatively, there is a relationship between the capture efficiency of a given microfilter that has a particular pore size distribution and the size of the particles suspended in the incoming solution. In general, three particle size ranges can be considered, namely:

I. Particles that are larger than the largest sized filter pore,

II. Particles that are significantly smaller than the smallest sized filter pore, and

III. Particles of intermediate size.

The filter will capture all Class I particles which are captured by mechanical entrapment. Such retention has been described as absolute retention in that the retention level will theoretically be unaffected by the test or process conditions, as long as the filter in integral. Class II particles are so small that their capture efficiency is low irrespective of process conditions. Class III particles may or may not be captured efficiently depending on the specific operating conditions.

This argument is supported by the results of a number of experiments that examined the retention of closely sized synthetic latex spheres suspended in DI water by various microfilters of different nominal pore size rating as a function of sphere size, loading, and operating conditions.[18,19]

2.2.5 Hydraulic Considerations: Liquid flow rates and processing times are a major concern in the production of DI water. The flow rate of a liquid DI through a microporous filter can be described by the Kozeny-Karman equation which describes the viscous flow of a fluid through a porous bed,[20] and which can be expressed as follows:

$$(1) \qquad G = \frac{AD_o^2 \epsilon \triangle P g_c}{32 \mu}$$

where:

G = Liquid volumetric flow rate

A = Filter Area

D_o = Capillary equivalent diameter

$\triangle P$ = Differential pressure across a filter

ϵ = Filter porosity

g_c = Gravitational constant

μ = Liquid viscosity

Fluid flow rate is directly proportional to differential pressure, filter area and porosity, and the square of the average pore diameter. It is inversely proportional to fluid viscosity. For any filter operating under nominally constant conditions, the flow rate will tend to decrease with time, because capture of particles by the filter effectively results in a reduction in the effective diameter of a pore and in filter porosity. Typical flow rates for pure water through representative commercial microporous filters are presented in Figure 9.

Pressure drop/flow rate characteristics are particularily important for the fine (e.g., <0.22 μm) filters used for DI water filtration. As shown in Table 5 there is a significant pressure drop penalty in going from an 0.2 μm filter to a smaller pore size. Brunswick and Nuclepore both claim that their cartridges offer less flow resistance than competitive products. Brunswick bases its claim on the asymmetry of its membrane which has a graded pore size distribution, and thus results in a higher membrane permeability. The higher flow capacity per pressure drop of the Nuclepore cartridges is due mainly to a higher filtration area per cartridge module.

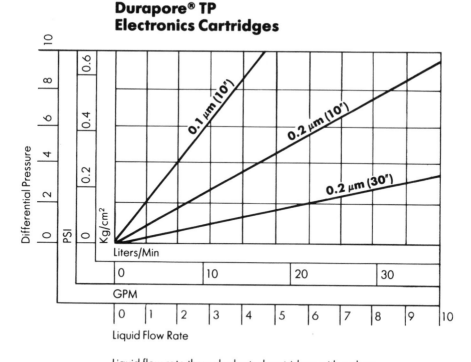

Liquid flow rate through pleated cartridges with a clean
wetting fluid having a viscosity of 1 cps. in housing.

Figure 9: Typical flow rate of water through microporous filters.

The useful life of a microporous membrane filter is limited by plugging of
the cartridge filter. This is a process specific parameter that depends on the level
of contamination present in the feed to the filter.

2.2.6 Filter Induced Contamination: Microporous membrane filters can
themselves be a source of contamination of the downstream liquid stream.
They may contain water soluble materials which could be extracted into the
filtered DI water stream, or they may be the source of particulate material.
Water soluble materials may include both inorganic or organic compounds
that can originate from any of the materials used in a cartridge assembly. Particles may include manufacturing debris, poorly bound substrate and membrane particles, and shards of the filtration media at sealing interfaces.[22]

To minimize the problem of filter induced extractables, all filter manufacturers have made an effort to use inert materials of construction, and to
thoroughly rinse and condition the finished product in what they consider
to be "pure water" before it is packaged and delivered ready for direct on-line
use. A recent investigation by Stewart[23] indicates however that there are considerable differences in the level of water soluble extractables in as-supplied

filters offered by different manufacturers. The results presented in Table 6 were obtained by analyzing samples of 16 liters of ultrapure water that had been circulating for 2 hours through a loop which contained a 10 inch filter cartridge. Manufacturer G had obviously achieved a more efficient final rinse of product than did the other manufacturers as of the time these tests were performed.

Table 6: Water Extracts from Manufacturers 10″ 0.2 μm Filters (ppbW)

Manufacturer	TOC	Na	K	Zn	Fe	Al	Si
	. . . . Limit of Detection						
	5	(0.02)	(0.02)	(0.01)	(0.02)	(0.2)	(0.5)
A	500	0.94	0.15	0.04	–	–	–
B	(a) 5,780	0.42	0.08	0.10	0.07	–	–
	(b)1,530						
C	1,080	11.0	0.08	0.03	–	–	–
D	1,300	2.86	0.07	–	–	–	–
E	610	1.01	0.07	0.11	0.03	–	–
F	190	0.71	0.02	0.04	0.05	–	–
G	100	0.97	0.15	–	–	–	–

Source: Reference 23.

Grant, Peacock and Accomazzo[24] recently reported on studies performed at Millipore to determine the particle shedding characteristics of microporous membrane filter cartridges in high purity water. A schematic of the test system is shown in Figure 10. The test procedure consisted of the following steps:

(1) Background check

(2) Cartridge insertion and system start-up

(3) A two hour steady-flow flush (@ 12 ℓ/min), followed by two hours of pulsed flow which was created by momentarily stopping the bulk of the flow through the cartridge every fifteen minutes.

(4) An additional two-day steady-flow flush, followed by a two hour pulsing cycle, and

(5) If there was significant shedding during Step (4), an additional three-day steady-flow flush, followed by a two hour pulsing cycle.

The tests were performed on a minimum of three samples of four commercial filter cartridges specifically recommended for the filtration of deionized water. These cartridges are described in Table 7. The major differences between the cartridges tested were the membrane type (polysulfone, charged nylon, or PVDF membrane) and cartridge configuration (pleated or stacked disc cartridge).

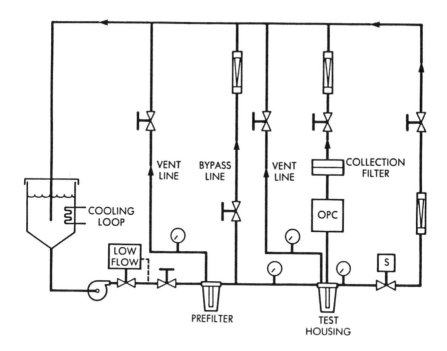

Figure 10: System flow schematic.[24]

Table 7: Properties of Filter Cartridges Evaluated in Shedding Tests

Cartridge No.	Membrane Type	Design Type	Materials of Construction
1	Dual layer Asymmetric 0.1 μm Polysulfone	10 inch pleated	Polyester membrane supports, Polypropylene cage, core and end caps Epoxy adhesive
2	Single layer 0.1 μm charge	10 inch pleated	Polyester membrane supports and end caps Polypropylene cage and core
3	Single layer 0.1 μm modified PVDF	10 inch pleated	Polyester membrane supports Polypropylene cage, core and end caps
4	Single layer 0.1 μm modified PVDF	5 inch Stacked Disc	Polysulfone disc membrane support

Source: Reference 24.

The downstream particle counts measured during the initial 2 hour flushing period for each of the cartridge types tested are presented in Figure 11. The effects of pulsing on cartridge performance are summarized in Figure 12. These results indicate that some microporous membrane filters can be a source of particulate contamination. In all cases, the downstream particle count was initially higher than the background level. For three of the four filters tested, the downstream particle count decayed to the background level within two hours under steady state flow. The one exception was the polysulfone pleated cartridges which continued to shed particles even after five days of flushing. Two of the four filters continued to shed particles during pulsing even after five days of flushing. Figure 13 shows that particle shedding for the nylon filters was increasing. This may indicate an inherent instability of materials in the filter structure. The particles shed by the polysulfone membrane were determined to be polysulfone spheres emanating from the membrane itself. The particles shed from the nylon filter, although visible, were not identified by the energy dispersive x-ray (EDX) analysis method indicating material with atomic numbers below fluorine.

Cartridge configuration appears to be a significant factor in terms of filter shedding under transient flow conditions. When pulsed, all the pleated cartridges shed significantly more particles than did the stacked disc cartridge. The superior performance of the stacked disc PVDF cartridge under transient flow conditions may be interpreted as being due to the greater mechanical support offered the membrane by the plate and frame assembly, which results in less membrane flexure than probably occurs with a pleated cartridge under the stresses caused by the transient flow conditions.

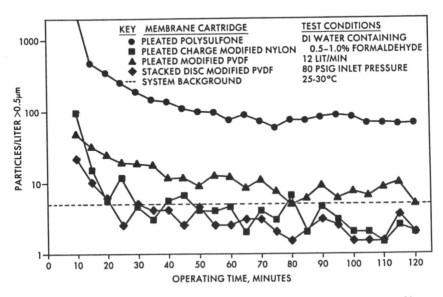

Figure 11: Comparison of cartridge shedding during steady-flow flush.[24]

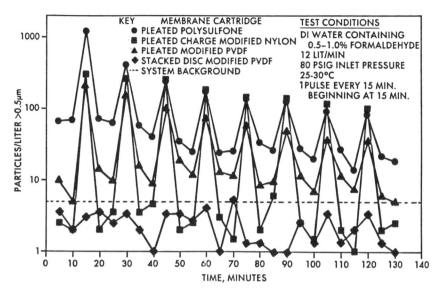

Figure 12: Comparison of cartridge shedding during initial pulsing cycle.[24]

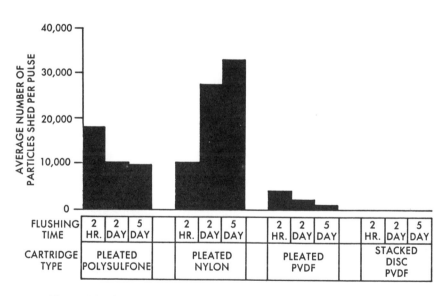

Figure 13: Summary of pulsing effects on cartridge performance.[24]

2.3 Ultrafiltration Membranes

2.3.1 Introduction: Ultrafiltration (UF) is a process of separation where-by a solution (or suspension) containing a solute (or a suspended colloidal material) that is significantly larger than the molecular dimension of the solvent (e.g., water) is depleted of solute by being forced, under pressure, to flow through a suitable membrane. The term ultrafiltration refers to a membrane with a pore size range greater than that of a reverse osmosis (RO) membrane but smaller than that of a microporous membrane, typically from 1 nm to 0.1 μm. When referring to pore size, the ultrafiltration industry usually states retention efficiency in terms of nominal molecular weight cutoff (NMWCO) rather than in geometrical terms. This practice is based on the use of known molecular weight indicator solutions of varying molecular weight to establish the nominal cutoff of ultrafiltration membranes. The above pore sizes correspond to nominal length values (NMWL) of approximately from 10^3 daltons to 10^6 daltons.

Ultrafiltration systems are being used to remove from DI water some extremely small contaminants that can pass through standard microporous filters. These contaminants include colloids (particularily silica), organics, and bacterial by-products (pyrogens). These contaminants can create a haze or a film on a wafer surface which can interfere with subsequent processing.

Ultrafiltration membrane filters used in DI water distribution systems typically have a NMWL rating of 100,000 daltons or less (approximately equivalent to a pore size rating of 0.01 μm or less) as indicated in Table 8. This table summarizes the characteristics listed in technical articles and in the sales literature of various suppliers for ultrafiltration membrane filtration products recommended for the filtration of DI water in the electronics industry.

2.3.2 Material Properties: Most ultrafiltration membranes operate on a size exclusion basis in that the solute or suspended colloid is removed from the liquid stream because it is larger than the effective size of the membrane pores. These ultrafiltration membranes are made from various polymers, most notably polyacrylonitrile, polysulfone and, more recently, polyvinylidene fluoride. They consist of a very thin (0.1 to 1.5 μm) dense "skin" which has an extremely fine pore structure of controlled pore size which opens to a much thicker (50 to 250 μm) substrate layer that has a much larger, open cell structure, as shown in Figure 14. One supplier, Asahi Chemical Company, offers a polyacrylonitrile ultrafiltration membrane with a sandwich structure that has two functional membranes on either side of a porous core.[25] The thin surface layer is the active element that effects separation while the substrate provides the mechanical strength required for the membrane to be used in practice. By using this configuration, the length of the active small pores is minimized and the pressure drop required to achieve a reasonable flow rate is not excessive.

A somewhat different operating principle is utilized in an ultrafilter offered by the Pall Corporation. This ultrafiltration cartridge combines a positively charged microporous nylon membrane with a negatively charged nylon membrane. Suspended colloidal materials are removed by electrostatic attraction to the oppositely charged membrane.[26]

2.3.3 Hydraulic Considerations: Ultrafiltration based on size exclusion is a surface filtration process. For effective ultrafiltration to occur, it is necessary to overcome or prevent concentration polarization, the accumulation of colloidal material being removed on the membrane surface. If left undisturbed, concentration polarization restricts solvent transport which markedly reduces filtration capacity. For this reason ultrafiltration is usually carried out in such a way that the flow path of the inlet fluid is parallel to the membrane surface.

Table 8: Characteristics of Commercial Ultrafiltration Cartridges Recommended for Deionized Water Service in the Electronics Industry*

DESCRIPTION			
MANUFACTURER	Asahi	Asahi	Brunswick
FILTER NAME	Kasei	Kasei	Ultrafilter
	FCV-3010	FCV-3010	BTUF100K
MWT CUT-OFF RATING, daltons	13,000	50,000	100,000
RATED PORE SIZE, µm			0.006
CARTRIDGE CONFIGURATION	Hollow Fiber	Hollow Fiber	Pleated
HEIGHT, IN	44	44	10
DIAMETER, IN	3.5	3.5	2.7
FILTRATION AREA, FT²	53.0	53.0	5.5
MATERIALS OF CONSTRUCTION			
MEMBRANE	ACR+	ACR+	PS
HOUSING			
VERTICAL SUPPORTS			PET
CORE			PS
SLEEVES			PS
END CAPS			PS
SUPPORT DISCS			
O-RINGS			S
POTTING ADHESIVE			PU
PERFORMANCE PROPERTIES			
INTEGRITY TESTABLE			YES
BUBBLE POINT TEST VALUE			
PSI			
MAX. DIFFERENTIAL PRESSURE			
PSIG	40	85	80
@ TEMPERATURE, O C	40	40	20
MAX. BACK PRESSURE			
PSIG			
@ TEMPERATURE, O C			
MAX. OPERATING TEMPERATURE, OC			
CONTINUOUS EXPOSURE	40	40	
INTERMITENT			
SANITIZABLE			YES
WATER FLOW, ΔP(PSI)/GPM	2.5	4.1	4.5
PERMEABILITY, GPM/ΔP-FT²	0.007	0.005	0.040
RESISTANCE RECOVERY TO 18 MEGOHM, GAL			16
EXTRACTABLES, MG/FT²			3
PARTICLE RETENTION - LATEX BEAD PASSAGE			
DIAMETER OF SPHERE QUANTITATIVELY REMOVED, µm			
PRESSURE DROP, K_L, PSI			

*Key to Table 8 is on page 241.

(continued)

Table 8: (continued)

DESCRIPTION MANUFACTURER FILTER NAME	Millipore Ultragard PTHK UB540	Millipore Ultrastak	Mitsubishi Rayon LMF 0610
NAMT CUT-OFF RATING, daltons	100,000	100,000	
RATED PORE SIZE, μm	0.006		0.05
CARTRIDGE CONFIGURATION	Spiral	Flat Plate	Hollow Fiber
HEIGHT, IN	12	3	4.3
DIAMETER, IN	3	8 x 15	1.8
FILTRATION AREA, FT²	12	10	5.6
MATERIALS OF CONSTRUCTION			
MEMBRANE	PS/PE	MPVDF	PE
HOUSING			PC
VERTICAL SUPPORTS	PP/PET		
CORE	PS		
SLEEVES	FEP		
END CAPS		PS	
SUPPORT DISCS		PS	
O-RINGS	EPR	EPR	
POTTING ADHESIVE	EPX	TP	URE
PERFORMANCE PROPERTIES			
INTEGRITY TESTABLE	YES	YES	YES
BUBBLE POINT TEST VALUE			
PSI			57
MAX. DIFFERENTIAL PRESSURE			
PSIG	100	50	30
@ TEMPERATURE, O C	50	25	45
MAX. BACK PRESSURE			
PSIG	5	5	
@ TEMPERATURE, O C	50	25	
MAX. OPERATING TEMPERATURE, OC			
CONTINUOUS EXPOSURE	50	35	45
INTERMITENT			
SANITIZABLE		YES	YES
WATER FLOW, ΔP (PSI)/GPM	10.0	2.4	0.4
PERMEABILITY, GPM/ΔP–FT²	0.008	0.042	0.459
RESISTANCE RECOVERY TO 18 MEGOHM, GAL	7		
EXTRACTABLES, MG/FT²			
PARTICLE RETENTION - LATEX BEAD PASSAGE			
DIAMETER OF SPHERE QUANTITATIVELY REMOVED, μm		0.05	0.05
PRESSURE DROP, K_L, PSI			

(continued)

Table 8: (continued)

DESCRIPTION			
MANUFACTURER	Nitto Denko	Nitto Denko	Pall
FILTER NAME			Ultrafilter
	NTU-3150-S4	NTU-3050-C3	NDZE
MWT CUT-OFF RATING, daltons	50,000	20,000	
RATED PORE SIZE, μm			0.04
CARTRIDGE CONFIGURATION	Spiral	Hollow Fiber	Pleated
HEIGHT, IN	40.0	42.4	10
DIAMETER, IN	3.9	3.5	
FILTRATION AREA, FT²			8.5
MATERIALS OF CONSTRUCTION			
MEMBRANE	PS	PS	NYL/PET
HOUSING		PVC	
VERTICAL SUPPORTS	PP/PES/PPO		PES
CORE	PVC		PP
SLEEVES	EPR		PP
END CAPS			PP
SUPPORT DISCS			
O-RINGS		S	
POTTING ADHESIVE	URE		TP
PERFORMANCE PROPERTIES			
INTEGRITY TESTABLE			YES
BUBBLE POINT TEST VALUE			
PSI			
MAX. DIFFERENTIAL PRESSURE			
PSIG	143	43	80
@ TEMPERATURE, 0 C	40	40	50
MAX. BACK PRESSURE			
PSIG			
@ TEMPERATURE, 0 C			
MAX. OPERATING TEMPERATURE, OC			
CONTINUOUS EXPOSURE	40	40	50
INTERMITENT			125
SANITIZABLE			YES
WATER FLOW, ΔP(PSI)/GPM	4.2	1.9	4.5
PERMEABILITY, GPM/ΔP-FT²			0.026
RESISTANCE RECOVERY TO 18 MEGOHM, GAL	600.0		
EXTRACTABLES, MG/FT²			4
PARTICLE RETENTION - LATEX BEAD PASSAGE			
DIAMETER OF SPHERE QUANTITATIVELY REMOVED, μm			0.04
PRESSURE DROP, K_L, PSI			180

(continued)

Table 8: (continued)

DESCRIPTION		
MANUFACTURER	Romicon	Romicon
FILTER NAME		
	GM80-HF 53	PM10-HF 53
NMWT CUT-OFF RATING, daltons	80,000	10,000
RATED PORE SIZE, µm		
CARTRIDGE CONFIGURATION	Hollow Fiber	Hollow Fiber
HEIGHT, IN	43	43
DIAMETER, IN	3.0	3.0
FILTRATION AREA, FT²	53.0	53.0
MATERIALS OF CONSTRUCTION		
MEMBRANE	ACR+	PS
HOUSING	PS	PS
VERTICAL SUPPORTS		
CORE		
SLEEVES		
END CAPS		
SUPPORT DISCS		
O-RINGS		
POTTING ADHESIVE	EPX	EPX
PERFORMANCE PROPERTIES		
INTEGRITY TESTABLE		
BUBBLE POINT TEST VALUE		
PSI		
MAX. DIFFERENTIAL PRESSURE		
PSIG	25	25
@ TEMPERATURE, 0 C	45	75
MAX. BACK PRESSURE		
PSIG	20	
@ TEMPERATURE, 0 C	45	
MAX. OPERATING TEMPERATURE, 0C		
CONTINUOUS EXPOSURE	45	75
INTERMITENT		95
SANITIZABLE		YES
WATER FLOW, ΔP(PSI)/GPM	4.5	3.8
PERMEABILITY, GPM/ΔP-FT²	0.004	0.005
RESISTANCE RECOVERY TO 18 MEGOHM, GAL		
EXTRACTABLES, MG/FT²		
PARTICLE RETENTION - LATEX BEAD PASSAGE		
DIAMETER OF SPHERE QUANTITATIVELY REMOVED, µm		
PRESSURE DROP, K_L, PSI		

(continued)

Table 8: (continued)

Key to Table 8

ACE+: ACETAL COPOLYMER
ACR+: ACRYLIC COPOLYMER
EPR : ETHYLENE-PROPYLENE RUBBER
EPX: EPOXY
MPVDF: MODIFIED POLYVINYLIDIENE FLUORIDE
NYL: NYLON
PBT: POLYBUTYL TEREPHTHALATE
PC: POLYCARBONATE

PES: POLYESTER
PET : POLYETHYL TEREPHTHALATE
PFA : POLYFLUOROALKOXY
PP : POLYPROPYLENE
PTFE: POLYTETRAFLUOROETHYLENE
PU : POLYURETHANE
S : SILICONE
TP : THERMOPLASTIC

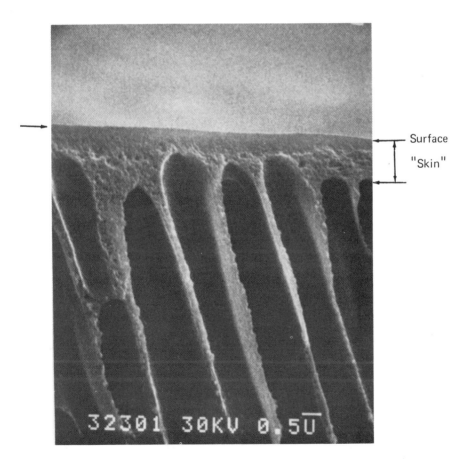

Surface

"Skin"

Figure 14: Photomicrograph of polysulfone UF membrane cross-section (100,000 NMWL).

The shear forces at the fluid/membrane interface resuspend the accumulated macrosolute into suspension, thereby significantly reducing concentration polarization. Cross-flow ultrafiltration systems produce two output streams: a purified filtrate and a retentate in which the macrosolute is concentrated, as shown in Figure 15.

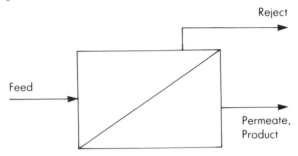

Figure 15: UF crossflow stream diagram.

Ultrafiltration membrane modules that operate in this mode are available in four different configurations: tubular, flat sheet (plate and frame), spiral wound flat sheet, and hollow fiber. Until recently, only the last two configurations have found extensive application in the purification of DI water. Hollow fiber UF membrane modules typically contain a plurality of parallel fibers that are bundled and sealed into a cartridge housing. Each fiber has a lumen of uniform diameter that is approximately 1 mm. Spiral type modules are more resistant to fouling than hollow fiber modules. However, the permeate can leave a hollow fiber module more readily, thus reducing equipment hold-up time, and increasing its dynamic response.[27] Millipore has recently introduced a stacked plate device that uses flat membrane and molded plastic plate supports, as shown in Figure 16. This device has minimum hold-up volume resulting in rapid rinse-up.

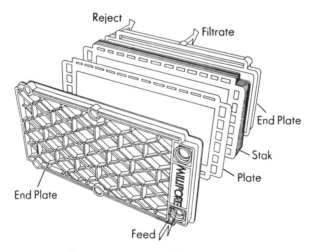

Figure 16: Ultrastak™ module.

In general, the permeability of cross-flow ultrafiltration membranes is much lower than that of microfiltration membranes. The permeabilities of the various hollow fiber and spiral ultrafilter membranes listed in Table 8 are all lower than 0.01 gpm/psi-ft^2, or approximately two orders of magnitude lower than the permeability of microfiltration membranes with a pore size rating of 0.2 μm. However, Millipore's new PVDF composite membrane, which consists of a thin UF membrane cast on a 0.2 μm PVDF microporous membrane (Figure 17) has a permeability of 0.04 gpm/psi-ft^2 when configured as a stacked plate device.

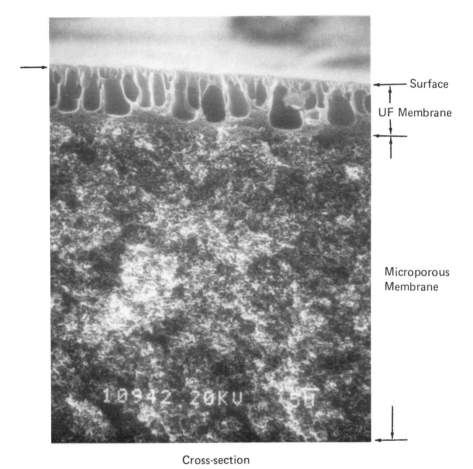

Cross-section

Figure 17: Photomicrograph of Millipore composite PVDF ultrafiltration membrane (100,000 NMWL on 0.2 μm microporous membrane).

The Brunswick, and Pall ultrafiltration cartridges listed in Table 8 are both tight pored pleated microfiltration membrane cartridges that are claimed to have the removal capabilities of cross-flow ultrafiltration membranes. The initial permeability of clean water for both these cartridges is approximately an order of magnitude higher than its initial permeability with the various hollow fiber and spiral cross-flow ultrafiltration membranes listed in Table 8. Similar claims are made for the Mitsubishi Rayon hollow fiber cartridge which operates in a flow through mode. However, no independent data exist which permits a comparison of the variation with time (or cumulative flow) of the permeabilities of these various ultrafiltration membranes. Cross-flow ultrafiltration membranes usually reach a steady state flow rate, whereas the flow rate of a dead-ended ultrafiltration membrane decays with time as its pores become blocked with captured particles. Cross-flow ultrafiltration membrane cartridges can furthermore be regenerated in-situ by either increasing the retentate flow rate or by back-flushing, while dead-ended ultrafiltration membrane cartridges can only be used once.

2.3.4 Contaminant Removal by Ultrafiltration: Ultrafiltration, even though it has been used for over a decade to effect the separation and concentration of biologically active organic molecules and of specialty polymers, is still a fairly novel technology in terms of its use in the purification of deionized water by the U.S. semiconductor industry. The ability of the Japanese semiconductor manufacturers to achieve a higher product yield than their overseas counterparts has been ascribed in part to their use of ultrafiltration to reduce the level of DI water contamination.[28]

As compared to microfiltration, there is relatively little published information on the contaminant removal efficiency of ultrafilters that has not been provided by their suppliers. Typical data are presented in Tables 9 and 10.

Table 9: Solute Retention of Various Commercial UF Membranes

Marker Compound	BtuFD-1M	Millipore 10^6	BtuFD-100K	Millipore 10^5	Amicon XM-100	Nuclepore 100,000
Dextran blue 1,000,000 MW	82	53	87	60	79	83
Immunoglobulin 150,000 MW	0	1	98	90	>99	>99
Bovine serum albumin 67,000 MW	0	0	4	9	79	24
Ovalbumin 43,000 MW	1	5	4	26	57	22
Myoglobin 17,800 MW	0	6	0	0	26	2
Vitamin B-12 1,200 MW	0	0	0	0	1	0

Source: Brunswick Technetics; Membrane Filter Products; San Diego, CA, October 1982.

Table 10: Performance Data—Pall NDZE Ultrafilter

Test Contaminant	Diameter (μm)	Efficiency (%)
Colloidal silica	0.021	>99.99
Latex spheres	0.05	100
Pyrogen*	0.001	>99.9997
T_1 phage particles	0.06	>99.9999
Mycoplasma	0.1–0.4	100 (T_R > 10^{12})**
Bacteria (Ps diminuta)	0.3	100 (T_R > 10^{12})**

*Pyrogens (bacterial endotoxins) are large molecules of various lipopolysaccharides. The test pyrogen used was *E. coli* endotoxin.

**T_R—Titre reduction is the ration of the number of influent bacteria to the number of effluent bacteria.

Source: Pall Ultra Fine Filter Corporation, Brochure UFS 850, July 1985.

A number of companies have reported that placing an ultrafilter up-stream of a 0.22 μm microfilter greatly extended the life of the microfilter, as represented by the data given in Figure 18.[22] Several in-plant studies have shown that there were significantly fewer particles on wafers rinsed with ultrafiltered than on wafers rinsed with 0.2 μm filtered DI water. Gaudet[30] reported finding 5 particles or less on 4 inch wafers rinsed in ultrafiltered as compared to 100 to 500 particles on wafers that had been rinsed with 0.2 μm filtered DI water.

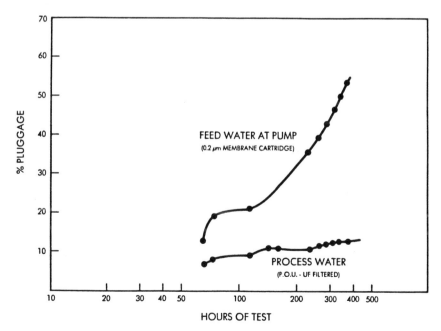

Figure 18: Plugging of 0.2 μm disc filter in DI water loop before and after ultrafiltration.[29]

2.3.5 Ultrafilter Induced Contamination: The comments made in Section 2.2.6 on microfilter induced contamination are believed to apply equally to ultrafilters. However, there is a paucity of data with regards to the shedding characteristics of ultrafilter membranes.

It should also be noted that the design of Millipore's most recently introduced ultrafiltration system (Ultrastak™) addresses the problem of particle shedding. By casting an ultrafiltration membrane on a 0.2 μm PVDF substrate, the product is integrity testable, bacterial retentive, and does not require a downstream filter. Because the composite filter is supported on a molded plastic plate, particle shedding characteristics are similar to those of the stacked disc microporous filters described in Section 2.2.6.

2.4 Reverse Osmosis Membranes

Whereas the use of reverse osmosis (RO) as a primary means of purifying incoming tap to near deionized quality is a well established technology, it has not been used to any significant extent in the U.S. to purify DI water in the distribution loop or at point-of-use. A major limitation of RO in terms of its use in distribution systems is that fairly high pressures, in excess of 200 psi, are required in order to obtain practical trans-membrane flux rates, while the pressure rating of DI water distribution systems in the U.S. is usually less than 125 psig.

There are indications, however, that in Japan so-called "loose" RO membranes are being evaluated as final filters in DI distribution systems as replacements for microporous membrane filters or ultrafilters. The incentive for using RO is that, in addition to being able to remove particulates, bacteria and pyrogens, is its ability to remove dissolved organics and that it can result in an increase in the resistivity of ultrapure water.[27,32]

3. SYSTEMS INTEGRATION AND PROCESS CONSIDERATIONS

3.1 Importance of Overall Process Design and Integration

Membrane filtration cartridges are just one element in the production of high purity deionized water. Unless they are properly integrated with the other components of a deionized water system, and unless the system is properly maintained and operated, the quality of the water delivered to the fabrication area may not be satisfactory, irrespective of the particle removal efficiency of the membrane filtration cartridges used.

Other factors which influence deionized water quality, and thus the apparent efficiency of membrane filters, include filter housings, distribution loop piping, other process elements that may be used in the distribution loop, choice of process conditions, and available methods of on-line analysis for waterborne contaminants.

3.2 Filter Housings

3.2.1 Requirements: The various filtration cartridges discussed above are all used in conjunction with a housing. Proper housing choice is critical if any membrane filter is to operate effectively. Key requirements of a housing include:[33]

(1) Adequate seal mechanism to eliminate fluid by-pass,

(2) Proper choice of materials and surface finish to prevent particulate generation, bacterial growth, and leaching of inorganic or organic contaminants,

(3) Internal design which optimizes fluid flow, allows the housing to be easily cleaned, and eliminates dead spots,

(4) Integrity test capability.

3.2.2 Design Considerations: A first design consideration is that the housing and the filter cartridge be compatible. It is imperative for the filter cartridge to fit in the housing without liquid by-pass, and that the sealing mechanism be such that the probability of obtaining a good filter/housing seal is very high. A second major design consideration is that flow path of the liquid through the cartridge be as direct as possible so as to minimize hold-up and eliminate any dead volume. The housing should also be compatible with the piping system used in the distribution loop with regards to fitting type, pressure rating and materials of construction.

3.2.3 Materials of Construction: Filter housings are made from a wide variety of materials of construction, including both polymeric and metallic materials. While passivated 316 stainless steel with a fine finish[33] is often used for filter housings in critical applications, for DI water service in the semiconductor industry, the current trend is to use housings molded from unfilled fluoropolymers. As discussed in the next section, these materials have been found to be among the more inert materials available.

3.3 Distribution Loop Design

3.3.1 Requirements: The purpose of the distribution loop is to provide water of specified quality to individual work stations in the fabrication area. It must be designed in such a way that DI water of acceptable quality is delivered at required volumes at constant pressure to any user station in the fabrication area irrespectively of the water demands of other user stations.

The major design consideration of any DI water system is that it not be of itself a major source of contaminants, such as particulates shed by the materials of construction, soluble contaminants extracted from these materials, bacteria which can flourish in DI water, and their decomposition products.

3.3.2 Microorganism Control: Microorganisms such as algae, fungi, and bacteria appear to be almost ubiquitous contaminants that find means to survive under even the most hostile conditions. The ability of bacteria to attach to surfaces can significantly enhance their ability to survive, especially in extremely oligotrophic (nutrient limited) environments where trace organic nutrients may become concentrated at solid-liquid interfaces. Bacteria are the primary organism that survive in deionized water systems, because of the large surface areas and nutrient limitations inherent in such systems. Despite all efforts to the contrary, it is impossible to keep a large volume water system totally free of bacteria (e.g., sterile). Given the proper environmental conditions, one bacterial cell in a sterile system could result in the generation of a fouling problem within a relatively short period of time.[34-36]

Bacteria are a form of self-replicating particulate contamination as well as a source of organic contamination. Each live microorganism is not only, in of itself, a comparatively large particle (>0.2 μm), but also, since it continuously sheds cell wall fragments, a source of many smaller particles (<0.1 μm). Upon death, as it decomposes into pyrogens and endotoxins, the organism remains a source of fine particle contamination.

While it is not possible to prevent bacteria from entering a DI water distribution system, it is possible to control their population with proper system design and operation.[34-39] Proper deisgn includes:

(a) The elimination of stagnant zones which provide bacteria favorable growth conditions and sufficient time to form large populations,

(b) The removal of reactive organic species that are potential nutrient sources from the system, and

(c) Provisions for sanitizing the equipment and contents of the DI water distribution system.

Proper operation entails establishing a bacterial monitoring program which routinely monitors bacterial levels in all equipment that use DI water in the fabrication facility, as well as ensuring proper operation of sanitizing equipment and the maintenance of good sanitation practices.

Design practices that tend to deny waterborne microorganisms the opportunity to lodge and reproduce within a distribution system include:

(a) Use of smooth inert materials of construction that are compatible with standard sanitizing agents, such as hydrogen peroxide or ozone,

(b) Careful joint construction, so as to eliminate crevices, threads or gaps where micro-organisms can accumulate,

(c) Maintenance of continuous DI water flow in the distribution system at all times, and providing adequate water velocities at all points in the system, and

(d) Elimination of dead legs in the distribution system.

3.3.3 Materials of Construction: While polyvinyl chloride (PVC) piping was used extensively in the past, the current trend in the semiconductor industry is to use fluorinated polymers for all piping, fittings, and equipment housings in the distribution system downstream of the central reverse osmosis (RO) system. Fluoropolymers currently used for DI water piping include "Halar" E-CTFE (ethylene-chlorotrifluoroethylene) (Allied Corp.), PVDF (polyvinylidene fluoride), and Teflon PFA (polyfluoroalkoxy) (E.I. DuPont de Nemours & Co., Inc.). Even though hardware made from these fluorinated polymers is expensive, these materials are used for the following reasons:

(a) Inertness and lack of reactivity towards DI water,

(b) High degree of oxidation resistance, even under sanitizing conditions where strong oxidizing agents and/or high temperatures are applied,

(c) Absence of additives and fillers,

(d) Ability to be heat welded, thereby allowing joints to be formed without the use of glue or solvents, and

(e) Good abrasion resistance.

Tests in high purity water have shown that PVDF piping, unlike that made of PVC, does not leach from its inside diameter.[40] This is thought to be attributable to the absence of plasticizers and pigments such as those used in the manufacture of PVC pipe,[41] and which have been found to be extractable in DI water.[42] Leaching of PVC piping may also "crater" the inside diameter of a pipe, providing areas out of the fluid turbulent flow lines in which bacteria may harbor and flourish.

3.4 Other Process Elements

A number of polishing units will usually be present in at some point of a DI water distribution loop. The purpose of these units is to remove trace levels of contaminants that eluded primary purification, or to remove contaminants that may be generated within the distribution system. Other than the filtration membrane systems already discussed, the units most commonly used are mixed bed ion exchangers and ultraviolet sterilizers.

The use of high purity mixed strong acid, strong base ion exchange resins in a single column has been the traditional method of eliminating the final traces of ionic material. Positive valence materials are exchanged for hydrogen on the cationic resin, and negative valence materials are exchanged for hydroxide on the anionic resin. The hydrogen and hydroxyl ions recombine to form water, thereby producing completely deionized water. The use of series mixed bed deionizers is quite common to guarantee the overall performance of the polishing stage. The resistivity of the water produced by mixed ion exchange beds is often in excess of 17 meg-ohm, indicating nearly total removal of all ionic species.

Mixed bed ion exchangers, however, do not remove organics, colloids, particles and bacteria which are also undesirable species, and indeed may increase the levels of certain of these contaminants.[43]

Ultraviolet light sterilizers are often used to control bacterial population in the DI water stream. Historically, low pressure mercury arc lamps which produce light with a wavelength of 254 nm have been used in this service.[44] The effectiveness of UV systems depends on the UV dose, wavelength, stream geometry and type of bacteria present. Properly applied and maintained, UV light units effectively eliminate planktonic bacteria suspended in the aqueous stream. However, adsorbed bacteria on pipe walls (e.g., biofilm) is not significantly affected by UV light because of its poor penetration.[36]

3.5 Process Conditions

The performance of a membrane filter will be influenced by process conditions which not only vary significantly from system to system, but also with time in any given system.

Filter challenge conditions continuously change with time, particularily if one considers flow rate variations, pressure drop and pressure drop fluctuations across a filter structure. One of the worst case situations for a filter is the challenge presented by a rapid change in flow rate and pressure differential, as may occur with the opening of a valve, and which can result in pressure pulses that exceed 600 psi/sec with a minimum peak of 60 psi.[37] The deleterious effects of pressure pulses on the shedding of particles from filters was discussed in the previous section, and is exemplified by the data presented in Figures 11 and 12. To the extent possible, as a consequence, sudden changes in flow rates should be avoided in DI water distribution systems.

A second process factor which will affect filter operation is temperature. The temperature of DI water systems can range from ambient ($15°$ to $20°C$) to $90°C$. An advantage of operating at higher temperatures is that the viscosity of water decreases with temperature, so that pressure drop for a given flow decreases with temperature. A disadvantage of operating at higher temperatures is that the maximum trans-membrane pressure ratings of these filters decreases significantly with temperature, so that the allowable flow rate through a filter element will decrease with temperature. As noted in Tables 5 and 9, the maximum allowable operating temperature can vary significantly with filter construction, with some filters rated only to a maximum temperature of $40°C$.

In a similar vein, the chemical resistance of microfilters can vary significantly with materials of construction. In DI water service, apart from extraction of organic compounds from the filter by DI water itself, and which should not occur beyond specified limits, the major concern is compatability of the filter elements with sterilizing agents used to control bacterial growth. These are usually powerful oxidizing agents, with hydrogen peroxide and ozone being preferred because they do not leave behind any deleterious decomposition products.

Due to the cost of purifying water, most wafer fabricators will reclaim at least some used DI water. Depending on the operations, reclaimed DI water will contain some undesirable impurities which will have to be eliminated before introduction into the distribution loop.

The performance of membrane filters also changes with time. In spite of the fact that membrane filter manufacturers now precondition their products by washing them extensively in DI water before shipment, introducing a new filter into a DI water circuit will usually result in a short term rise in contaminants. The usual measure of this transient period is the number of gallons of DI water that have to be flushed through a filter after installation before water resistance reaches a required operating level which is usually in excess of 17 meg-ohm.

Properties of membrane filters inherently change on a longer term basis, over a time frame of weeks and months. As particles that are removed from suspension accumulate on the surface and within the pores of dead-end microfilters, the porosity and permeability of these elements decrease with time. An intermediate benefit is that the capture efficiency of an aged filter is higher than that of a fresh filter because captured particles tend to reduce the effective pore size of the filter. However, pore blockage also results in reduced flow rate per unit pressure driving force. Sooner or later, depending on the cleanliness of the liquid stream, it becomes desirable to replace an existing filter cartridge with a fresh element.

3.6 Methods of Analysis

The purpose of including a membrane filter in a DI water process loop is to remove microcontaminants from the stream. A number of test methods have been developed to monitor the effectiveness of a filtration system other than using significant variations in device manufacturing yield which is an expensive approach to water analysis. A number of test procedures have been developed and which include methods of determining:

(1) Whether all the fluid stream is actually being filtered, e.g., "integrity testing," and

(2) The concentration of contaminants in the filtered fluid, including bacteria, suspended particles, and dissolved species.

3.6.1 Integrity Testing: An important feature of a membrane filtration system is its ability to be integrity tested in-situ before and after use. An integrity test will detect a damaged membrane, ineffective seals, or a system leak, thereby indicating whether any process fluid is not being properly filtered. A number of methods of integrity testing are used, with the most important being the Bubble Point Test and the Diffusion Test.

Bubble Point Testing — Membrane filters have discrete uniform passages penetrating from one side to the other which can be thought of as fine uniform capillaries. The bubble point test is based on the fact that liquid is held in these capillary tubes by surface tension and that the minimum pressure required to force liquid out of the tubes is a measure of tube diameter, as indicated by the following equation:

(2)
$$P = \frac{K4\phi\cos\theta}{d}$$

where:

P = Bubble point pressure

ϕ = Surface tension

θ = Liquid-solid contact angle

d = Pore diameter

K = Pore shape correction factor

A bubble point test is performed by prewetting the filter, increasing the pressure of air upstream of the filter, and watching for air bubbles downstream to indicate the passage of air through filter capillaries. The pressure at which a steady continuous stream of bubbles appears is the bubble point pressure. An observed bubble point, significantly lower than the bubble point specified for that particular filter, indicates a damaged membrane, ineffective seals or a system leak. A bubble point that meets specifications ensures the system is integral.

Diffusion Testing — In high volume systems, where a large volume of downstream water must be displaced before bubbles can be detected, a diffusion test can be performed instead of a Bubble Point Test. This diffusion test is based on the fact that in a wetted membrane filter, under pressure, air flows through the water filled pores at differential pressures below the bubble point pressure by a diffusion process that follows Fick's Law. But, in the large area filters used in high volume systems, it is significant and can be measured to perform a sensitive filter integrity test.[45] It has been found that applying pressure at 80% of the filter bubble point provides an effective test of filter integrity since there would be a dramatic increase in air (and water) flow at lower pressures, and which would be indicative of a damaged membrane, the wrong pore size filter, the absence of any filter, ineffective seals or system leaks.

3.6.2 Monitoring for Bacteria: Process DI water should be analyzed routinely for the presence of bacteria at various points in the purification system. An effective monitoring system will provide information about the efficiency of sanitization processes and warn of any sudden build up of bacteria in the system. Because they can flourish and multiply so rapidly, bacteria have been considered to be the most important available indicator of purified water quality.[35]

Several techniques for enumeration of bacteria are currently in use. These techniques include vial count assays, direct count epifluorescent microscopy, scanning electron microscopy (SEM) and biochemical techniques such as culture plates and bioluminescence assay.[35,46] The relatively low nutrient count of purified water is the major limiting factor in microbial growth and is an important consideration in the selection of microbial detection technique. Proper sampling is also important. Care should be taken to sample the true water stream and to eliminate valve or sampling port contamination.[47] If one is monitoring for a bacterial population of less than 10 organisms per liter, it is important that the sample be large enough to be statistically meaningful.

3.6.3 Monitoring for Particles: Automatic particle counters are now available that allow for accurate on-line, real time, continuous monitoring of the presence of suspended particles in a liquid stream. By using laser based sensors, streams can be monitored for the presence of particles as small as 0.3 to 0.5 μm in diameter.[48,49] These instruments allow for continuous, real time monitoring, and are valuable because they can detect even the transient presence of particles in filtered water. Their major limitations are that they can not distinguish between types of particles, or even between particles and bubbles, and that they can not detect particles smaller than 0.3 μm in diameter, even though it has been suggested that particles as small as 0.1 μm could be detected with an optimized laser light-scattering detector.[50]

Monitoring the filtration efficiency of fine microfilters (with a pore size rating of under 0.2 μm) or of ultrafilters can not currently be performed on line. Monitoring for the presence of particles smaller than 0.3 μm requires batch sampling and off-line analysis. A sample of the water stream is collected, and a known volume of the sample is passed through a flat membrane filter with a pore size smaller than the finest particle size of interest. The particles collected on the filter are then examined under a scanning electron microscope (SEM). SEM are now available that have a resolution of less than 5 nm.[51] In addition to particle size information, an SEM provides information as to particle morphology and composition.

3.6.4 Monitoring for Dissolved Species:
The measurement of the electrical conductivity is the established method of monitoring DI water quality. While this measurement is simple and rapid, and thus valuable and necessary, it is no longer sufficient. It is a limited measurement that only provides information as to total concentration of ionic species. It does not distinguish between ions, and it provides no information on uncharged species, such as particles, bacteria, and organic compounds. Different ionic species can now be detected off-line to ppb levels by ion chromatography.[52]

Total organic carbon (TOC) concentration levels as low as 20 ppb can now be monitored on-line.[53] The importance of this development is that it allows ultrafiltered water to be monitored on a real-time basis for a wide variety of organic compounds, including macromolecular colloids too small to be detected by any on-line particle size measurement method available.

3.7 Choosing a Membrane Filtration System

The purpose of this article is to provide persons who become responsible for selecting a membrane filtration system for a semiconductor deionized water facility with sufficient background information to allow them to make a reasonable and successful choice. This choice has to be considered within the context of not only the DI water system, but the overall manufacturing system, and involves the following issues:

(1) The first issue that has to be addressed is to understand the end-product manufacturing operation, and how it is affected by the purity of DI water used in the operation. This establishes the purity requirements of the DI water that is to be produced.

(2) The second factor that has to be considered is the purity of the water feed to the proposed filtration system. This establishes the challenge levels of the contaminants that have to be removed.

(3) The above information provides the basis for choosing the type(s) of membrane filter(s) and their size rating(s) that would be appropriate to system requirements. If the stream contains species that can not be removed with membrane filters, not only will other process equipment need to be specified and included into the system, but its potential impact on the filtration modules has to be considered.

(4) The required filter area has to be determined on the basis of the required output flow rates and available pressure differential, under both peak transient and steady state conditions. This establishes the size and number of filter elements required. If these appear to be excessive, it may be necessary to increase the system pumping capacity.

(5) Establish any process constraints on the choice of materials of construction that can be safely used in the proposed system.

These constraints also apply to all wetted parts of a membrane filter, and have to be taken into consideration in choosing among alternate candidate vendor products.

(6) Contact reputable filter vendors for technical recommendations and quotations.

(7) Procure candidate filters for in-house testing and evaluation before recommending any specific vendor's products for manufacturing use.

REFERENCES

1. Kopp, R., *Transcript 1985 SEMI Information Services Seminar,* p. 77 Newport Beach, CA (January 1985).
2. Bardina, J., *Microcontamination, Vol. 2* (1), pp 29, 78 (February/March 1984).
3. Monkowski, J.R., *Proc. 30th Technical Meeting Institute of Environmental Sciences,* pp 81–84, Orlando, FL (May 1984).
4. Logan, M.S., Kaminsky, E.B. and Kritzler, W.R., *Proc. 30th Technical Meeting, Institute of Environmental Sciences,* Orlando, FL, pp 152–159 (May 1984).
5. Bansai, I.K., *J. Env. Sciences, Vol. 28* (4), pp 20–23 (July/August 1983).
6. Gaudet, P.W., "Point of Use Ultrafiltration of Deionized Rinse Water and Effects on Microelectronics Device Quality," ASTM Special Technical Publication 850, Gupta, D.C., Editor (1984).
7. Monkowski, J.R. and Zahour, R.T., *J.T. Baker Chemicals for Electronics, Reprint 1* (1982).
8. Mistry, C., Gilliland, J., Koutalides, A. and Mandaro, R.M., *Proc. 45th Annual Meeting, International Water Conference* (1984).
9. Balazs, M.K. and Poirier, S.J., *Proc. 3rd Semiconductor Pure Water Conference,* p. 153–162 (1983).
10. "SEMI Suggested Guidelines for Pure Water for Semiconductor Processing," Semiconductor Equipment & Materials Institute, Mountain View, CA (1986).
11. Lonsdale, H.K., *J. Membrane Science, Vol. 10,* p. 81 (1982).
12. Porter, M.C., *Proc. 3rd. World Filtration Congress, Vol. 2,* p. 451 (1982).
13. Overbeek, E.J.W. and J. Th. G., "Theory of the Stability of Lyophobic Colloids," Elsevier Publishing Company, Amsterdam, Holland (1948).
14. London, F., *Z. Physik, Vol. 63,* p. 245 (1930).
15. Tolliver, D.L. and Schroeder, H.S., *Microcontamination, Vol. 1* (1), pp 34–43, 78 (June/July 1983).
16. Simonetti, J.A., Schroeder, H.G. and Meltzer, T.H., *Ultrapure Water, Vol. 3* (4), pp 46–48, 50–51 (July/August 1986).
17. "Microbiological Evaluation of Filters for Sterilizing Liquids," HIMA Document No. 3, Volume 4 (1982).
18. Goldsmith, S.H. and Grundelman, G.P., *Solid State Technology, Vol. 28* (5), pp 219–224 (May 1985).
19. Wallhauser, K.H., *Pharm. Ind., Vol. 41,* pp 475–481 (1979).

20. Kozeny, J., *Sitzber. Akad. Wiss. Wien. Math-naturw. Kl., Abt. IIa, Vol. 136,* pp 271–306 (1927).

21. "Membrane Filter Specifications," Millipore Catalog and Purchasing Guide, p. 6–7, Millipore Products Division, Millipore Corporation, Bedford, MA 01730 (July 1983).

22. Goldsmith, S.H., Barski, J.P. and Grundelman, G.P., *Microcontamination, Vol. 2* (1), pp 47–52 (June/July 1984).

23. Stewart, D., Microelectronics Technical Symposium on DI Water Contamination Control Technology, Sponsored by Millipore Corporation, Bedford, MA (May 20, 1985).

24. Grant, D.C., Peacock, S.L. and Accomazzo, M.A., Proc. 5th Annual Semiconductor Pure Water Conference, San Francisco, CA (1986).

25. Warashina, T., Hashino, Y. and Kobayashi, T., *Chem Tech., Vol. 15* (9), pp 558–561 (September 1985).

26. "The New Disposable Ultrafilter for Ultrapure Water," Pall Filtration News, Pall Ultrafine Filtration Corporation, East Hills, New York (Spring 1985).

27. Nakagome, K. and Brady, M.F., *Ultrapure Water, Vol. 6* (2), pp 34–38, (May/June 1986).

28. Motomura, H. *Microcontamination, Vol. 1* (3), pp 45–51 (February/ March 1984).

29. Accomazzo, M.A., Proc. 3rd Annual Semiconductor Pure Water Conference, San Francisco, CA (1983).

30. Gaudet, P., ASTM Special Technical Publication 850, Gupta, D.C., Editor (1984).

31. "Contamination Control of Semiconductor Process Fluids," p. 22, Lit. No. EC085, Millipore Corporation, Bedford, MA (November 1985).

32. Yokoyama, F., Proc. 3rd. Annual Semiconductor Pure Water Conference, San Jose, CA (January 1984).

33. Yaeger, S. Proc. 3rd Annual Semiconductor Pure Water Conference, San Jose, CA, pp 101–109 (January 1984).

34. Mittelman, M.W., *Microcontamination, Vol. 3* (10), pp 51–55, 70 (October 1985).

35. Mittelman, M.W., *Microcontamination, Vol. 3* (11), pp 42–58, (November 1985).

36. Mittelman, M.W., *Microcontamination, Vol. 4* (1), pp 30–40, 70 (January 1986).

37. Harned, W., *J. Env. Sciences, Vol. 29* (3), pp 32–34 (May/June 1986).

38. Rechen, H.C., *Microcontamination, Vol. 3* (7), pp 22–29 (July 1985).

39. Nickerson, G.T., *Ultrapure Water, Vol. 4* (4), pp 52–55 (July/August 1966).

40. Couture, S.D. and Capaccio, R.S., *Microcontamination, Vol. 2* (2), pp 44–47, 70 (April/May 1984).

41. Hanselka, R., Proc. 3rd Annual Semiconductor Pure Water Conference, San Jose, CA (January 1984).

42. McConnelee, P.A., Poirier, S.J. and Hanselka, R., *Semiconductor Intl., Vol. 9* (8), pp 82–85 (August 1986).

43. Frith, Jr., C.F., *J. Env. Sci., Vol. 27* (4), pp 22–26 (July/August 1984).

44. Collentro, W.V., *Ultrapure Water, Vol. 3* (4), pp 56-58 (July/August 1986).
45. Reti, A.R., *Bull. Parenteral Drug Assn., Vol. 31* (4), pp 187-194 (July/August 1977).
46. Smith, K.C., *Proc. 30th Annual Technical Meeting, Institute of Env. Sciences,* Orlando, FL, pp 138-140 (May 1984).
47. Green, B.L. and Haffenreffer, A.H., *Proc. 30th Annual Technical Meeting, Institute of Env. Sciences,* Orlando, FL, pp 132-137 (May 1984).
48. Goldsmith, S.H., Barski, J.P. and Grundelman, G.P., *Microcontamination, Vol. 2* (4), pp 47-52 (June/July 1984).
49. Peacock, S.L., *Proceedings 5th Annual Semiconductor Pure Water Conference,* San Francisco, CA (1986).
50. Knollenberg, R.G., *Proc. 30th Annual Technical Meeting, Institute of Env. Sciences,* Orlando, FL, pp 46-55 (May 1984).
51. "Scanning Electron Microscope, Model S 570," Catalogue EX-E612, Hitachi Corporation, Tokyo, Japan.
52. Houskova, J. and Chu, T., *Solid State Technology, Vol. 29* (5), pp 205-208 (May 1986).
53. Poirier, S.J., *Ultrapure Water, Vol. 2* (4), pp 30-32 (July/August 1985).

8

Monitoring Systems for Semiconductor Fluids

Robert G. Knollenberg

1. HISTORICAL PERSPECTIVE

One's nature is to first attack problems that are most apparent—not necessarily those that are potentially the most severe. The approach to monitoring and controlling microcontamination in VLSI semiconductor manufacturing is a case in point. Clean rooms for semiconductor manufacturing came into existence with early efforts in producing microelectronic circuits. Everyone could see the dust in ambient air, and thus, today we have VLF (Vertical Laminar Flow) rooms with tested 99.99% guaranteed removal of all particles larger than 0.1 μm. But few people are aware that our drinking water typically has significantly higher particle concentrations than the air we breath (on a particles per unit volume basis). Likewise, VLSI semiconductor plant deionized water (hereafter DI water) supplies exhibit enhanced microcontaminant concentrations over clean room air. We simply cannot easily see the particle contaminants suspended in liquids, not withstanding the fact that they are more difficult to remove. One might further speculate that process gases (being largely hidden from view) might also be easily contaminated without detection, as certainly would sourcing by those machinery, equipment, and apparatus essential to VLSI manufacturing. In fact, only in the last few years have semiconductor manufacturers become aware of particle sourcing and defect generation *from within* the processing equipment utilized on the manufacturing line. Returning to contamination in semiconductor process liquids, we conclude that attempts to control particulates have come recently not because it is a recent problem but simply because we have *now* recognized it for what it has always been—a significant problem in defect generation. The real impact of contamination control within semiconductor process liquids has yet to be realized. While direct correlation of yield improvements with clean room aerosol reduction has been well established, such positive correlation to use of low particulate process liquids has not. We simply know that the rooms cannot be further improved by quantum steps and that other sources are stronger candidates for increased control. Process liquids undoubtedly possess much higher levels of particulate contamination on a particles per unit volume basis and if used in significant quantity are likely primary sources. It is the monitoring of these contaminant levels that this chapter addresses.

The magnitude of contamination within process liquids has only recently been appreciated. Consider that Federal Standard 209B (Federal Air Classification Standard) is now 25 years old yet no standards exist for *any* liquid and the Purewater Conference (devoted to microcontamination in a single process field, i.e., water) has only recorded its fourth anniversary at this writing. Admittedly, significant efforts have recently been focused on the contamination and quality of DI water. DI because metallic cations are recognized killers in semiconductors; resistivity in the 15 to 20 megohm cm range has been a requirement for several years. Subsequent to resistivity came the monitoring of silica, total organic carbon, bacteria, and now finally suspended particles. One might speculate that the eventual thrust will be to measure total dissolved nonionized species—a difficult task to say the least. However, as the die size increases and the feature size decreases, more intense controls and monitors will undoubtedly be required. DI water specifications deemed acceptable to electronic users and manufacturers of DI systems are currently being studied and evaluated in the hope of developing standards. However, established standards take years to formulate and have a vanishingly small period of applicability in a rapidly advancing and ever changing technology.

But what about the ever expanding list of process chemicals. Process liquids in the form of acids, solvents, and photoresists were largely ignored while particle monitoring and control in air and DI water were advancing. Prior to 1980, process liquids were largely purchased according to grade and chemical analysis. However, manufacturers became convinced that point-of-use controls on chemical supplies were required for yield improvements. Process engineers on the VLSI manufacturing lines began investigating the quality of all materials. But in the process liquid area, there simply were no accepted standards or specifications. Users typically monitored their process liquids and controlled contaminants as best they could to enhance yields because no one knew what was "good enough." Overspecification was believed to be the safer route than exact specification in the absence of knowing exact requirements. It is highly probable that certain process liquids may already be cleaner than required. While this may be true today for certain process liquids, it most likely won't be for long.

Inherent in any control process is the ability to monitor. This chapter concerns itself with particle microcontaminant monitoring of all semiconductor process liquids. We first survey measurements that have been made to establish a baseline. We embark on a discussion of material problems and compatibilities of fluids with vessels, plumbing and measuring apparatus with particular attention paid to corrosive effects on viewing optics and sensors. The variety of available sampling methodologies are explored and tradeoffs in their application examined. Applicable measurement techniques are surveyed with detailed emphasis on optical light scattering methods. A thorough theoretical examination of the optical behavior of liquid suspended particles is given. System configurations and applications and analysis complete the formal presentations in this chapter.

2. SURVEY OF MEASURED LEVELS OF MICROCONTAMINATION IN SEMICONDUCTOR PROCESS LIQUIDS

Before we can talk intelligently about liquids as being "clean" or "contaminated," it is important to determine appropriate size ranges of interest. There is a well recognized need for submicron sensitivities for semiconductor process liquids but what about the upper size limits? Clearly, a large particle may be more damaging than a small one unless it settles out in a process or washes off more easily, and while it is desirable to measure every particle—no matter how large or small—obviously compromises must be made.

Most manufacturers of liquid particle counters design instruments to size as small as possible (generally, submicron) and limit the larger size to two orders of magnitude above the lowest submicron threshold by simply counting all particles larger than this upper limit. Thus, a liquid particle counter may have a 0.5 to 15.0 μm, 0.3 to 7.5 μm, or other similar size range divided up into a few or as many as 15 size class intervals including a final size class for particles greater than the uppermost class limit.

The use of liquid particle counters in the submicron size ranges for semiconductor process liquids has been a recent phenomenon as previously addressed. Only after many other sources had been nearly totally contained, did the need to control, define, and monitor particulate contamination in process liquids gain appreciation. The typical procedure for contamination control in liquids was based on reliance of point-of-use filters in the process liquid lines. A point-of-use filter with pore size of 0.2 μm or smaller is typically used in DI water installations. Corrosive process liquids such as photoresist and acids or solvents were frequently not dispensed through a filter but used as received from the supplier or from a process vessel of unknown cleanliness. Today, the value of final filtration is becoming more and more recognized as necessary for quality production and yield improvements.

We can summarize several reasons for the apparent disregard of contaminants in liquid systems. First, corrosive liquid cleaning processes with available filters have only recently become available. So have particle counting and sizing instruments for the submicron size range. Third, the standards and specifications needed to define required liquid cleanliness levels were, and still are, nonexistent. Attempts made in the past to compile meaningful specifications for a wide variety of liquids have no widespread acceptability. Now that new and better methods of process liquid filtration coupled with means of verifying submicron cleanliness levels are available, the information needed to develop useful specifications can be hopefully obtained.

A usable contaminant specification for process liquids must consider the tolerances of the product to be protected. For instance, particle size ranges might be stated on a basis related to the minimum geometry of the device, exposure to the liquid being produced, and projected defect introduction span. For instance, a 10 μm layer of photoresist applied to 100 cm^2 of a wafer surface amounts to 0.1 ml of fluid. If only 100 particles larger than a critical threshold per 100 ml are present, there will only be one sensitive particle per every 10 wafers—a low level of potential defect generation. Specifications must be written so that *available* monitoring and control equipment can be used. Very little information is currently available on the particle size distributions that can be expected in new process liquids of varying types, the reduction in contamination by filtration, or the increase in contamination introduced by processing steps. This problem is further complicated by the needs for chemical analysis of major impurities and dissolved species in particular, which when combined with suspended particulate contaminants provide a clearer overall picture of the apparent chemical grade of material. (Consider that 0.0001% dissolved solids allows for a mass equivalent to 10^6 one micron particles per ml.) In the absence of standards, we can only discuss measurements in terms of specific numerical concentrations above a certain size threshold or within specific size intervals. Such measurements serve only as a *relative* baseline and cannot be regarded as absolute since, as we shall see later, the measured size is only an estimate of real particle size.

Table 1 is a partial listing of commonly used semiconductor process materials and refractive index data. The literature is devoid of any extensive survey of particle contaminaton in "as received process" liquids. Articles by Dillenbeck[1,2] provide an indication of the potential problems involved in measurements of particle counts larger than 0.5 μm: Two different brands of acetone, costing the same, varied by factors of 40, and two different brands of isopropyl alcohol and photoresist varied by factors of 30. Such differences might at first appear to reflect different manufacturing process controls were it not for Dillenbeck's observation that found samples of the same vendor material, but of different lot, varying by a factor of over 100! Dillenbeck found concentrations varying from a high of over 25,000 ml (>0.5 μm) in sulfuric acid to a low of 8 ml (>0.5 μm) in isopropyl alcohol. While not trying to explain all possible reasons for such differences, he was able to eliminate the bottle as the possible contamination source for the high count cases. His experiments showed that the lack of adequate rinsing could only account for 100 ml (>0.5 μm)—a possible factor in the clean liquids but not a substantial contributor to the highly contaminated liquids. (We will later show that such data have a natural sizing bias depending on the liquid refractive index, and as such, data cannot be easily compared from chemical to chemical without normalization.) Furthermore, certain process liquids were filtered at point-of-use just prior to measurement (and as expected, were cleaner) while others were not likely filtered by anyone. Regardless of the filtration procedure, more viscous fluids have poor filtration characteristics while less viscous fluids filter well.

Because of the obvious pitfalls in any such survey, it may be appropriate to start with measurements of nonsemiconductor grade chemicals and examine them. To that end, we have provided representative measurements performed at Particle Measuring Systems, Inc. (PMS) for nine *reagent* grade chemicals supplied by local vendors, two photoresists supplied by national manufacturers and DI water supplied from PMS's system. The measurements were performed with a laser light scattering instrument, PMS model IMOLV, having a 0.3 to 7.5 μm size range. To avoid bottle-to-bottle and batch-to-batch variations, samples from different lots were mixed together to represent an average. These data are presented grouped into solids, solvents, and photoresists in Table 2 and Figure 1.

**Table 1: The Refractive Indices of Commonly Used Semiconductor
Process Materials**

Gases[*]	Refractive Index
Nitrogen	1.00
Oxygen	1.00
Argon	1.00
Hydrogen	1.00
Hydrochloric Acid	1.00

Liquids	
Hydrofluoric Acid (50%)	1.29
Ammonium Fluoride (40%)	1.33
Ammonium Hydroxide (30%)	1.33
Methanol	1.33
Water	1.33
Acetone	1.36
Hydrogen Peroxide (30%)	1.36
Trichlorotrifluoroethane (Freon TF)	1.36
Acetic Acid (Concentrated)	1.37
Methyl Ethyl Ketone	1.38
2-Propanol	1.38
Butyl Acetate	1.39
Nitric Acid (50%)	1.40
Hydrochloric Acid (38%)	1.41
Dichloromethane (Methylene Chloride)	1.42
Phosphoric Acid (85%)	1.42
Potassium Hydroxide (50%)	1.42
Sulfuric Acid (80%)	1.43
Sodium Hydroxide	1.44
Trichloroethane	1.48
Trichloroethylene	1.48
Toluene	1.49
Tetrachloroethylene	1.50
Xylene	1.50

Solids	
Silicon	3.90
Germanium	3.50
Carbon	2.00
Silicon Dioxide	1.46

[*]The refractive indices of these gases is non-zero only in
the fourth decimal place at STP.

The data are presented with measurements of liquid refractive index. As we will later see, particles are better detected by optical particle counters the more they differ in refractive index from the liquid media. As is observed in the table, contamination levels are relatively high, although not substantially higher than reported by Dillenbeck. (Bottle-to-bottle and lot-to-lot differences were also observed.) The size distributions of Figure 1 are of particular interest in that they show similar trends (increasing numbers with decreasing size) but with varying slopes. DI water, toluene, and trichloroethylene have much lower slopes than the other liquids. The DI water was point-of-use filtered from PMS's supply and while very clean, not much cleaner than reagent grade trichloroethylene and toluene. The 2-propanol was surprisingly contaminated implying little or no controls prior to packaging. Of the acids sulfuric was much more contaminated than the other three. The photoresists were nearly as contaminated as the sulfuric acid sample. Most of the more viscous liquids showed higher concentrations of particles. There is no obvious correlation with refractive index. We will further elaborate on this data later and reveal interesting effects of dilution. For now, we will be satisfied with knowing that apparent concentrations may be anticipated varying from a few to tens of thousands of particles per ml at sizes above $0.3 \mu m$.

Table 2: Summary of Particle Microcontamination Measurements in Semiconductor Process Liquids

LIQUID	MEASURED TOTAL NUMBER $>0.3\mu m$ per 100 ml	REFRACTIVE INDEX
Solvents		
DI Water (Filtered)	2.02×10^2	1.33
Acetone	7.08×10^4	1.36
2-Propanol	9.81×10^5	1.38
Trichloroethylene	7.04×10^2	1.48
Toluene	7.40×10^2	1.49
Xylene	6.42×10^4	1.50
Acids		
Hydrofluoric (50%)	4.2×10^4	1.29
Hydrochloric (Conc.)	1.76×10^5	1.43
Nitric (Conc.)	3.48×10^4	1.44
Sulfuric (Conc.)	3.27×10^6	1.45
Photoresists		
Vendor A	9.7×10^5	1.40
Vendor B	2.2×10^5	1.41

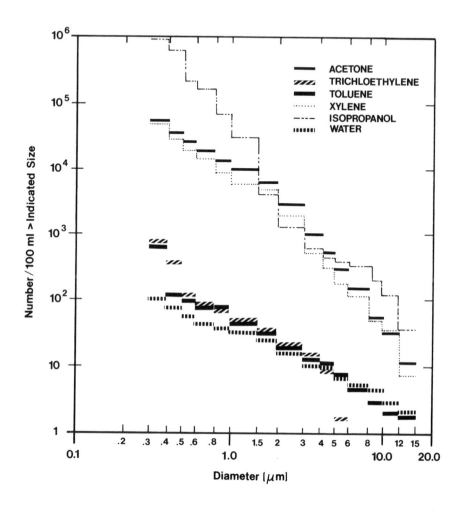

a

Figure 1: Measured size distributions in semiconductor process liquids. (a) solvents, (b) acids, (c) photoresists. All of the above measurements were performed without prefiltration. The instrument used was a PMS IMOLV sensor with a 0.3 to 15 μm size range.

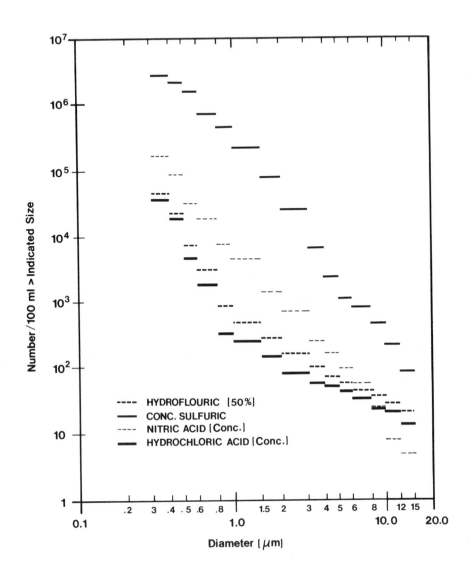

b

Figure 1: (continued)

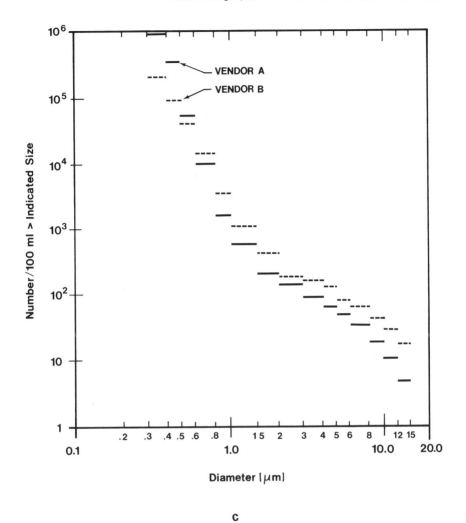

c

Figure 1: (continued)

3. MATERIAL PROBLEMS IN LIQUID MEDIA

Regardless of the monitoring system employed, it is necessary to construct the wetted portions thereof of materials that are capable of handling the liquid media at moderate pressures and perhaps temperatures without leaking, dissolving, or reacting with the process liquid media. When one looks over the list in Table 1, there are some notable "bad actors," e.g., hydrofluoric, nitric, and sulfuric acid. In most cases it is necessary for an instrument to be compatible with any of the process liquids on the list. In those cases where an instrument is dedicated to only one or several of them, the material selection problems are substantially reduced. For instance, a variety of inexpensive plastics are compatible with hydrofluoric acid but dissolve in xylene. Common glasses are compatible with almost everything except hydrofluoric acid. Stainless steel in the form of the 316L alloy will be suitable for cold acids (except hydrofluoric) but not hot ones. Gold, although prohibitively expensive, is suitable for all pure acids but not certain acid pairs. Probably, the most universally used material in chemical apparatus outside of Pyrex® glass is tetrafluoroethylene (TFE) and other fluorinated hydrocarbons with trade names such as: Teflon, Halar, Kel-F, Kal-Rez, SyG-F, PFA, and PVDF. For all but high temperatures and pressures, such materials are adequate, but they all tend to cold-flow and are dimensionally unstable at temperatures above 100°C. If welded together in tubing assemblies, they will not likely leak but when used in mechanical joints with other materials, most likely *will* after several temperature cycles or after prolonged pressure applications. Many manufacturers of apparatus using these materials tend to favor constructions with stainless steel exteriors, such as cylindrical bands around valve joints to contain the dimensionally unstable material. This should be considered as good practice but will not preclude leaks from developing under the higher temperatures and pressures encountered.

The requirement of optical windows for optical monitoring instruments creates two additional problems. The windows must be compatible with the liquid and appropriately sealed. There are no fundamentally new problems here. Glass in the form of fused silica, BK-7 (borosilicate), or Pyrex are suitable window materials for all but hydrofluoric acid and can be sealed with Teflon "O" rings to body parts manufactured from TFE, 316L stainless steel, or other selected material. For hydrofluoric acid, glasses are not suitable and only two crystalline materials have been in use—sapphire and $MgFl_2$. Sapphire is the harder of the two but has a much higher refractive index. A high refractive index can be disadvantageous for certain applications. $MgFl_2$ has one of the lowest refractive indices of any crystalline or glass material and is the most commonly used antireflection coating material in optical thin-film deposition work. Coatings of $MgFl_2$ are slightly soluble in hot concentrated hydrochloric acid and 50% nitric acid, but the crystalline $MgFl_2$ shows negligible recession in tests conducted at PMS with the materials in Table 1. Both sapphire and $MgFl_2$ are birefringent and the optical axis of the crystal should be oriented along the viewing path to avoid spurious optical effects.

An optical counter sizing particles in corrosive liquids typically has two windows (one on the condensing side and another on the collecting side) defining limits of an illumination path over which particles are viewed, detected, and sized. The fluid is constrained in the other two directions by cell walls. In cases where these optical cells define very small volumes, problems relating to material dimensional instability must again be addressed. Such problems are alleviated when the viewing cell is fabricated as a capillary as described in Section 8.

In concluding this section, we should mention that many of the considerations applied to material selection have their basis in safety requirements. No one wants to see a leak develop in a pressurized hot acid fill-line! However, there is an additional concern about contamination of the process liquid from dissolution of wetted materials. In semiconductor processing, cations from the dissolution of metals are particularly problematic. Thus, the industry has been steadily eliminating metal tubing. DI water lines have gone from PVC to PVDF and PFA (or TFE types). Gas lines have in many cases been upgraded from copper to stainless steel (304 then to 316L).

4. SAMPLING METHODOLOGIES

There are several different methods used to sample particulates in liquid media. First, consider that a single semiconductor plant may use as much as 100,000 gallons of DI water a day and at the same time only 20 gallons of hydrofluoric acid. The DI water arrives in a 3" main line and is distributed all over the plant branching out to point-of-use equipment while the hydrofluoric acid may come in one liter bottles. It is desirable to be able to characterize *both* materials equally well. Clearly, the DI water must be monitored on a continuous basis to obtain a representative sample. Samples of the HF acid may be taken for analysis at any time prior to its intended use. These two extreme examples introduce us to choices in sampling methods. Phrases such as: in-line, on-line, off-line, extractive batch, "in-situ," and volumetric all have specific meanings when applied to liquid sampling. In many cases there is confusion about the proper terminology, and manufacturers unfortunately use different terminology interchangeably. The following definitions will be adhered to in this chapter:

In-Line—Refers to a monitoring sensor placed in the fluid line in which all of the fluid flowing through the line passes through the sensor. Usually refers to continuous monitoring by an "in-situ" sensor (see Figure 2).

On-Line, Off-Line—They both refer to situations where a small portion of the main flow is separated and routed through a sensor. The sampled liquid is then either expended to a drain or returned to the main line (see Figure 3).

Extractive Batch—A sample is taken from an off-line set up or out of a container for batch analysis (see Figure 4).

"In-Situ"—Refers to sampling without extraction, intrusion, or disturbing the liquid in its existing state. Applies primarily to in-line sampling where a small (generally less than 1%) but representative portion of the total flow is actually sampled on a continuous basis.

Volumetric—Refers to sampling wherein the total flow passing through the sensor is analyzed for particle contaminants.

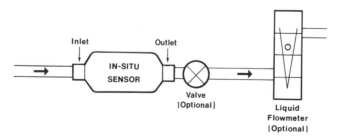

Figure 2: In-line sampling configuration. "In-situ" sensors are required for in-line measurements for all but the lowest of flow rates.

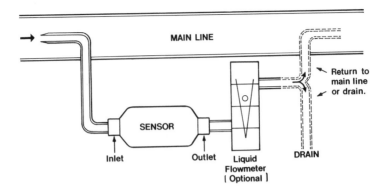

Figure 3: On-line/off-line sampling configuration. Off-line sampling and on-line sampling refer to any process where a sample flow is diverted from a main line through a sensor. Volumetric or "in-situ" sensing can be used depending on flow rate. Spent samples can be returned to the main line or diverted to a drain.

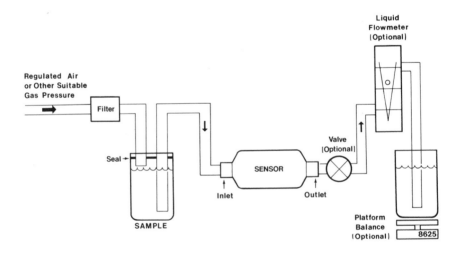

Figure 4: Batch sampling configuration. The above configuration is designed for unidirectional flow to minimize potential system contamination. It is easily configured for automation since the volume is continuously monitored, and samples may be subdivided into small aliquots.

From the above, it must be made clear that "volumetric" and "in-situ" define different classes of sensors each of which could be used in either of the first three listed *sampling methods* depending on the flow rates involved. Volumetric sensors having submicron capabilities only sample up to 100 to 200 ml/min^{-1}, while "in-situ" sensors might require a minimum of a liter per minute with no fundamental upper limit. Thus, while an "in-situ" sensor with a flow rate of 100 ml/min^{-1} might be of the "in-situ" or "volumetric" type, an application with 100 liters per minute would necessarily be limited to an "in-situ" type.

For "on-line" (off-line) setups either type of sensor can be used depending on whether the sample line is dimensioned to provide a large or small portion of the total flow. If a small portion of the total flow is found desirable, a volumetric sensor is recommended, while the opposite is true for a larger flow. One problem here with a volumetric sensor is that it is impossible to achieve iso-kinetic flow at the inlet with the very small orifice required to limit the sample flow to approximately 100 ml/min^{-1} within lines having any reasonable flow rates. Also, the time to flush out contamination collecting in a "dead line" might be prohibitively long if the sample line is operated on a discontinuous basis.

Simple setups for "in-line" and batch *analysis* measurements were shown in Figures 2 and 4. The "in-line" flow set-up of Figure 2 samples a fixed percentage of flow but again requires a knowledge of total flow over a sampling period or flow monitoring. In many applications, the total volume usage is monitored or otherwise known and can be applied to coincident contamination data to reduce it to concentration values. Otherwise, a flow measurement in some form is required if concentration data is desired. Some "in-situ" sensors have the ability to measure particle velocity directly which can be used to compute flow rates and thereby directly obtain concentration data. However, velocity data can only exist when there are particles, and no velocity data is obtained for the cleanest of liquids. Figure 4 is entitled unidirectional which refers to the case when the flow always passes through the sensor in the same direction. Because there is no recirculation or flow reversal, the sensor has the least probability of self-contamination. We will see shortly that more flexible setups have greater tendencies to self-contaminate. The arrangement in Figure 4 shows the liquid placed into a pressure vessel and filtered air used to force the liquid through the sensor. The advantage of pressure over vacuum (applied at sensor outlet) is that cavitation of the liquid can be precluded and more viscous material can be forced through the sensor at suitable flow rates provided by the higher pressure differentials. The recommended sensor in such application is invariably a volumetric sensor since the sample aliquot is reasonably small, and all of it *can* be measured. To obtain concentration, one must measure a known volume which can be accomplished by manually monitoring the transfer or by using a continuous monitor in the form of a flowmeter or platform balance. Both devices provide continuous flow monitoring capabilities; however, there is no flowmeter available to handle HF acid while no such restriction applies to the balance. The balance has the added advantage that most have RS-232 outputs affording convenient computer processing.

Within the last couple of years, instrument firms have begun to offer what are called *corrosive liquid tank samplers* which have many desirable features for contamination monitoring in semiconductor process liquids. Figure 5 depicts the operation of such a device used with a volumetric sensor. Here a sample is drawn from any open vessel into a burette by a syringe pump which is connected to the

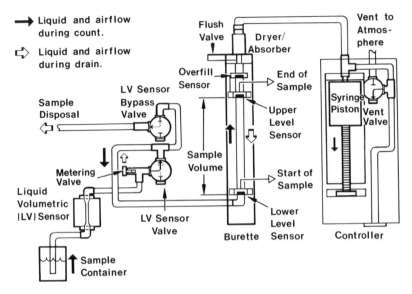

COUNTING DURING FILL MODE

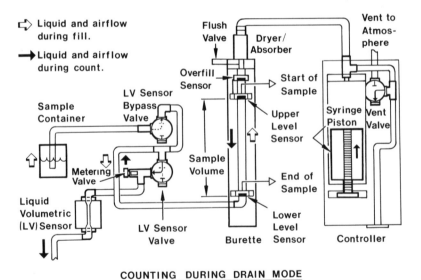

COUNTING DURING DRAIN MODE

Figure 5: Corrosive liquid tank sampler configuration. Samples may be taken during the fill cycle (vacuum) with minimal self-contamination but at slower rates than during the drain cycle (pressure). The principle advantage of such devices is the ability to sample from any open container.

burette via an air line. The syringe produces vacuum or pressure depending upon its venting and direction of travel. Particle counting can occur during the time the burette fills (vacuum) or during the time it drains (pressure). The set-up offers a lot of flexibility in handling all process liquids regardless of container size be it a large tank or a small beaker but has some distinct disadvantages. The system easily self-contaminates unless the count-during-fill mode is utilized exclusively and flow rates are not constant due to the air gap between the sampled liquid and piston of the syringe pump. Attempts to utilize wetted syringe pumps have yet to be realized in product offerings.

5. OVERVIEW OF MEASUREMENT METHODS

The approach to particle sizing is most often limited to methods that are familiar to the user embodying his particular expertise. We can site acoustic, electrostatic, electrolytic, magnetic and optical methods as all being successfully applied to particle size measurements, although not all in liquid media. Because of the need for submicron measurements in semiconductor process liquids, most nonoptical techniques have sensitivity problems. In fact, the only *optical* techniques suitable for continuous monitoring are single particle techniques. Single particle optical methods embody imaging, light scattering, or light extinction. Ensemble optical techniques (where more than one particle is observed simutaneously and particle size information is measured from deconvoluted bulk scattering measurements) include dynamic light scattering and Fraunhofer diffraction. Scattering is the removal of light from a beam via reflection, refraction, and diffraction. An extinction process is the inverse of scattering where one measures the transmitted beam and measures the light lost. Extinction and scattering are equal in the absence of absorption.

The reason that ensemble methods are not applicable to microcontamination measurements is quite simple: In these applications, the probability of more than one particle being present in an illuminated field is vanishingly small. Ensemble methods are applicable to particle products in the form of slurries or suspension but not to dispersed particle populations observed individually as rare events.

With regard to nonoptical methods that are also single particle techniques, there remains to be discussed the electrolytic and acoustic methods. The popular Coulter[3] counter which revolutionized blood cell counting 25 years ago operates as a particle volume sensing device in which the particle displaces a carrier electrolyte and is sensed as a charge in fluid resistivity or conductivity within a small orifice. It is ideal for biological work because isotonic solutions (1% NaCl) are ideal biological cell suspending media and provide the needed conductivity range. However, 18 megohm cm DI water, strong concentrated acids, and nonconductive hydrocarbon solvents are not within the range of reasonable conductivities/resistivities for operating such a device at any of the particle sizes of interest to microcontamination. With regard to acoustic devices that have been used for microcontamination measurements, they must also operate as single particle devices. While they can detect the presence of ensembles of submicron particles there is considerable debate as to their efficacy in observing submicron particles within a definable volume.[4] They are "in-situ" devices by nature and the volume sensed is much more difficult to bound and to monitor uniformity of response. They have largely been limited to measurements in DI water.

We are thus primarily concerned with single particle optical methods embodying imaging, extinction, and scattering. The above factors being recognized one must further come to grips with the basic optical detection process. Except for imaging *a particle is much more difficult to detect and size in a liquid than in a gas.* The process of "oil immersion" in microscopy increases resolution by substituting the air interface between the imaged object and the microscope objective with a higher index liquid media permitting the collection of higher spatial orders of diffracted light—thus increasing resolution. One can take advantage of this fact when sizing particles in liquid media using imaging methods other than microscopy. But the presence of the higher liquid refractive index that increases resolution via imaging reduces the amount of scattering by a particle thus making detection via collected scattered light or extinction more difficult. To understand exactly why this is true is well beyond the scope of this chapter; however, a certain level of understanding is desirable.

Fundamentally, particles of size comparable to the wavelength (λ) of illuminating radiation remove (via scattering) an amount of energy less than that intercepted on their cross section, the disparity increasing as particle size decrease. In liquids this loss of scattering efficiency occurs at a larger size than in air. The relevant factor here is the contrast in refractive index (the ratio of the particle refractive index to that of the medium) $n_c = n_p/n_m$ which for air since $n_m = 1$ is numerically equal to n_p but for a liquid is not. If a particle has the same refractive index as the liquid medium it is invisible and thus not detectable via light scattering, imaging, or extinction. Likewise a 1 μm water ($n_p = 1.33$) droplet in air with an index contrast of 1.33 produces the same signal as a 1 μm sphere of high index glass ($n_p = 1.77$) in water since $n_c = 1.33$ in both cases. Since the index contrast is always higher for aerosol than hydrosol, greater signals are realized for aerosol with less size variation due to particle refractive index changes. Thankfully, liquids in general have lower refractive indices than solids, and the size versus signal relationship is very steep in the submicron region permitting acceptable sizing tolerances for a wide range of particles refractive indices. Another result of the lower index contrast is that light scattered by particles in liquids is more strongly forward scattered. Thus, while less light is scattered, the bulk of it is somewhat easier to detect by conventional optical systems.

Microscope imaging systems necessarily require filtration as a first step for sample collection. A suitable housing must be arranged to hold a membrane filter with apparatus to control the flow and time the sample. Care must be taken to preclude contamination during the assembly/disassembly and microscopic counting and sizing process. There is obviously no real-time data availability nor time-line data history. Sizing is generally less accurate than with particle counters and optimization of sampling requires some prior knowledge of concentration. There is one ultimate advantage, it is currently the only way to obtain morphological and chemical constituent information on contaminants. One can also take samples to an electron microscope and obtain data on sizes below the resolution of optical particle counters.

With regard to *single particle extinction counters,* the following points can be made. At first glance it would appear that an extinction measurement must provide greater detectability since it includes all of the light possibily scattered. In fact, everything being equal, the extinction signal is larger than the light scattering signal. However, an extinction measurement must be distinguished from the random fluctuations (noise) of the entire illuminated transmitting beam while a light scattering measurement by design excludes the transmitted beam. Considering beams of the same size (cross section), a light scattering arrangement can generally detect signals an order of magnitude below that offered by extinction. For years extinction sensors have been used in larger particle sizing applications (hydraulics and pharmaceuticals) but have generally pressed no lower than 1 to 2 μm. Thus, manufacturers of such sensors have necessarily switched to light scattering to extend into the submicron range demanded by the semiconductor industry. In the following sections, we will limit ourselves to the treatment of instruments using light scattering techniques. Table 3 includes current instruments offered by manufacturers capable of submicron single particle size measurements in liquid media.

Table 3: Currently Available Instruments for Submicron Single Particle Size Measurements in Semiconductor Process Liquids

MANUFACTURER	MODEL NO.	(ADVERTISED) MINIMUM DETECTABLE SIZE	NOMINAL FLOW RATE	LIGHT SOURCE
Volumetric Instruments				
Pacific Scientific Menlo Park, CA	346	0.4μm	100ml/min	HeNe Laser
Particle Measuring Systems, Inc. Boulder, CO	IMOLV	0.3μm	100ml/min	HeNe Laser
Climet Instruments Redlands, CA	CI-221	0.5μm	Not Specified	Filament Lamp
Met One Grants Pass, OR	201	0.5μm	100ml/min	Laser Diode
"In-Situ" Instruments				
Particle Measuring Systems, Inc. Boulder, CO	LPOU/CLPOU	0.3μm	5ℓ/min[*]	HeNe Laser
Particle Measuring Systems, Inc. Boulder, CO	LIL	0.3μm	5-100g/min[*]	HeNe Laser
Particle Measuring Systems, Inc. Boulder, CO	PLILP	0.3μm	1-20ℓ/min[*]	HeNe Laser

[*]Sample cells compatible with a variety of flow rates available.

6. DETAILED DESCRIPTION OF LIGHT SCATTERING RESPONSE FOR LIQUID SUSPENDED PARTICLES

Theoretical Considerations

In this section we will use some theoretical expressions to calculate the expected light scattering response of particles in liquid media. These will determine ultimate lower sizing limits, provide estimates of calibration response characteristics, and illuminate effects of variations in both fluid and particle refractive index.

The basic processes by which particles remove (scatter) energy from a beam of light include reflection, refraction, diffraction, and absorption (see Figure 6).

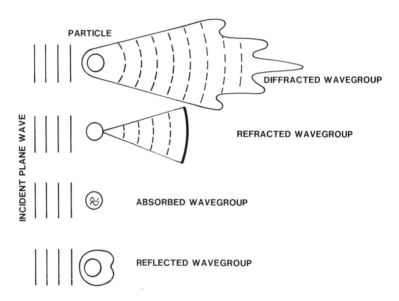

Figure 6: Wavegroups contributing to light scattering. The above wavegroups are always present. Although it is impossible to accurately separate scattered light into each at submicron sizes, many scattering features are easily traced to the individual processes.

A particle whose size is many times the wavelength of illumination is described as in the geometric optics range. It removes an amount of energy equivalent to twice its cross-sectional area (approximately $1/2 D^2$). One areal cross section is removed by the sum of reflection, refraction, and absorption and one areal cross section via diffraction. The computation of the magnitude and direction of the scattered light can be calculated via geometric optics using optical ray tracing and bulk particle properties but is nontrivial in computational magnitude. When the particle approaches the wavelength in size, rays can no longer be traced through a particle and theoretical solutions must treat the particle in closed analytical forms necessitating complicated field theory solutions (Mie region[5]). However, when the particle size become *much* smaller than the wavelength, considerable approximations are possible which have resulted in simple analytical expressions for the light scattering process (Rayleigh region[6]). Here, the scattered response expression reveals a volume squared (or D^6) as well as a λ^{-4} dependency.

Scattering cross sections are employed in dealing with the interactions of matter and radiant energy. In the field of light scattering, angular and total scattering cross sections are basic concepts which lead to several coefficients and expressions having great practical utility. The angular scattering cross section of a particle $\sigma_p(\theta)$ is defined as that cross section of an incident wave, acted on by the particle having an area such that the power flowing across it is equal to the power scattered by the particle per steradian at an angle θ. The total scattering cross section σ_p of a small particle is defined as that cross section of an incident wave, acted on by the particle having an area such that the power flowing across it is equal to the power scattered in *all* directions. A comparison of this definition with the one given for $\sigma_p(\theta)$ inciates that σ_p is equal to the integral of $\sigma_p(\theta)$ over 4π sr. The customary units for σ_p are cm^2, and when the incident flux has unit irradiance, the total amount of flux scattered in all directions is equal to the numerical value of σ_p.

The scattering efficiency factor Q_{sc} is defined as the ratio of the total scattering cross section (area of the wavefront acted on by the particle) to the geometric cross section of the particle. For a sphere this latter cross section is just $\pi D^2/4$. Thus, the scattering efficiency factor of a small sphere, in the size range considered here, can be expressed as:

(1)
$$Q_{sc} = 4\sigma_p/\pi D^2$$

We now see that an important aspect of scattering is the ratio of particle size to wavelength rather than the absolute value of either, and Q_{sc} is thus the fraction of the cross-sectional area of the particles that acts on the incident wavefront. Thus, particles that are very small relative to the wavelength are inefficient scatterers ($Q_{sc} \ll 1$), but the efficiency rises rapidly as the fourth power of particle size to where at sizes of λ, approach 200% or more, and $Q_{sc} \cong 2$.

As a practical matter, particles behave as Rayleigh scatterers at sizes below about 0.2 μm for the most used 633 HeNe laser wavelength. This is about the lower threshold for current liquid particle counting technology; thus we need to utilize Mie theory for computations at currently measurable liquid microcontaminant sizes. From Mie scattering theory for a polarized wave of wavelength λ traveling in media of index n_m incident on a sphere of diameter D and refractive index n_p, the cross section for a given solid angle Ω is given by:

(2)
$$\sigma\Omega = \frac{\lambda^2}{4\pi} \int_\Omega \left\{ |S_1|^2 + |S_2|^2 \right\} \sin\theta \, d\theta,$$

where $S_1(\lambda, n_p, n_m, \theta)$ and $S_2(\lambda, n_p, n_m, \theta)$ are the Mie scattering amplitude functions corresponding to light polarized with electric vector perpendicular and parallel to the plane of scattering. The angular integration is over the solid angle subtended by the light-collecting optics. The refractive index in the more general case of Mie solutions is a complex quantity $n = n_r - in_k$, where n_r is the real portion containing the refractivity, and n_k is the imaginary portion reflecting possible absorption. The details of computations using (2) are beyond the scope of this writing; however, computations are readily accomplished today using small computers.

Most manufacturers of optical particle counters use polystyrene latex (PSL) microspheres as the primary calibration particles. These microspheres have a refractive index of 1.586 at 633 nm increasing to about 1.60 at the blue end of the spectrum. Figure 7 shows the results of computations of the total scattering cross sections for PSL in air and water. PSL is ordinarily totally transparent.

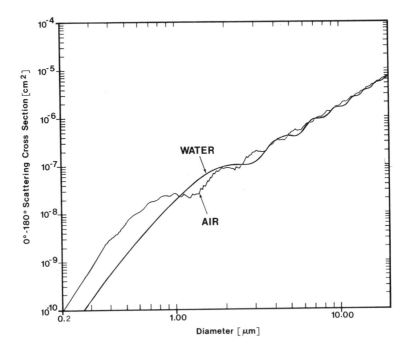

Figure 7: Total scattering cross sections for PSL spheres in air and water. These theoretical response waves are for PSL spheres in air and water (n = 1.585) and a wavelength of λ = 633 nm.

However, the effects of possible absorption by a water immersed particle with the same real refractive index as PSL but having various amounts of absorption would be as shown in Figure 8. In both figures there are several noteworthy features. The slopes become parallel at sizes below about 0.2 μm; the smallest particles scatter more light when they are absorbing and the submicron particles scatter more light in air than in water.

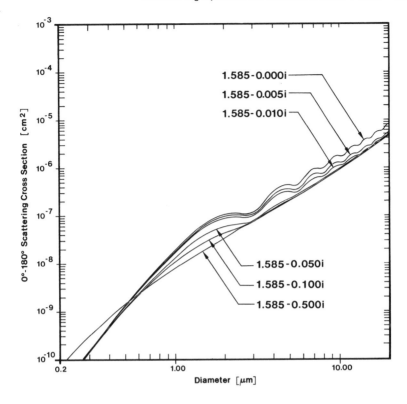

Figure 8: Total scattering cross sections for absorbing particles having a real refractive index equal to PSL in water. Computations are for a wavelength of $\lambda = 633$ nm.

In the above computations we have been comparing total scattering cross sections. In any practical instrument one has to consider the collection of all scattered light as impossible, although a large percentage of the light scattered can be realized. Figure 9 compares computations for a $10°$ to $60°$ collecting solid angle with the total scattering cross section for a refractive index at 1.586 at 633 nm wavelength. The reader can readily observe that the pair of curves are very similar. As the Rayleigh region is approached, the curves are parallel. In the Rayleigh region the slope is approximately D^6 independent of the collecting geometry. Also, the $10°$ to $60°$ collecting geometry maximizes the percentage of light collected in the critical submicron size range.

Thus far we have only treated one particle of a specific refractive index in one liquid (water). We previously mentioned that the appropriate parameter is the index contrast which is the ratio of the particle refractive index to that of the liquid. Figure 10 illustrates how latex particles of various size would scatter in liquids of various refractive index. Thus, a one micron PSL sphere in xylene (n = 1.50) would scatter the same amount of light as a 0.63 μm PSL sphere in water (n = 1.33) and a 1.58 μm PSL sphere in sulfuric acid (n = 1.43) is equivalent to a 1.0 μm PSL sphere in hydrofluoric acid (n = 1.29). Clearly, the higher the liquid refractive index the smaller the apparent particle size. But what about particles that have refractive indices lower than or equal to that of the liquid? Figure 11 illustrates the problem for several particle sizes. It is observed that the particle may scatter almost the same for an amount higher as for an amount lower in refractive index, but the effect on estimating the true particle size is apparent. A bubble in water (n_{Rel} = 0.7) scatters the same as a particle of n = 1.7; i.e., more than a latex sphere.

Index contrast can also exist in another form: The particle may have the same real component as the liquid but a different imaginary component. Figures 12 and 13 reveal the effects of varying absorption by the liquid and varying absorption by the particle respectively. In both figures n_c(Real) = 1. It may be surprising to think that a particle can absorb and induce scattering whereas if it did not absorb, it would not scatter at all. However, if one looks at metals which have real indices near unity but nearly infinite imaginary index components, it helps to understand the problem; metal particles do not transmit (refract) light at all but are highly reflective.

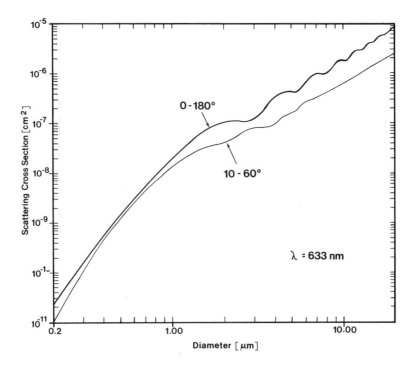

Figure 9: Comparison of total scattering cross sections and a 10° to 60° collecting geometry for PSL in water. Computations are for λ = 633 nm. Note the high percentage of light collected in the submicron size range for this geometry.

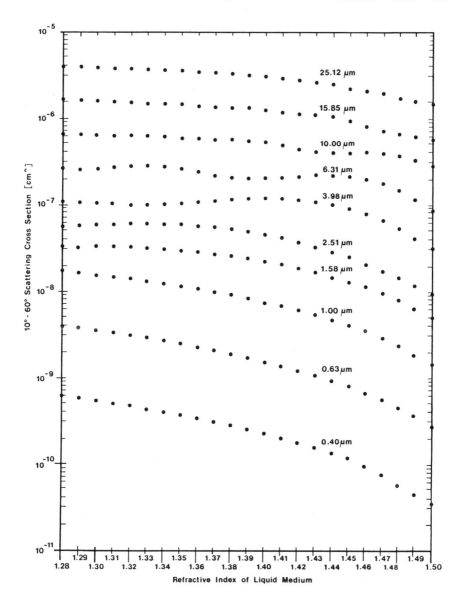

Figure 10: Scattering cross sections for PSL in liquids of varying refractive indices. Computations are for a 10° to 60° collecting solid angle and λ = 633 nm.
The above curves graphically demonstrate the loss of size sensitivity as the liquid
refractive index increases. Compare with tabulated values for various liquids in
Table 1.

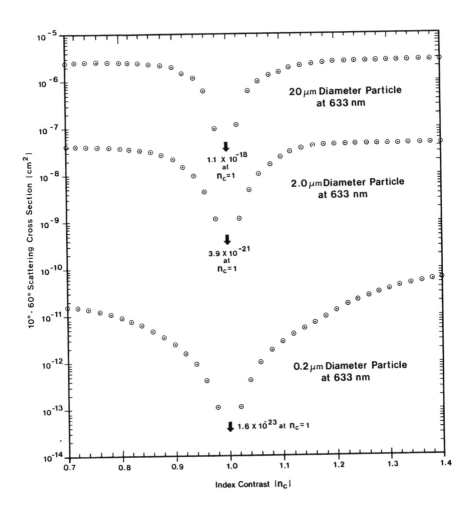

Figure 11: Scattering cross sections as function of index contrast. Computations are for a 10° to 60° collecting solid angle and λ = 633 nm. When n_c = 1 the non-zero values computed are the result of computational limits.

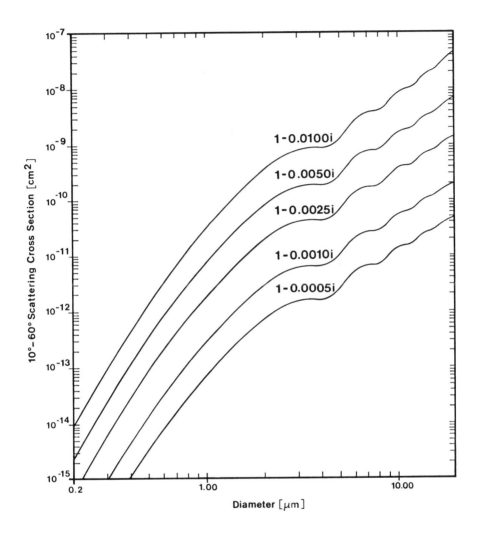

Figure 12: Scattering by a transparent particle in an absorbing liquid. These computations assumed perfectly matched real refractive indices [n_c(Real) = 1] of particle and liquid. The collecting solid angle was 10° to 60° and λ = 633 nm.

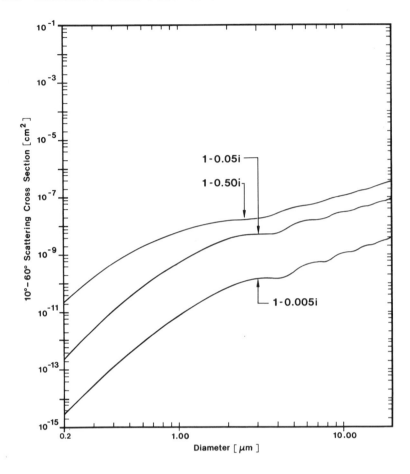

Figure 13: Scattering by an absorbing particle in a transparent liquid. These computations assumed perfectly matched real refractive indices [n_c(Real) = 1] of particle and liquid. The collecting solid angle was 10° to 60° and λ = 633 nm.

7. DEFINING THE LOWER LIMIT OF PARTICLE SIZE USING LASER LIGHT SCATTERING

At this point in our discussion we have predicted the response into the submicron size range but have not defined an optical system capable of making such measurements. In order to define what is required, we need to consider signal-to-noise ratio for various configurations and exactly what constitutes the minimum detectable particle size.

A particle can be assumed to be detectable if in a great majority of sizing opportunities there is a high success ratio (e.g., 9 out of 10 times). Furthermore, background noise should introduce negligible false counts. Background noise is defined as any unwanted signal but is ultimately limited to background light in all optical particle counters. This background light comes from two primary sources; stray light scattered from optical surfaces and scattering from liquid media carrying the particle through the viewing optics. In most cases, bulk liquid media scattering becomes the fundamental limiting factor. In order to reduce media scattering, one must reduce the viewing volume under observation. Outside of the practical problem that restricting the viewing volume reduces the potential volumetric sampling rate, there are also fundamental limits as to how small a volume can be illuminated. Knollenberg[7] showed that such a volume would have dimensions of about 1 μm^3, and appropriately 10% of the scattered light could be collected. The total scattering cross section of 1 μm^3 of water is approximately $10^{-17} cm^2$ ($10^{-5} cm^{-1}$). (This estimate is inferred from Reference 8 and comes from measurements commonly made in the field of chemistry. In principle one could use Rayleigh or Mie scattering to compute these values; however, it requires a knowledge of the detailed clustering of molecular aggregates and is difficult to parameterize.) Any laser is just capable of putting all of its energy into this volume wherein its total cross section is $10^{-8} cm^2$ (1 μm^2). Thus, in the limit the effective media scattering cross section per unit of laser beam cross section over a 1 μm pathlength is 10^{-9}. Thus, if one watt laser were utilized, approximately one nanowatt of power would be scattered by 1 μm^3 of the bulk liquid media. An immediate problem is that it will be impossible to distinguish a photon that a particle scatters from that by the bulk liquid necessitating that the particle scatter more photons in its viewing time than the photon noise associated with the bulk liquid scattering.

The problem here reduces to a statistical photon counting problem wherein the variations in any particular count are proportional to the square root of the count itself (classical statistical noise). The noise associated with the bulk liquid scattering will thus be proportional to the square root of the bulk liquid scattering signal. In the case of a one microsecond viewing time, the noise (statistical) associated with observing 10^{-9}W flux is 3×10^{-10}W. The one watt laser produces $1.0 W \mu^{-2}$ or $10^8 W cm^{-2}$. Consequently, a particle of $3 \times 10^{-17} cm^2$ would generate a signal equivalent to the liquid ledia generated noise. This corresponds to a of about 350Å in size and is the resulting theoretical limit sought. However, with current optical systems, we can only collect about 10% of the scattered light, and this reduces the signal level by 10 or S/N by 3.3 times and raises the minimum detectable size to about 400Å. Also, the viewing volume of 1 μ^3 is too small to be of practical interest. Scaling the viewing volume to a more practical size of 100 μ x 100 μ x 100 μ results in a minimum detectable size of 750Å or just slightly less than 0.1 μm. (See Reference 7 for details of similar calculations in gases.)

In the above analyses it was assumed that the light source was a thermally quiet one. That is the only fluctuations produced by the light source were those generated by the random generation of photons and the statistical bunching of them. To obtain the maximum sensitivity from an instrument, a laser is generally required to produce the needed energy density. Lasers are not thermally quiet sources but are rather noisy at times due to mode competition between the ever changing axial modes. Such noise manifests itself as sinusoids of changing amplitude and frequency. The quietest lasers are single frequency devices where mode competition is eliminated. However, even single frequency lasers typically have 50% greater noise than an equivalent thermal source of equal power (e.g., a dc white filament lamp). Aside from their greater noise and also being more costly than white light sources, lasers offer numerous advantages when used for sizing particles. The most important considerations and advantages of laser light sources are:

- High energy density
- Ability to filter
- Ideal collimation characteristics
- Light source stability (dc)
- Long lifetime (15,000 to 20,000 hours versus 1,000 to 2,000 hours)
- Improved optical system design (no chromatic aberrations)
- Detectors matched to a single wavelength

The most important of the above is the higher energy density which makes available lower size sensitivities approaching 0.1 μm. Because of the higher noise of lasers, there is no ultimate advantage in using them for extinction measurements unless they are single frequency devices. Their primary advantage comes about in light scattering instruments wherein the background light (both stray and bulk liquid media generated) is suppressed to negligible levels. In the instrument descriptions that follow, laser light source devices will be described in detail.

8. DESCRIPTION OF INSTRUMENTS

Instruments that are currently used to monitor contaminants within semiconductor process liquids are largely laser based using light scattering to size particles that can be categorized as belonging to one of two possible classes: those that measure remotely and sample a small portion of the total volume passing through *("in-situ" instruments),* and those that sample all of the volume passing through *(volumetric instruments).* Both instruments have characteristics which allow them to be optimally used under different circumstances. Either instrument can logically view only a small illuminated volume if maximum sensitivity is desired (which is required for microcontamination measurements). Therefore, a volumetric instrument must have a highly restricted flow to allow all of the fluid to pass through the illuminated view volume. Typical dimensions of the minimum restriction are submillimeter enabling maximum fluid flows of only 100 to 200 ml/min. An "in-situ" instrument on the other hand has no required flow restrictions since the particles are viewed remotely and the relevant viewing volume can be established by optical parameters (fields-of-view, depths-of-field, etc.) rather than physical boundaries.

Volumetric Instruments

The necessity to size all particles within a small volume of liquid in a single pass dictates the use of volumetric sampling. This is the most common sampling method in use and many procedures have developed surrounding its use. Optical systems for two liquid volumetric (LV) sensors manufacturered by PMS are shown in Figure 14. They both use a laser illumination source and a flow cell constructed to optical tolerances. Sizing is accomplished by pulse height analysis of the detected light signals produced by flowing particles. The sample cell is generally constructed of metal or fluid compatible synthetic and is fabricated to leave walls defining a flow cell of 1 to 2 mm length and 0.25 to 0.5 mm width. It is capped on both ends with windows with the laser optical path coinciding with the longest dimension. The cell is constructed as a mechanical sandwich, typically assembled from six parts (including two windows) as shown in Figure 15. The sandwiched parts can be either accurately lapped or sealed with gaskets or cemented to obtain leak-free assemblies. Windows can be either 'o' ring sealed or cemented. When operated as a light scattering cell, the assembly requires that surfaces A and B be highly reflective so that the light scattered by particles near the far side of the flow cross section (being multiply reflected) is collected with the approximate same efficiency as a particle at the near side of the flow cross section. The interfaces can be finished to optical flatness permitting disassembly and reassembly without the use of gaskets or sealants.

As shown in Figure 14, the LV sensor can be either an extinction or light scattering device. PMS manufactures a laser extinction device, while Pacific Scientific offers numerous white light extinction devices. As is generally the case, light scattering permits increased sensitivity. However, for sizes 2 μm and larger, the extinction devices give better resolution. This is due to the difficulty in collecting light equally from particles passing through the center versus the far reaches of the cell. With extinction there is only the requirement to make beam loss measurements and thus assure that the particles are uniformly illuminated. With regard to the latter, we should mention that the laser beam in either case is focused in only one dimension to about 40 to 50 μm using a cylindrical lens. The other dimension is approximately 1 mm and essentially collimated. When the beam enters the cell walls truncate it to the dimension of the sample cell minor width (0.25 to 0.5 mm). The truncated Gaussion laser beam diverges only slightly (due to slit diffraction) resulting in approximate collimation through the sample volume.

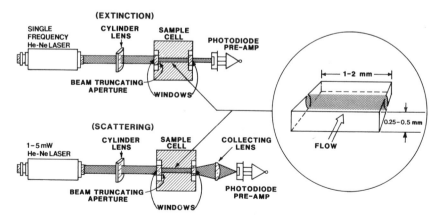

Figure 14: Volumetric sensors. The cells used for extinction or scattering differ in light collecting geometries but are otherwise identical.

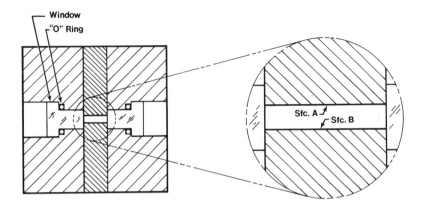

Figure 15: Conventional volumetric cell construction assembly.

The scattering signals generally vary as r^2 except at submicron sizes as shown in Figure 9. In general, the size resolution of the LV sensor is better than what one might first suspect considering the far reaches of such a sample cavity and the problems of maintaining uniform illumination. However, the flow has a strong gradient across the flow cell and fewer particles pass along the sample cell walls than through the center fortuitously biasing the measurement volume in favor of regions of preferred uniform illumination.

There are four manufacturers of the above light scattering LV sensors. Only Climet's instrument device is a nonlaser device (see Table 3). All tend to use similar materials, i.e., Kel-F or Halar for body parts and crystalline sapphire or $MgFl_2$ for windows. Three of the instruments size down slightly below 0.5 μm. The lower threshold of 0.5 μm is well above the 0.3 μm lower sensitivity or "in-situ" sensors. This is true even though both instrument types have similar viewing volume sizes using the same laser powers. The reason for the loss in sensitivity in the LV sensors is that the fluid/vessel wall interfaces generate fairly high levels of stray light—at least an order-of-magnitude above that of the liquid media. The stray light establishes a noise background level from which the light scattered by individual particles must be differentiated. Obviously, if the noise background is greater than the particle scattering signal, the particle cannot be measured. With "in-situ" type instruments the particles are illuminated and viewed through windows whose fluid interfaces can be removed far enough from the illuminated "view volume" to be out of the depth-of-field, and their light scattering noise contribution is of a negligible nature.

PMS has recently described a new volumetric cell design in which the interfacial stray light can be minimized to such a low level that size sensitivities of the nearly ideal "in-situ" methodologies can be duplicated. The background light can be reduced by nearly 100x below that of the prior cells allowing the lower limit of sizing to be extended from 0.5 μm down to 0.3 μm. (The ratio of scattering signals for 0.5 μm and 0.3 μm is approximately a factor of 10.)

In the newer technology cells, the mechanical sandwich is replaced with three optical glass parts which are bonded together as shown in Figure 16 forming an integrated micro-optical cell. The three parts are: (1) a capillary tube,

(2) a window, and (3) a lens. The fluid flow is constrained to the capillary, and all interfaces are glass-fluid. As an instrument assembly, this integrated micro-optical liquid volumetric (IMOLV) sensor functions as illustrated in Figure 17.

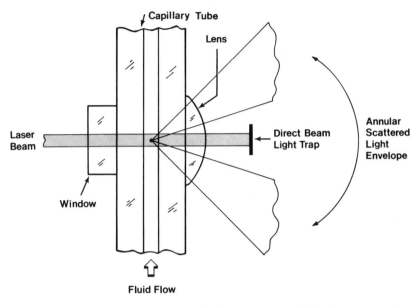

Figure 16: Integrated micro-optical liquid volumetric (IMOLV) sensor cell construction. This cell allows for the collection of light from 10° to 60°.

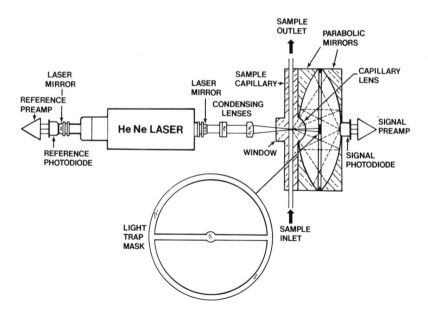

Figure 17: Integrated micro-optical liquid volumetric (IMOLV) sensor optical system.

A laser is used as the source of illumination and is astigmaticly focused using cylinder lenses to cover the capillary cross section as a thin cross section of illumination. The transmitted light is absorbed by a central mask dimensioned to block all normally transmitted light, but not to appreciably reduce the collectable scattered light. The scattered light from particles is reimaged onto a photodetector and becomes the signal for particle size analysis. The function of the entrance window is to move the entering glass-air interface away from the centrally viewed volume of the capillary. The exit lens has the same purpose plus one additional important function; the radius of curvature is selected to coincide with the distance from the center of the "view volume." This allows for rays emanating from scattering particles to be undeviated passing through the glass-air interface upon exiting the lens, and the particles, as viewed from the detector, appear to be in the same positions as if the glass were absent. The scattered light can be collected by a variety of optical systems. For maximum collection of scattering signal, a parabolic mirror pair at 1:1 conjugates is utilized. This optical system for collecting scattered light generates a large solid angle ($10°$ to $60°$) and the necessary central mask used to block transmitted light can be conveniently positioned between the mirrors.

The primary advantage of the IMOLV cell is its greatly reduced level of scattered light. Since the fluid-glass interface produces a reflective loss calculated according to the well known Fresnel formula:

$$\text{(3)} \qquad \text{Reflectivity} = (n_{GI} - n_{FI}/n_{GI} + n_{FI})^2$$

where n_{GI} is the refractive index of glass and n_{FI} is the fluid refractive index, the reflection losses become zero when the fluid and glass interfaces are equal but are also extremely small in nearly all cases for common liquids (where for liquids the range of n is approximately 1.3 to 1.5 and for glasses 1.45 to 1.55). In the former cells the wall materials used have extremely high imaginary refractive index components and high reflectivities.

The scattering losses occurring at interfaces are also proportional to the index contrast between fluid and confining vessel walls. For instance, a small glass imperfection at a fluid interface will scatter much less than an identical defect (in size and shape) at a metal interface wetted with the same liquid. If the two materials at an interface have identical refractive indices, such imperfections also produce zero scattering. Thus, with great care in cell construction, volumetric sensors can be designed to approach the same low stray light contribution as "in-situ" devices. However, the capillary in the IMOLV needs to be free from defects and contaminants that scatter light whereas the "in-situ" sensor is essentially immune to window/fluid interfacial conditions.

"In-Situ" Liquid Instruments

There are two basic "in-situ" techniques in use, both use laser based darkfield imaging systems. (Currently PMS is the only known manufacturer of submicron "in-situ" sensors.) The first system to be described involves a particle trajectory analysis scheme using complementary masks to define particles traversing a preferred sample volume and can be described with the aid of Figure 18. A high resolution laser optical system collects particle scattered light and reimages it at a selected magnification within a dark field. The transmitted laser beam is dumped at a central stop on the first lens element. A beam splitter produces two image planes for two detectors. The reflected image prism face is masked with a vertical slit to block central transmission during transit. The other detector prism face is unmasked. The masked beam splitter derives two signals which, in conjunction with double pulse height analysis, provide a means of determining if a particle's position is in the desired sample volume. The relative size of the sample volume cross section with respect to the laser beam is depicted in Figure 19. The sample cross section includes only the region near the center of the laser beam and is noticeably diamond shaped. The center of this diamond cross section coincides with the object plane of the collecting optics. The points of this diamond cross section define the limiting depth-of-field. Both the width of the sample cross section and the depth-of-field vary inversely with the magnification used in the collecting optics.

The functioning of the sample volume definition process can be explained as follows: For small particles (essentially point objects), the image size is linearly related to the numerical aperture of the collecting optics and is given approximately by: image size = N.A. x displacement from object plane. It is apparent that only images that are near the object plane *and* the center of the sample volume form images with light concentrated on the opaque slit on the masked detector. One normally uses a gain ratio of masked aperture detector to signal aperture detector of >1 for best noise immunity. Particles whose pulse amplitudes seen by the masked aperture detector are greater than those seen by the signal aperture detector are rejected. The diamond shaped sample area (product of the sample volume width and one-half of the depth-of-field) results. Sizing is accomplished by pulse height analysis of the unmasked signal detector.

This method of trajectory analysis has two primary advantages. First, it is possible to localize as small a region as necessary to achieve uniform illumination by increasing magnification. A second result of the use of this method is that stray light can be controlled by magnification and depth-of-field parameters. This type of system has been used to localize regions of a laser beam less than 20 μm wide. Thus, with a highly focused HeNe laser and 1 mW output power, one can achieve 0.2 μm sensitivities in liquid media with solid state detectors. The primary disadvantage lies in the very small sample cross section that results and lower alignment stability which have resulted in current offering of only 0.3 μm sensitivity.

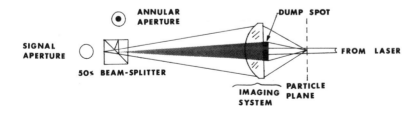

Figure 18: Simplified optical system for "in-situ" sensor.

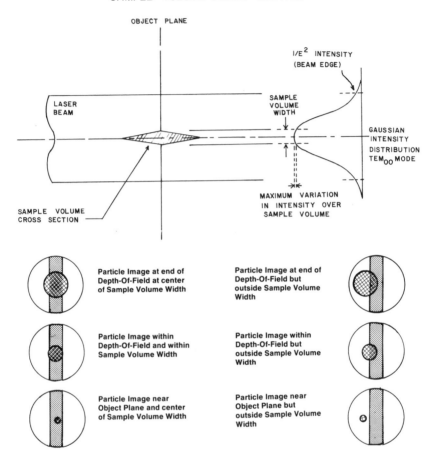

Figure 19: Sample volume definition and particle positions and image sizes for "in-situ" system in Figure 18. The diamond-shaped sample volume results from analysis of image trajectories.

The second type of "in-situ" sensor uses a polarization scheme to define the valid sample volume (see Figure 20). Here a polarized laser output is split into "S" and "P" polarized states. These beam components are shaped with lenses to produce beams of circular (S) and elliptical (P) cross sections in the viewing volume. The smaller circular beam (S) is aligned so that when particles traverse both beams, they must have crossed the center of the elliptical beam (P) as shown in Figure 21. This isolates the preferred sample volume in one dimension only. The other dimension (depth-of-field) truncates naturally because the circular beam coverages and diverges more rapidly than the elliptical beam near the focal plane. When the collected "S" and "P" scattering signals are compared, the acceptable sample volume is defined by those particles whose "S" scattering exceeds "P" scattering. Sizing is accomplished by analysis of the "P" scattering component.

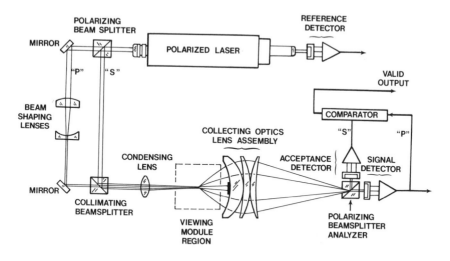

Figure 20: Polarized optical system for "in-situ" measurements used in PMS PLILP and LIL instruments.

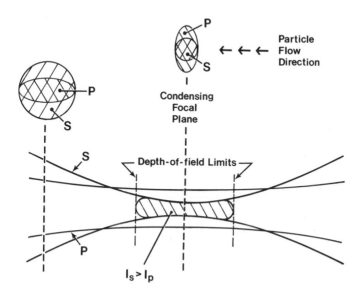

Figure 21: Sample volume definition using optical system of Figure 20. Particles are defined as acceptable when they pass through cross-hatched region where the intensity of "S" scattered light is greater than "P" scattered light.

The above polarization scheme has some advantages over the complementary masking scheme. First, acceptable particles are defined by an acceptance criteria rather than rejection criteria. The coincidence of both "P" and "S" components improves the noise immunity markedly. Secondly, the sample volume is inherently larger and independent of magnification. Thirdly, the imaging optics are not critical since the quality of images is not a factor. Fourthly, since there are two independent beams that can be juxtaposed, it is possible to devise a velocimeter that is more accurate than when using a single beam as in Figure 18.

Both of the "in-situ" type of systems can be utilized for "in-line" or point-of-use type measurements. PMS has, in fact, developed three sensor families using the "in-situ" techniques described. These are the liquid-in-line (LIL), liquid-point-of-use (LPOU) and portable liquid-in-line probe (PLILP) sensors. The LIL and PLILP sensors use the polarization scheme. Photographs are shown in Figure 22. The PLILP is unique in that a windowed viewing module is used to isolate the sample volume from the sensor optics as shown in Figure 22a. Variations in liquid refractive index are compensated for by having windows of compensating thicknesses. In general, viewing modules of $\frac{1}{4}$" to minimum of 2" internal sample flow dimension are available in four ranges of liquid refractive index compatibility. The LPOU, PLILP, and LIL give best performance in the submicron size range. Because the scattering signal varies so strongly in size, high resolution is achieved; however, the signal level is still about an order of magnitude lower than with aerosol. Figure 23 shows typical calibration data using latex and glass spheres in water.

All of the liquid sensors available today can be purchased with a few or many size resolution channels. A particle counter is generally defined as one that has eight or fewer size class intervals while a particle size spectrometer is one that has 15 or more. The inference in the distinction is that a proper definitive spectrum requires more size classes for defining spectral features. In general, data acquisition can be hardware or software driven, and complete computer control and automation is available from device manufacturers.

9. APPLICATION PROBLEM AREAS

In this section we will discuss the test procedures and data treatment for microcontamination measurements in process liquids. Particular attention is first given to test procedures. All test procedures must be carried out carefully to avoid erroneous data which can result from either incorrect measurement techniques or incorrect liquid handling procedures. Standard procedures and test methods are required whether sampling from containers, main liquid lines, point-of-use dispensers, or system reservoirs. We then turn our attention to data processing and interpretation. In a preceding section, we discovered that particle size measurements in liquids are subject to considerable uncertainty due to variations in index contrast. Absolute size measurements are difficult to obtain and require special sample preparations. We, thus, finally turn our attention to the different task of data interpretation.

Operating Procedures

The operating procedures for both in-line measurements and batch analysis

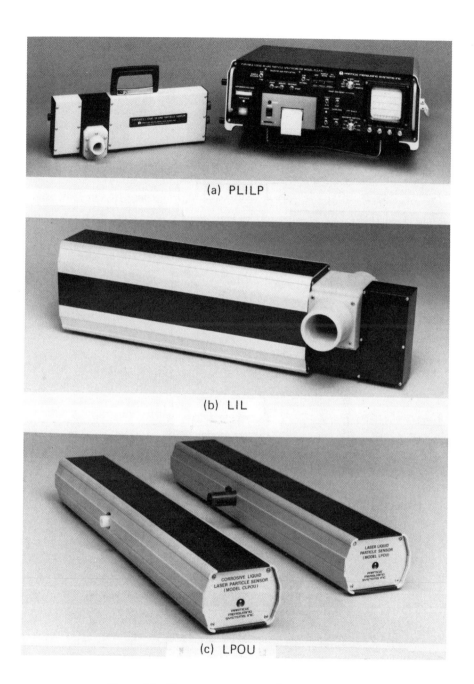

(a) PLILP

(b) LIL

(c) LPOU

Figure 22: Photographs of PMS "in-situ" sensors.

of sample aliquots must take into consideration the following:

Sample size

Sample flow rate

Sampling frequency

Sample delivery system

Post collection, storage, and transport effects (batch sampling only)

Consistent data may only be obtained after careful consideration of the above, followed by the definition in detail of all steps to be followed, and the apparatus used for liquid handling.

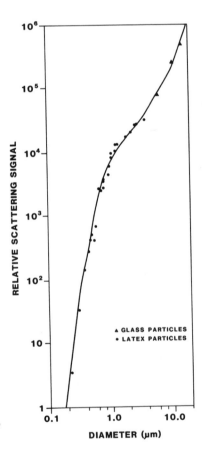

Figure 23: Calibration data typical of "in-situ" sensors. Data are for a 10° to 45° collecting solid angle.

Unless information on particle morphology or composition is required, sub-micron contamination measurements are invariably made with optical particle counters. While laser particle counters are satisfactory for particles larger than 0.3 μm diameter for obtaining size and concentration data, correct calibration and operating procedures must be followed to ensure that valid data are generated. In the area of process liquid handling methods, one should bear in mind that existing techniques have primarily been adopted for the requirements of the medical and pharmaceutical product manufacturing industries. There the emphasis has been on sterility, and although high quality filters are generally used, some residual inert particle loading is found acceptable. For example, only particles larger than 10 μm are to be counted in parenteral fluids as specified by the *U.S. Pharmacopoeia.* Obviously, the procedures in liquid handling methodologies in use for the pharmaceutical industry require significant refinements before being applied to semiconductor process liquids. We are now ready to examine several problem areas which arise in typical applications of microcontamination measurements in semiconductor process liquids.

Container Cleaning

A significant amount of background contamination arises from the storage containers in which semiconductor process liquids are provided as well as the sample containers used to transfer liquids for batch analysis. Popular container materials are glass and plastics such as polyethylene, polypropylene, or fluorinated hydrocarbons. Notwithstanding that chemical manufacturers carry out good cleaning practices on containers, containers and their closures are still a potent source of microcontaminants. It presently appears that glass is a preferred container material over plastic although not necessarily preferred over fluorinated hydrocarbons. The exact source of these contaminants are far from being well understood, but it appears that the levels of contamination increase with time of storage in plastic containers but not in glass containers.

Several cleaning cycles are generally used in sampling containers. There are typically two or three solvent-surfactant high shear rate flushings followed by rinsings with clean, DI water. Drying may not be necessary when the final rinse liquid is completely miscible with the process liquid to be sampled unless the containers are to be stored before use. Water-detergent cleaning is generally sufficient to remove contaminants present in semiconductor process liquids.

Regardless of the container material type, the preferred closure is a fluorocarbon-lined cap. Reliable sealing with minimum debris generation is easier with this material than with any other type of gasketing. In general, container closures should be cleaned in the same way as the containers. It is a good practice to clean sample containers and closures in batches large enough to allow two or more of the containers to be used for background measurements by filling with clean liquid (most likely DI water) and measuring background contaminant levels.

Possible Contaminants

Most semiconductor process liquids are not affected by as many species of contaminants as found in DI water systems, e.g., bacterial contamination. Except for alcohols, most semiconductor process liquids are either sufficiently oxidizing or reducing to destroy microorganisms. In spite of the high degree of cleanliness in DI water supply systems, bacterial growth is a common problem. While the nutrient supply is admittedly low, certain bacteria can survive and grow over a wide range of temperature and pH. Single bacteria are typically a few tenths of a micron in size; however, colonies of bacteria are often found typically growing in the enhanced nutrient levels trapped by filters, sometimes growing in sufficient numbers to clog filters. Once formed, they are difficult to remove and may require system disassembly, cleaning, or replacement if flushing with biocides and ultraviolet light control measures prove inadequate.

Artifacts

The generation of particulate artifacts due to improper material handling is one of the greatest problems in measuring submicron particulates in process semiconductor liquids. To permit satisfactory measurements of submicron particles, procedures proven satisfactory for much larger particles may have to be refined, modified, or abandoned. For instance, the simple flexing of a transport line can release many submicron particles even if the lines have been well flushed. These may be generated as a result of sloughing of surface filaments as a result of stress concentration or be released as previously deposited particles due to the disruption of binding forces. Sloughing is a particular problem when plastic lines become chemically crazed. It is a recommended practice to avoid flexing of sample lines just before or during a sampling operation otherwise sloughed material will positively bias the measurements.

Valves are another source of contamination artifact generation. The dead volume and dead lines associated with valve operation provide accumulation zones for microcontaminants. Also, valves must have surface-to-surface sliding parts generating wear particles. When sampling at point-of-use stations, valve dead lines must be flished until steady state levels of contaminants may be measured. This may require several tens of minutes for inactive systems. It is recommended that the number of valves in a sampling system be kept to a minimum and their adjustments be few. Another source of particulates that can be generated by valve operation results from the hydraulic jump (pressure pulse associated mostly with valve closure) which can rattle and shake lines over considerable distances.

Another source of artifact generation is through the trapping of airborne contaminants when using open containers. Containers, tanks, or sample storage vessels must be protected from environments that contain particulate debris of any kind. This is generally not a big problem in semiconductor contamination areas since most of the analysis is performed in clean rooms or under clean hooded benches. In general, common sense will dictate when samples become contaminated. It is a useful practice to actually expose open containers to the lab atmosphere as a method of determining the magnitude of this artifact transport mechanism.

One deleterious effect of measuring reasonably high levels of contaminants in liquids is that of contamination of the viewing windows of a particle counter. These attached scatterers will increase the background light and can generally raise the noise level to the point where noise is counted in the most sensitive size channels of the instrument. As previously discussed, this is a greater problem with volumetric systems than with "in-situ" systems. However, in either case, the optical system will from time to time require cleaning of the internal wetted optical surfaces. The manufacturer's recommendations should be followed when cleaning is required.

One has to keep in mind that any problem that occurs with instruments designed to sample larger particles can occur with submicron instruments, but the problem will be aggravated with sensitivity to smaller particle sizes. For example, extreme care must be taken to ensure that rinsing liquids are miscible with the liquid being sampled. Otherwise, some emulsion droplets will occur. These homogenized droplets may be undetectable until submicron sensitivities are applied. Trapped in cavities immiscible liquids may be sourced for much longer periods of time than anticipated.

A final problem area that must always be considered in examining any low count particle levels is that of ensuring that a data set is statistically valid. One should recognize that the natural variation of any population is proportional to the square root of N, where N is the total sample size. Once the population is of such sufficient size that the natural variation is acceptable, comparisons can be made from sample to sample. It is important also to recognize that the statistical base is generated by the raw data counts and not by some concentration value expressed in engineering units with artifically large significant figures. Finally, it is a good practice to make background measurements and subtract them from the data samples when either artifact or noise are suspected false count contributors. It is generally sufficient to operate the instrument without flow while making a background count. However, the liquid *must* be clean.

Data Interpretation

We now come to the most difficult task of interpreting the levels of contamination measured in various process liquids. We saw that in Figures 10 and 11 that the apparent size measured by an optical particle counter is not an absolute size but to a first order is in fact a strong function of the disparity between refractive index of the particle and the process liquid. We also know that all instrument calibrations are based upon a suspension of latex microspheres in water. Latex is a higher index material than many contaminants, while water is one of the lowest index process liquids. Thus, the latex/water binary generates a relatively strong response. Examining Figure 10, it is clear that a 0.5 μm latex particle in water scatters the equivalent of a 1.0 μm particle in sulfuric acid. Furthermore, a quick glance at Figure 1 reveals that the number of particles of 1.0 μm size as compared to 0.5 μm is typically several factors of 2. It goes without saying that it would be difficult to compare levels of contamination between fluids of different refractive index. A final complication is revealed in Figure 11, which shows that a particle could be of higher or lower refractive index than as compared to the suspending liquid and generate an identical light scattering response. So how are we to interpret our measurements?

At the present time there are two approaches to be followed: (1) measurements can be made and compared on a *relative basis* as long as they are from the same liquid or a liquid of equivalent refractive index, (2) measurements can be made on an *absolute basis* by diluting a process liquid with a *low index clean solvent* to achieve refractive index values comparable to water. The second approach requires some additional explanation. What in essence we are recommending is that the index contrast of the liquid/particle binary be reduced to that of a water/particle binary via dilution. This in turn largely reduces the index contrast problem to that for which the particle counter was calibrated (i.e., a liquid index equal to that of water). To accomplish a satisfactory dilution, one first must have available solvents much cleaner than the liquid under test. Fortunately, DI water is the cleanest liquid ordinarily found and several of the low index solvents such as methanol and acetone are also more easily cleaned up to provide suitable solvents where water is unsuitable (e.g., for diluting xylene).

In accepting dilution as a possible means of correcting for the lowered index contrast of contaminants in high index liquids, an assumption is made that the contaminants typically have higher refractive indices than liquids. Such an assumption is generally valid as long as one is talking about solid particles and low index liquids. There are only a few solid materials which have refractive indices below 1.4; magnesium fluoride is 1.37 and cryolite is 1.35. (Because of their low refractive indices, these materials are useful in optical coating work for their use in antireflective coatings.) However, the refractive indices of most chemical solids average between 1.5 and 1.6, including most plastics, glasses, and a host of inorganic compounds. Materials such as silicon, carbon, and germanium have much higher refractive indices (3.9, 2.0, and 3.4 respectively) but, aside from carbon, would only be anticipated in process reservoirs. Thus, the bulk of possible solid contaminants have refractive indices close to the upper end of the liquid range, and it is imperative to recognize that apparent *contamination levels in these high index liquids are invariably underestimated.* As an example, we diluted sulfuric acid by a factor of 40x with filtered DI water. Since the sulfuric acid was nearly 10,000 times more contaminated than our DI water (see Table 2 and Figure 1), we could simply neglect the contribution from the water. One would expect a huge reduction in counts, but in fact there was virtually none. Considering the 40:1 dilution ratio, there indeed was a large enhancement of apparent contamination. These results in Figure 24 show the potential errors in measurements taken without considering the relative refractive index contrast effects. The H_2SO_4 tested was in fact several times more contaminated than it appeared in its concentrated state.

Following the above experiment, we can finally interpret the results given in Figure 1. *All of the higher index process liquids were more contaminated than the measurements truly revealed.* In particular, the toluene, xylene, and trichloroethylene low values are obviously too low, while the acetone and 2-propanol are sufficiently close to water in refractive index to give comparable measurements. The acids, with the exception of hydrofluoric, are also underestimates of true microcontamination levels. So are the values for the pair of photoresists, although less so. On an absolute basis the sulfuric acid was the most contaminated followed by the photoresists, nitric and hydrochloric acids, high index hydrocarbon solvents, hydrofluoric acid, low index hydrocarbon solvents, and water, in that order. It is noteworthy that the higher index liquids were, as a whole, (2-propanol excepted) more contaminated. Why? Probably not because of any optical property at all, but simply because higher index liquids are generally more viscous—and yes, more difficult to filter. It is thus important to understand the true nature of all light scattering particle counters. They all measure an optical equivalent size. In liquids lack of respect for this fact can quickly lead one to erroneous conclusions regarding microcontamination measurements.

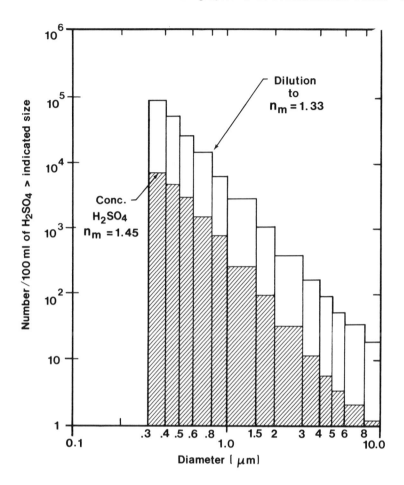

Figure 24: Effects of dilution on measured particle size distribution in sulfuric acid. This important experiment demonstrates that high index liquids are invariably much more contaminated than measurements in their concentrated state can reveal. The data can be interpreted to yield an estimate of the refractive index of the particle contaminants. A value of n_p = 1.55 is compatible with both data sets.

REFERENCES

1. Dillenbeck, Keith, *Microcontamination,* Vol. 2, No. 6 (December, 1984).
2. Dillenbeck, Keith and DeVore, Steve, Proceedings – Institute of Environmental Sciences, pp. 127-134 (1985).
3. Coulter Instruments, Hialeah, Florida.
4. Gatlin, Larry A. and Nail, Steven L., *Journal of Parenteral Science and Technology,* Vol. 35, No. 5, pp. 227-230 (September-October, 1981).

5. Van de Hulst, H.C., *Light Scattering by Small Particles,* p. 8, General Publishing Company, Ltd., Toronto (1981).
6. Rayleigh, Lord, *Phil. Mag.,* Vol. 41, 1871, pp. 107-120, 274-279.
7. Knollenberg, Robert G., *Journal of Environmental Sciences,* pp. 32-47 (January, February 1985).
8. *Handbook of Optics,* Chap. 15, Optical Scoiety of America, McGraw-Hill (1978).

9

Particles in Ultrapure Process Gases

Gerhard Kasper and H.Y. Wen

INTRODUCTION

The Challenge to High Technology

Particle control is accepted as an integral part of any serious effort to develop contamination sensitive technologies, today. While this type of impurity may occasionally receive a disproportionate amount of attention relative to other important factors, there can be no doubt that heightened concern over particles produced significant technological improvements for ultraclean gases.

At the same time we must acknowledge that several problems surrounding particulate contamination remain to be addressed, including central issues such as the relationship between contamination and product yield. We are generally unable to predict or even estimate the influence of a given source of particles arising at a certain point of a gas distribution system although, conversely, the failure of a microelectronic curcuit is often traceable to damage from a particle during production.

This may have to do at least in part with the fact that particles as a discrete type of impurity follow their own laws. Whatever the cause, the situation is unsatisfactory from an economic point of view because it tends to foster disproportionate measures in one area while neglecting another. Among the most valuable benefits of better insight into the relationships between cause and effect may be a *systems concept* for contamination control in gases, that places particles alongside with gaseous impurities and all other process design criteria. Such an approach would help optimize a given piece of equipment or production step by deliberately weighing the risks of various measures.

While this ambitious goal may not be in reach yet, the present review hopes to make a small contribution by pointing out the well developed scientific and technological base which needs to be tapped and activated to deal with particles in ultraclean gases.

301

Challenge for Particle Technology

The need for totally particle free gases represents a challenge to aerosol research, as well. Not that this relatively young discipline would have failed to accumulate a respectable body of knowledge and know-how since its emergence around the turn of the century. However, particle technology has traditionally had to deal with comparatively very dirty gases containing large numbers of particles, where the problem posed itself as one of diluting excessive concentrations rather than searching for elusive events in pure gases.

To fathom the new challenge, consider a pair of idealized examples. Figure 1 is symbolic for the steep decline in particle concentration from the ambient air we breathe daily, down to the levels required for a successful VLSI production. A very crude measurement of total particle concentration in normal room air (without smokers present and not in a very polluted environment), made with a condensation nuclei counter, typically yields on the order of 10^5 particles/cm^3 equivalent to roughly 10^9 per ft^3. In state-of-the-art VLSI grade nitrogen the same instrument will typically detect around 2 to 3 particles/ft^3 or the equivalent of 10^{-4} per cm^3. In either case, all particles in the size range of about 0.01 μm up to several μm are included.

As we make this formidable step of 8 or 9 orders of magnitude, particle technology enters what has been termed the "reentrainment dominated regime" where literally everything becomes a source of particles.[1] This will necessitate in the following chapters a reevaluation of past knowledge and a discussion of new strategies to handle measurements at very low particle concentrations.

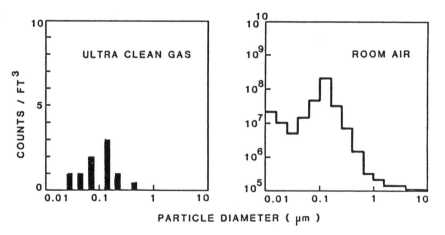

Figure 1: Two conceivable size distributions of particles, from a high-purity gas (left) and from normal room air (right), to symbolize the enormous difference in particle concentration between the two systems.

Contents and Structure of This Chapter

This review is not to be misunderstood as a compendium of particle science and technology. Instead of a futile and preposterous attempt to cover the entire field, we shall concentrate on those issues, where interesting progress was made or where new perspectives can be gained. Nevertheless, the article is not a loose sequence of selected problems; it is structured around three closely related topics concerning "particle free" process gases. Namely (1) the kind of particles to expect, (2) what kind of analysis equipment to use in looking for these particles, and (3) how to eliminate them effectively.

Consequently we will begin with identifying some typical sources of particles. Without attempting a complete enumeration of all possibilities, a few illustrative examples will be presented to familiarize readers with elementary mechanisms of particle generation which under the right circumstances could lead to a source of harmful particles. Furthermore, we shall see that each formation mechanism has a typical range of particle sizes in which it is most active. The shape of this size distribution can sometimes be used as a fingerprint to pinpoint one source or at least to exclude certain others.

The discussion of sources will lead us to the choices of equipment available for particle analysis in clean gases, including both on-line counters as well as off-line techniques for particle identification. Much of the article revolves around trading-off equipment characteristics and limitations specific to very low particle concentrations, because little of this information has so far been documented and organized. The material is not confined to routine gas monitoring situations ("less than 10 particles/ft^3 above 0.2 μm") but includes always the wider need to help identify troubling contaminant sources.

Finally a brief review of filter performance in general and what we can expect from high quality filters in producing particle free process gases will be outlined. Naturally this discussion will focus on membrane filters while leaving the fibrous filters mostly to the sections on clean rooms in other parts of this handbook.

PARTICLE SOURCES AND FORMATION MECHANISMS

Since the identification of particle sources is a key to contamination control, we will have to review the basic processes of particle formation which could be of importance to compressed gases and clean rooms. An understanding of the causes of particle generation is not only useful in avoiding problems, but it helps also in choosing the right analytical technique for identifying an existing source. Each category of sources tends to produce particles in a characteristic size range. Determination of this size range often permits a preliminary judgement on the possible nature of a contamination problem much like a "fingerprint," without resorting to cumbersome identification of chemical and physical properties of individual particles.

It is impossible to foresee all conceivable particle sources, nor is it desirable in this context to cover such a wide field exhaustively. This can be left to excellent reviews of the subject which are available in the literature.[2-4] Much rather, the topic will be approached by briefly describing the most widely found mechanisms of aerosol generation, followed by a few typical case studies. We will concentrate on the characteristic size range (Figure 2) and, where useful, on the relative concentration of particles produced by these basic mechanisms, while avoiding most questions of identification of individual particles (by morphology, composition, etc.).

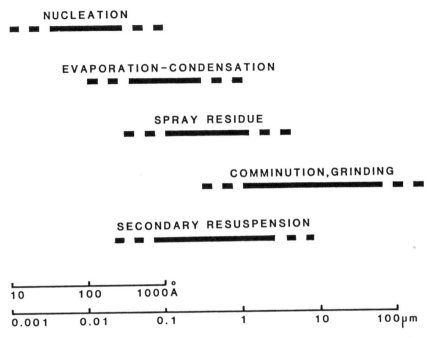

Figure 2: Characteristic size ranges for various sources of particulate contamination.

The Primary Mechanisms

"Gas-to-particle conversion" embraces a wide range of processes in which physicochemical reactions in the gas phase lead to particles. The term originates from environmental chemistry where it is known for a long time that atmospheric particle pollution (e.g. "smog") has gas phase precursors such as NO_x, SO_2, O_3, etc. which together react in complex sequences, usually along with energy from photons (sun light) to form discrete particles. The particles in their embryonic stage are extremely small, often containing no more than a few hundred molecules clustered around an ion, but are usually very abundant. (In the presence of an already existing particle population the gas-to-particle conversion may be heterogeneous, i.e., no new particles are formed, but the existing ones grow. However, this case is not terribly important for particle free gases.)

The typical prerequisites for gas-to-particle conversion thus include one or several species of vapor molecules or radicals, which are ready to nucleate via a chemical reaction that produces a supersaturated species, perhaps in the presence of ions. The homogeneous reaction typically produces large numbers of very fine particles with diameters in the order of magnitude of 100 Å. The total mass of these particles is negligible.

Evaporation-condensation mechanisms are among the most abundant in all environments. The process starts by evaporating a given material from a heated surface, followed by recondensation of this vapor in a cooler region further away from the heat source. Any substance is capable of producing particles in this way. Liquids, metals, salts, plastics, even refractories may evaporate and re-condense if the appropriate temperature range is reached. Very often it is not the bulk of a surface which evaporates, but a thin layer of more volatile contamination .

Obviously there are numerous possibilities in a semiconductor processing environment for aerosol formation by evaporation-condensation. An often quoted example is the hot plate in a laboratory from which a thin film of ubiquitous hydrocarbons evaporates to form a dense cloud of very fine particles, usually on the order of a few hundred angstroms in size. Heated metals are also capable of producing metallic or oxide particles even below their melting tem-peratures. Such particles are usually smaller than 0.1 μm.

The typical particle size range for evaporation-condensation is 0.01 to 0.5 μm under the conditions we will mostly be concerned with. Particle concen-trations are usually high (before dilution). Particles may either be spherical or crystalline.

Spray-drying processes generate particles by concurrent atomization of a liquid and evaporation of the droplets so formed. When a droplet evaporates, it leaves behind a residue particle composed of all nonvolatile organic and in-organic contamination dissolved in the bulk fluid.[5]

The size of a residue particle is related to the original spray droplet diameter by $(C_v)^{1/3}$, where C_v is the volume concentration of contaminants in the liquid.

Liquid sprays are often formed deliberately in a rinsing process or acci-dentally by splashing or bubbling a volatile liquid, thereby forming fine sattel-lite droplets. Another well known case is the secondary aerosol formed when dispersing polystyrene latex particles.[35]

The particle size of main concern for spray drying processes is roughly 0.05 to 1 μm. (Much larger residue particles are unlikely to be formed from the clean fluids considered here.)

Mechanical disintegration or grinding will generally produce the largest particles, with sizes typically above 1 μm up to hundreds of μm. Such particles occur when objects rub against each other, either deliberately (such as brushes of an electric motor, a break, etc.) or accidentally (as in the case of worn bearings). Textile fibers, organic material from human skin, hair, etc. are also generated in this way.

Such particles are usually too large to form an aerosol by themselves. How-ever, they may get carried along in a gas stream of sufficient strength.

Case I: Outdoor Atmospheric Aerosol. Although outdoor aerosols should have little in common with high purity process gases, this case offers a very good opportunity to illustrate the foregoing discussion. Figure 3 is an idealized, typi-cal size distribution of atmospheric aerosol in a moderately polluted urban environment.[6]

The "nucleation mode" contains particles formed by gas-to-particle con-version and by processes of the evaporation-condensation type (combustion, automobile exhausts, etc.). Their number concentration is generally high in the presence of a source. Particle sizes are typically below 0.05 μm.

The "coarse mode" contains particles formed by mechanical disintegration (brake pad dust, tire abrasion, soil particles, etc.). Their sizes range from a few μm up to hundreds of μm. However they are relatively few in number.

The "accumulation mode" was so named because it is thought to contain primarily aged particles which have been carried over from the nucleation mode by coagulation (i.e., clustering together of particles to form a bigger particle) or other growth mechanisms.

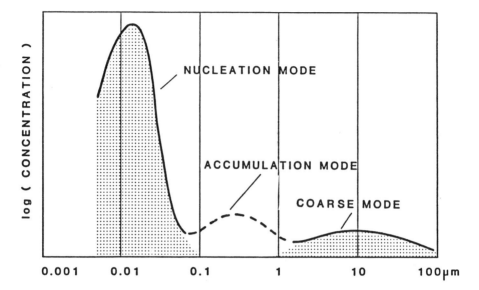

Figure 3: A schematic size distribution of atmospheric particles, displaying three modes associated with different particle sources.

Case II: A Heated Metal Tube. The experiment described in the following figures was made with an internally electropolished but poorly cleaned stainless steel tubing system under alternating conditions of room temperature and heating with an electrical heating tape. The gas flowing through the system was N_2.

Measurements were made with a laser-optical particle counter (LPC) of 0.2 μm sensitivity and a condensation nuclei counter (CNC) which detects particles from roughly 2 to 0.01 μm. (See the discussion in later sections.)

The average particle level of the "cold" system fluctuates around 20 particles/ft^3. Figure 4 shows the effect of a gradual increase in temperature to 200°C on the particle concentration and size distribution. The LPC records no significant change in particle concentration above the level found in the cold system. The CNC detects a dramatic rise in particle concentration, strongly hinting at an evaporation condensation type of source with particle sizes well below 0.1 μm.

Figure 5 shows a concurrent increase in moisture level in the gas with the increase in particle concentration during the heating cycles. Although H_2O itself is probably not the particle source, it serves as an indication that other species (e.g. hydrocarbons) are also evaporating and perhaps forming particles. The decrease in peak particle concentration with consecutive heating cycles indicates further that the particle source is slowly being depleted.

Figure 6 finally shows a size distribution of these particles measured by an electric differential mobility classifier (DMC; see description in a later section).

The distribution indeed peaks in the size range which is characteristic for the evaporation-condensation mechanism.

Secondary Generation by Particle Resuspension

Resuspension, also called "Reentrainment" means that particles are dislodged from the surfaces of a duct, valve, etc. and carried along with the gas stream. Reentrainment is not really a primary generation mechanism (i.e. no new particles are formed) but it acts as such because it changes the original particle concentration and size distribution in the gas.

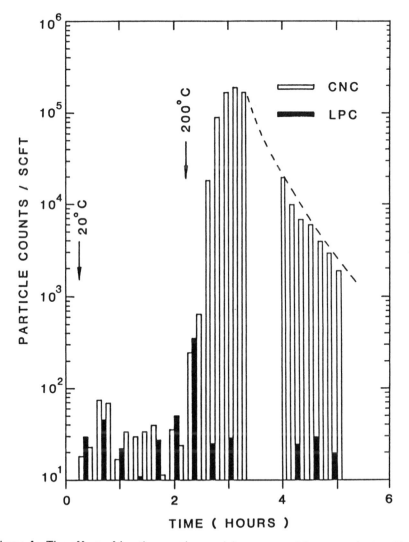

Figure 4: The effect of heating on the particles generated by a metal tube. The increased particle concentration is registered only by the CNC, not the laser particle counter.

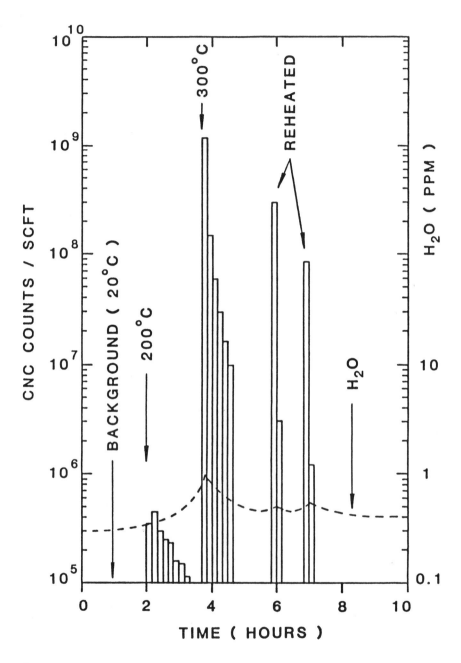

Figure 5: The correlation of particle and moisture generation from a periodically heated metal tube. The peak particle concentration diminishes with each heating cycle.

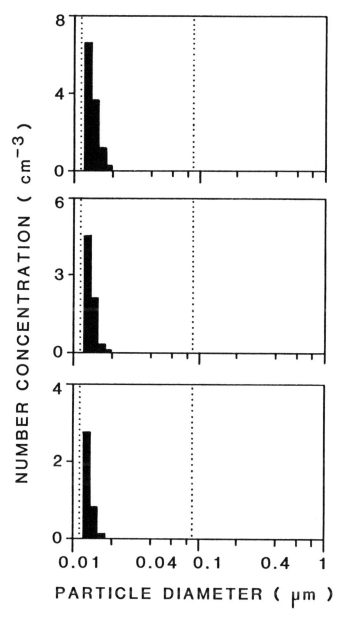

Figure 6: Size distribution of particles generated by heated metal tube, recorded by a differential electrical mobility classifier.

Particles are dislodged primarily by mechanical shock or by the shear action of the gas. The latter is a strong function of the flow velocity as illustrated in Figure 7. The data were generated by varying the flow rate through a thin capillary tube used normally for purposes of chromatography. The inner surfaces of the tube were "specially cleaned" by the manufacturer. The measurements show that such a tube is a virtually infinite source of particles. The most noteworthy point of Figure 7 is the sharp onset of particle reentrainment with gas velocity.

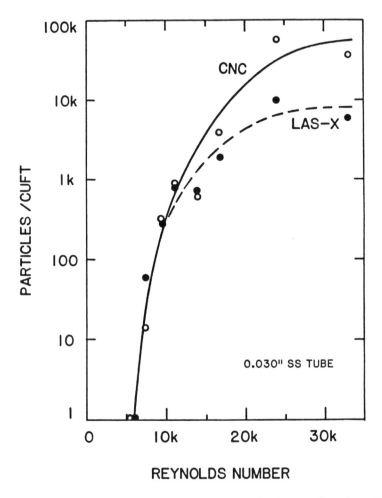

Figure 7: Particle reentrainment from a thin metal tube as a function of flow through the tube.

Another aspect of particle generation by reentrainment from surfaces is the importance of relatively minute amounts of contamination. The following example will show that a tiny spot of particulate-containing "dirt" can change the particle concentration by a factor 100 during a period of one to several days. Measurements were carried out in three phases. First, a perfectly clean gas supply system for N_2 was tested and determined to generate on the order of 1 particle/ft^3 or less, measured independently by a laser particle counter (LPC) and condensation nuclei counter.

This perfectly clean gas line was later contaminated accidentally by a tiny spot of grease from a tube fitting, which resulted in a dramatic rise in particle concentration (shown in Figure 8). The elevated level of particle counts decreased only very slowly and would not have reached the original "clean" level for several days. After the contaminated spot was discovered and removed, however, the counts immediately (i.e. in the first hour of measurements) returned to below 1 particle/ft^3.

A detail to note in Figure 8 is the fact that CNC and LPC both record virtually the same particle levels during the contamination episode which means that the size of particles generated by the reentraining source was mostly above 0.2 μm.

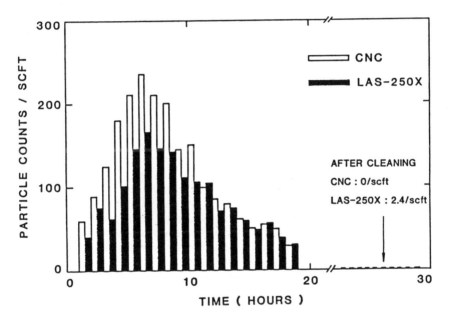

Figure 8: Particles generated over a long period of time by a contaminated section of a sampling tube. The majority of particles is larger than 0.2 μm and thus recorded by both the CNC and the laser counter.

SELECTION CRITERIA FOR ON-LINE PARTICLE ANALYSIS EQUIPMENT

This section outlines and defines the essential technical characteristics to specifically look for in choosing particle analysis equipment for high-purity gas applications. By speaking of on-line particle analysis we shall confine ourselves to devices that measure particle concentration and/or size continuously, quasi in "real time." Not covered are chemical analysis and off-line techniques such as microscopic methods that first require taking a sample which is then analyzed elsewhere at a later point in time.

The Particle Size Range of Interest

According to the survey of particle sources in the preceding chapter there really is no fundamental limit to the particle size range of hypothetical contamination sources. "Particles" can be found anywhere from molecular dimensions up to grains of sand. Although aerosol science has developed a variety of techniques to cover this size range, it would be rather uneconomical to cover the entire spectrum as a matter of routine.

Fortunately, there are some practical constraints that help define our problem. For example, the probability of finding particles much larger than about 1 μm in a clean gas line is usually quite small, because coarse particles are easily removed by most cleaning procedures or by filtration. Coarse particles with aerodynamic diameters above about 2 or 3 μm (depending on conditions) are also rather difficult to sample, i.e. to carry through the sampling tube(s) into the counter, without sizeable losses due to inertia and sedimentation. Considerable efforts and precautions are usually required to obtain meaningful data beyond a few microns with a particle counter. Due to our earlier argument that coarse particles are usually very rare in a well engineered high-purity gas line, it seems reasonable to accept the practical operating limit of approximately 2 or 3 μm imposed by the sampling system as a working basis for the remainder of the article.

Nevertheless, one has to be cautious in drawing the inverse conclusion that a gas is free of coarse particles at a given point because the counter shows no record. There are often indirect signs, such as the shape of the size spectrum, or supplemental analysis techniques that can be used to confirm the presence of coarse particles, if necessary.

We must now define the lower end of the particle size spectrum that should be covered in our review. Semiconductor manufacturers widely use the "One Tenth Rule" which suggests a limit at one tenth of the critical dimension of an integrated circuit. This typically brings us into the range of about 0.05 μm to 0.5 μm. It is difficult, however, to define exactly what the critical dimension of a complex integrated circuit is, considering that the most advanced MOS technology is now using gate oxide layer thicknesses of only a few hundred Angstrom units while horizontal spacings may easily be more than one order of magnitude larger. To account for this as well as for future developments, we have to extend our survey well beyond the magic number of 0.1 μm, down to 0.01 μm, where we encounter the next practical limit of particle counting technology. In doing so, we must clearly keep in mind that analytical methods and solutions for ultrafine particles are not generally "off the shelf" techniques and may require experience to operate and interpret the data.

The Particle Concentration Range

The second point to consider in the selection of analysis equipment is the generally very low particle content of high-purity gases. Levels are typically below 100 particles/ft^3 and often in the range of only one or two particles per ft^3. In the presence of major contamination sources they may go up to 10^6/ft^3 equivalent to under 100/cm^3.

This limits our choice to single particle counters, i.e. devices capable of individually detecting and possibly sizing each incoming particle. Other types of devices which must integrate over an ensemble of particles to derive a measurable signal such as charge sensors or extinction photometers are unsuitable. The usefulness of sequentially stepping size classifiers such as electrical mobility analyzers or diffusion batteries is also limited (see discussion later on).

The Choices

This leaves us with two types of on-line single particle detectors. The principal tool above ca 0.1 μm is the light-scattering particle spectrometer (LPS). Detection of significantly smaller particles is possible with a condensation nuclei counter (CNC).

There are of course other types of single particle detectors that have been commercialized, such as the so-called Aerodynamic Particle Sizer (trade name of TSI, Inc.) with an operating range above ca 0.5 μm, whose detection ranges are too far at the fringe of our area of interest to be included here.

Detection Limits and Counting Efficiency

The counting efficiency (or "detection efficiency") of a particle analyzer is defined as the probability of detecting a particle that was actually contained in the aerosol stream originally entering the device. Counting efficiency is one measure of quality for proper representation of particle contamination by a counter.

Figure 9 shows the counting efficiency as a function of particle size for a fictitious but rather typical counter. The operating range is bounded by the upper and lower detection limits, i.e. the size limits above or below which particles are no longer detected by the analyzer. Such limits are imposed either by the physics of the detection process (for example when signals become too weak) or by the fact that some sort of loss mechanism may prevent particles from entering the sensing volume of the detector ("inlet losses").

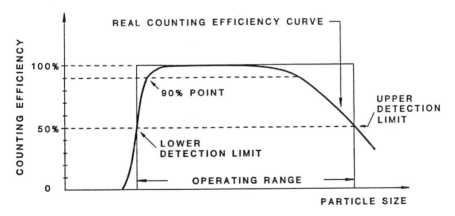

Figure 9: A schematic counting efficiency curve of a particle counter.

Such inlet losses occur to a varying degree in all counters. They depend on design and flow conditions of a sampling system and are of course a function of particle size. We shall discuss this point in more detail in later sections and concern ourselves here only with a rule of thumb that is useful in quickly judging a situation. For the particle sizes that we are interested in, namely from roughly 0.02 to 2 µm, losses are not terribly important except at the fringes of that range. Losses will gradually increase both toward large particles (typically above 1 µm) which are deposited in the sampling and inlet system by inertia and sedimentation, as well as toward very small particles due to diffusion.

Detection limits are seldom sharp boundaries. Most often they take the form of gradual drops in detection efficiency (Figure 9). The "detection limit" thus becomes a matter of definition and can vary from application to application depending on what percentage of losses is tolerable. Common choices are the 50% or the 90% point of the efficiency curve.

We shall discuss counting efficiency in more detail in conjunction with specific instruments.

Internally Generated Counts

Internally generated counts are defined as counts which are not generated by particles entering the instrument from outside. Although it may sound absurd, this phenomenon is still one of the key limitations of particle counters for high purity gas applications.

Internal counts can result from actual particles dislodged from duct walls within the instrument, e.g. by mechanical shock or gas purging action. They can also be the result of an improperly adjusted lower threshold relative to the amplifier noise for the electrical pulses from the photodetector that converts the light pulses from particles. A fourth possible source are radioactive discharges within the photodetector (an established phenomenon in ultrasensitive electrometers) or cosmic radiation induced discharges. Depending on the source, internal counts can come in bursts, they can be randomly distributed over time, or they can follow some other functional relationship.

Remedial measures are possible in part, depending on the source of noise. For example, counts from actual particles can be eliminated by appropriate cleaning; some electrical noise can be filtered by electrical shielding, etc. Pulses from cosmic muons or stray particles within the instrument can be eliminated by pulse width analysis (e.g. in the HP-LAS or the LPS 525 from PMS, Inc.) and other design features. Unless special noise rejection systems are incorporated into the counter (such as the LPC 101, LPC 110 or the LPC 525 from PMS) it is generally not possible to suppress the majority of noise counts. We must expect some level of background noise which tends to increase with the sensitivity of optical particle counters.

Irrespective of its cause, internal noise amounts to a certain count rate per time which in turn converts into a concentration per volume of sample gas, expressed e.g. as counts/ft^3. Since these internally generated counts cannot be distinguished from real particles during normal operation, they can make measurements at ultralow particle concentrations impossible.

The phenomenon of internally generated counts in aerosol counters has only been discovered very recently with the advent of ultraclean gas technology. Most of its basic aspects are not new, though, and quite well understood in other domains, such as low level ionizing radiation measurements.[7]

There are different ways of determining the level of internally generated counts of an aerosol instrument. In some cases it is possible to close off the aerosol inlet and run the instrument on internal sheath gas circulation mode. Otherwise, an "absolute" filter may be attached to the particle inlet. In this case one has to ensure, however, that the filter is truly "absolute" or that the level of particles originating from it is significantly below the level of internal noise one is trying to measure. The analysis of the filter can itself be a very challenging task.[8]

In the remainder of this section we shall discuss some examples, using data from two laser particle spectrometers (LAS-X 0.12 μm and LAS 250-X of PMS, Inc.) while leaving specific equipment data for later on. The measurements were made after blocking the sampling inlets so that the instruments were operating on internally circulated flow only. All counts recorded in this way are internally generated, although possibly from different sources. Figure 10 shows histograms of typical hourly counts for LAS-X (top) and LAS-250-X (bottom).

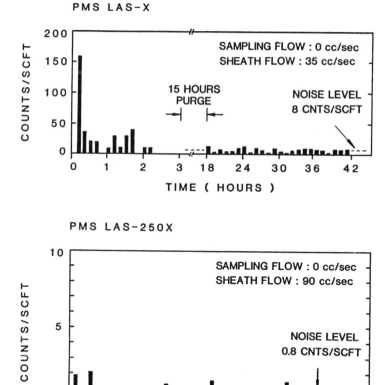

Figure 10: Time series of internally generated counts for two particle counters.

Normally such measurements are made after count rates have stabilized at their lowest possible level to insure that the "true" noise level is measured. (It is at times very difficult to determine whether steady state conditions have been reached.) However, the upper Figure 10 initialy shows a continuous decrease of counts over several hours and only after further purging do the counts settle down at an approximately constant time average corresponding to about 8 counts/scft. This initial decay results from particles released slowly off the walls of the counter inlet system which had been contaminated during a previous run with room air. The data were included in the figure to illustrate how the ducts within a counter display a "memory" for prior history. In cases of prolonged exposure of a counter to dirty gas it can take up to a week before the internal noise settles down to its original level.

From here on we will make an important distinction. As opposed to (total) internally generated counts as they were defined at the top, we shall speak of *background counts* when we actually mean those counts which do not originate from internal particle contamination. This distinction makes sense because the background counts are relatively constant and independent of time and location compared to internal contamination which may vary over several orders of magnitude depending on how clean the device has been kept.

The examples of Figure 10 show that, once it has stabilized, the internal count level of an instrument averaged over a day remains quite constant. The counts from one sampling period to the next, however, show a seemingly random distribution: If we choose an interval width of, say, 10 min, some intervals will contain no count and some will have several. Table 1 gives a frequency distribution of background counts for a typical instrument. It was derived from a time series of about 350 consecutive 10 min internals after ensuring that the counter was virtually free of internal particle contamination. The table shows a peak for the distribution at 1 count/interval.

Table 1: Frequency Distribution of Background Counts of a Laser Particle Counter (LAS-X) in a Series of 347 Consecutive Time Intervals of 10 Minutes Each

Counts/time interval:	0	1	2	3	4	5	6
Number of Occurrences:	90	133	80	33	9	2	0
Decumulative Frequency [%]:	100	74	36	13	3	0.6	0

The obvious reason for this variability of background counts (i.e. of counts that are caused by electrical or nuclear interference, not by particles) is that they are recorded when electrical spikes riding on the amplifier noise exceed the lower signal threshold. Without going into the details of such phenomena, we will mention only some of the practical and fundamental consequences for particle analysis in very clean gases.

First, it is a rule of thumb that the more sensitive counters are also more susceptible to noise, simply because they usually operate closer to the noise level in order to detect the smallest particles. (Powerful lasers are expensive.) For the same reason, condensation nuclei counters are relatively free of noise since they are designed to detect liquid droplets with the comfortable size of 10 μm.

Second, the background level of an instrument is bound to change when readjusting its threshold. Such readjustments are sometimes required during maintenance, e.g. to compensate for an aging light source. The noise level of a counter should therefore be checked after each service.

More fundamentally, the distribution of pulses in time, such as in Table 1, can be used to determine sources of noise. One of the most powerful tools is graphic analysis by a Poisson plot such as Figure 11. Very nearly random sequences of rare events (radioactive decay, extraterrestrial high energy particles striking the photo diode, etc.) should be Poisson distributed and appear as straight vertical lines on this specially prepared graph paper. Any skewness, bulges, etc. of the distribution curve would indicate that the pulses come in bursts, are combinations of different sources and so on.

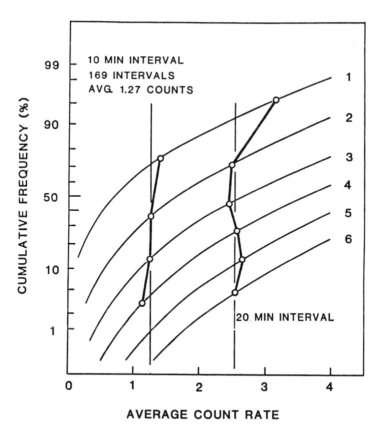

Figure 11: Cumulative frequency distribution of internally generated counts for a laser particle counter (LAS 250-X) plotted on Poisson graph paper.

Figure 11 is a Poisson plot of the data from Table 1 which had been obtained by closing the aerosol inlet of the counter and running it on its internal circulation flow. The fact that the distribution is quite vertical means that the source of noise was fairly random.

A final point concerns only those particle detectors capable of size discrimination. Since background counts are recorded whenever a noise spike exceeds the threshold, the frequency of events is highest in the lowest bin of the multichannel analyzer.

This is illustrated in Table 2 where the distribution of (mostly) background counts has been recorded for the first 7 channels of a highly sensitive laser particle spectrometer with very good size resolution. In case of a counter with broader size internals, all these noise counts would be concentrated in channel 1, of course.

As a consequence of the fact that noise is usually recorded only in the smallest size range, one can generally conclude that events in higher size classes are the result of particles.

Table 2: Distribution of Background Counts/Ft^3 in the First 7 Narrowly Spaced Channels of a Sensitive High Resolution Laser Particle Spectrometer (LAS-X)

Time Interval [h]	Channel 1	2	3	4	5	6	7
1		3	1	2		1	1
2	1			1			
3		1	2	2			
4	2	1					
5	3	1					
6	3	1	1				1
7	3	3	2		1		
8				1			
9	1	3		1	2		
10	1	1		2		1	
11	1			1			
12	1	3	1	1	1		
13		2	1				1
14		1		1			
15	1		1	1		1	
16	4			1		1	
17	1	3	2	1			
18	3	2	1	1			
19	2	1	2		1		
20	2	2		1			
21		1		1			
22	3					1	
23	3	2			1		
24	3	3					
TOTALS	38	34	14	18	6	5	3

LIGHT-SCATTERING PARTICLE COUNTERS

The section intends to elucidate key performance characteristics for high-purity gas applications and some of the trade-offs that need to be made in choosing the right LPC. Some points are simplified or not touched at all because they have been covered elsewhere in the literature.

Operating Principle

In operating an LPC, particles are passed one by one through a beam of light (either from a laser or an incandescent lamp). Each particle creates a pulse of scattered light which varies in intensity ("height") with the wave length of the light, the angle of observation relative to the illuminating beam, the shape and size of the particle and its refractive index. Many excellent books and articles are available on this subject, e.g. by Kerker[9] and Van de Hulst.[10]

An LPC measures the pulse intensities and sorts them into bins ("size classes") of a multichannel analyzer by comparing with an internal calibration curve. Thereby a distribution of number concentration versus particle diameter is obtained.

The crucial calibration curve of a pulse height versus size is established for all commercial equipment using transparent polystyrene latex spheres. When spheres of differing refractive indices are measured they are of course not classified by their true size (unless a special calibration curve is established) but by their "optical equivalent diameter." The same happens with particles that are not perfectly spherical. Naturally, this creates some uncertainty in the "particle diameter" established by an LPC.

A wealth of information is available today on the size response of optical counters, indicating that virtually every device has certain irregularities in its calibration curve for some sort of particles. A common rule of thumb is that the overall worst case uncertainty in particle size can be bounded by a factor of $2^{\pm\frac{1}{2}}$ for a properly designed and functioning instrument.

It is appropriate, however, to place this size uncertainty in perspective. For the application concerned it is seldom critical to distinguish between particles of, say 0.2 μm and 0.28 μm in size. Often it is more important to have an exact measure of concentration. Many commercial instruments already take this into account by having only few and very wide size bins.

Key Characteristics of Commercial Optical Particle Counters

Table 3 contains selected characteristics for some of the instruments that were available to the US market over the past few years. The table is representative for the range of technical capabilities on the market at the time this article was revised. However, being included does not constitute an endorsement and vice versa. Furthermore, rapid technical advances are taking place in the technology of optical counters which will relativate the following section without however rendering it basically incorrect. Equipment with nominal detection limits above 0.3 μm was ignored.

The three properties listed for each instrument are *nominal detection limit* (closely related to counting efficiency, but not the same), *sampling flow rate* and *average background level*. The importance of the lower detection limit goes without saying. The sampling flow rate is of equal importance at very low particle concentrations because it determines the time required to obtain statistically valid information about the gas. The table shows roughly an inverse relationship between smallest detectable particle size and sampling rate. For example, the LAS-X model designed for 0.12 μm takes about 1½ hours to sample one ft^3 of gas; for the LAS-X version capable of reaching 0.09 μm this time increases to 5½ hours. Such differences can be critical when operating in very clean gases.

A similar relationship exists between detection limit and average background level. As the counters become more sensitive, they tend also to become more susceptible to noise. Counters at the 0.1 μm level typically have a background count of about 10 counts/ft^3. At that level it becomes important to decide what particle concentrations are to be expected in the gas since such a counter is obviously not suitable for ultraclean environments with 10 or less particles/ft^3 of gas.

Table 3: A Sampling of Light Scattering Particle Counters for High-Purity Gas Applications

The list was compiled mainly from available information on tested instruments and thus represents mostly equipment available to the market in North America. The list is representative for instrument capabilities in general, but not complete. Inclusion does not constitute endorsement or vice versa.

Brand/Model		Light Source	Nominal Lower Size limit (μm)	Sampling time for 1 ft^3	Average level of background counts (counts/ft^3)
PMS	LAS-X	Laser	0.09	5¼ hrs	ca. 30
PMS	LAS-X	Laser	0.12 (Mode 3)	1¼ hrs	8-10
			0.17 (Mode 2)		0
PMS	LAS-X	Laser	0.10[a]	3/4 – 1¼ hrs	4-8
Hitachi	TSI 400	Laser	0.1	1¼ hrs	1.3
Royco	5200	Laser	0.1	10 min	10-14 [c]
PMS	LPC-101	Laser	0.1	10 min	5-15 w/o NRC
					2 with NRC
Climet	CI-6400	Laser	0.1	10 min	not available
PMS	LPC-110	Laser	0.1	1 min	not available
Met One	205	Laser	0.16	1 min	1
PMS	LAS-250X	Laser	0.2	10 min	1
PMS	LPC-525	Laser	0.2	1 min	0 with NRC
Climet	CI-6300	Laser	0.2	1 min	not available
Royco	5100	Laser	0.3	1 min	0
PMS	LPC-555	Laser	0.3	1 min	0
Climet	CI-8060	Wht.Lt.	0.3	1 min	0
PMS	HP-LAS	Laser	0.3	(50 min) [b]	0

NRC = noise rejection circuitry

(a) Newer version.

(b) Typical value, sampling rate is a function of pressure.

(c) Prototype data; may be reduced in future.

Overall, Table 3 would indicate that the optimal combination of instrument characteristics lies around 0.2 μm with a flow rate of at least 0.1 ft^3/min and a noise level of less than 1 per ft^3. There are some fine points, however, concerning the difference between nominal and actual detection limits which will be discussed below. The following sections of this chapter intend to provide some insight into the underlying causes for these relationships and the necessary trade-offs between conflicting requirements. Explanations and comments will also be given for some of the obvious exceptions to the rule contained in Table 3.

Smallest Detectable Particle Size Versus Sampling Flow Rate

According to the physics of light scattering by spheres initially developed by Lord Rayleigh (1881ff), Gustav Mie (1908) and Peter Debye (1909), the intensity of the light pulse scattered by a particle passing through the light beam of a counter decreases rapidly with the particle's size. In the vicinity of 0.1 μm the decrease is proportional to r^6 where r is the particle radius.

Technically speaking this means that every slight gain in sensitivity requires substantially higher illumination intensities. Particle counters using white incandescent light are practically limited to between 0.3 to 0.5 μm. With powerful lasers it is possible to detect particles down to about 0.05 μm. The first such instruments were described in the literature in 1976.[11,12] In order to do so economically, many commercial devices make use of the higher intensities of the laser cavity.

In addition to the costliness of lowering the detection limit, some very sensitive laser counters suffer from low sampling flow rates as reflected in Table 3.

One physical reason for this is the "Rayleigh background," i.e. the background of light scattered by the gas molecules present in the sampling volume along with the particle, against which the increasingly weaker pulses of small particles must be detected. While the pulse height decreases rapidly with size, the Rayleigh background remains only a function of gas pressure. In order to improve the signal-to-noise ratio, it is necessary to collect as many photons emitted from the particle as possible. Thus the particle must pass the laser beam slowly, i.e. at reduced flow velocity.

The Rayleigh scattering background sets a limit of about 0.05 μm at atmospheric pressure, but the signal-to-noise ratio worsens with increasing gas pressure. This is important for high-pressure counters. For the much denser liquids the practical limit lies in the vicinity of 0.2 μm.

Another point to be mentioned here without further discussion is that the gas flow in active cavity lasers leads to a beam intensity modulation which also tends to worsen the effective signal-to-noise ratio of such a counter (Knollenberg, private communication).

Background Counts Versus Sampling Flow Rate

The basic definitions and explanations about internally generated counts and background counts were given in the previous chapter, where the discussion was purposely held general enough to include all kinds of single particle counting equipment. However, most of the examples concerned LPCs.

Table 3 contains a list of background count data compiled mostly in our laboratory. The listed values are thus derived from one or a few available counters of a given type. Furthermore, the values can change somewhat due to readjustments during instrument service (see previous section) and thus represent typical levels.

An obvious point made in the table is that background noise becomes important when stepping from 0.2 to 0.1 μm instruments. This can have to do with the fact that the 0.1 μm devices are more sensitive to noise, for example because they may have larger photo diodes picking up more cosmic muons, or because they operate closer to the amplifier noise level. For example, the LPC 101 and the LAS-X are both 0.1 μm instruments. While the LAS-X was found at various times and with two different instruments to have about 8 to 12 counts/scft, the LPC 101 models evaluated had count levels on the order of 30. The likely explanation for this is (1) that the LPC 101 has a larger detector diode which is also very sensitive to EMI and (2) that it has a small signal-to-noise ratio as a result of the elevated flow rate. (However, recent versions of the LPC 101 are now equipped with a noise rejection logic capable of eliminating very short pulses such as muon spikes; consequently the noise level has dropped to around 2).

Increased sensitivity is not, however, the only cause for higher background levels of some counters. Another practical reason becomes evident by comparing the LAS-X and the LAS 250-X which are both very similar in hardware design. Table 4 shows a comparison of actually measured noise levels for the two instruments. We see that (due to their similar design) the two instruments pick up almost the same number of noise counts per unit time, 5.1 for the LAS 250-X versus 5.04 for the LAS-X in its most sensitive mode (0.12 μm). However, due to the 10 fold difference in sampling flow rates, this converts to a 10 fold difference in background counts per scft, namely 0.85 versus 8.4.

Table 4: Average Levels of Internally Generated Counts for Two Laser Particle Spectrometers

	LAS-X		LAS-250X
	MODE 3 (0.12 μm)	MODE 2 (0.17 μm)	
Internally generated counts: (with aerosol inlet closed)			
[counts/hr]	5.04	0	5.1
[counts/ft^3]	8.4	0	0.85
Remeasured after factory service:			
[counts/hr]	7.56	0	-
[counts/ft^3]	12.6	0	-
Count level with "absolute" filter at aerosol inlet:			
[counts/hr]	-	-	7.6
[counts/ft^3]	-	-	1.3

Table 4 also illustrates some points made earlier:

(a) Concerning the influence of readjustments of the signal threshold on the background level: a factory readjustment raised the average count from 8.4 to 12.6 for the LAS-X.

(b) The fact that most noise counts are registered in the lowest size channel(s) of a counter: when the LAS-X is operated in its less sensitive mode 2 (0.17 μm), there is virtually no noise.

Counting Efficiency and Detection Limits

As shown earlier in Figure 9, the counting efficiency curve determines the useful operating range of an LPC in terms of particle size and, specifically, its effective upper and lower detection limits.

A counting efficiency curve such as the one shown in Figure 9 is usually obtained by comparing the instrument in question with a reference device for a range of particle sizes. A series of such measurements is shown in Figure 12 for various optical counters.[13] The data were obtained from careful measurements with aerosols of PSL spheres, with solid inorganic spheres and with organic spheres generated by a vibrating orifice aerosol generator.

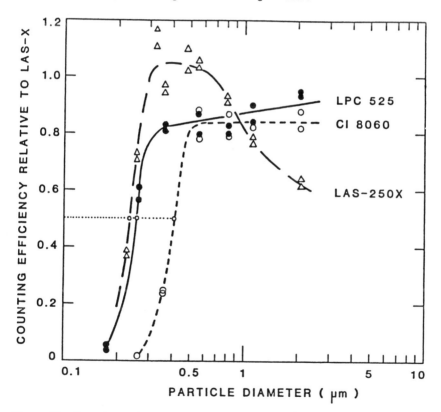

Figure 12: Counting efficiencies of 3 commercial optical particle counters relative to a LAS-X (from Wen and Kasper[13]).

The reference instrument in all cases was an LAS-X (PMS, Inc.) with 0.12 µm capability. This type of device (Royco 226/236) had earlier been investigated as part of a study by Gebhart et al.[14] who determined that its counting efficiency between 0.1 and 1 µm was virtually 100% (Figure 13). The LAS-X thus served as a secondary standard.

The curves of Figure 12 show that among the instruments evaluated by us, the laser counters reach their specified nominal lower detection limits within about 0.05 µm while the white light counter falls short by over 0.1 µm, always comparing 50% points, of course. All instruments represented were either new or recently serviced by the manufacturer. In addition, supporting measurements such as pulse height analyses were carried out to avoid misalignments and other common causes for erroneous data. In all likelihood, the curves are thus representative for equipment performance. It appears that the performance levels on which manufacturers of LPCs base their specifications are not always reached in careful evaluations. In fact, a shortfall of 0.1 µm or less for the lower detection limit as in Figure 12 seems modest compared to some of the results reported by Gebhart et al. (Figure 13; note that all counters in this figure are rated at 0.3 µm, except for Royco 226/236 and DAP 2000, which are 0.12 and 0.5 µm, respectively.)

Concerning the upper detection limits suggested by Figure 12, we see that except for the LAS 250-X, all counters detect 50% to 100% of incoming particles up to a few microns. (One must keep in mind that the curves in Figure 12 are relative to a LAS-X which itself has a 50% point around 3 µm!)

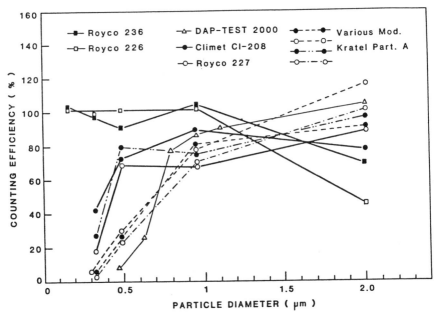

Figure 13: Counting efficiencies of several commercial optical particle counters incl. Royco 226/236 resp. LAS-X (from Gebhart et al.[14] with permission of the authors).

An investigation into the rather unusual performance of the LAS 250-X[13] showed that its accelerated loss of counting efficiency toward larger particles was caused by the design of its inlet and could be avoided. Details of this are given in Reference 13.

Special Comments on High-Pressure Counters

High-pressure counters are designed to operate at pipline pressures up to about 10 bar (150 psi), either as in-line instruments such as the HP-LAS (PMS, Inc.) or as extractive sampling counters (Climet 208 P or PMS LPC 101-HP).

The sampling rate of in-line counters depends on the pressure which has to be measured separately to convert particle counts per time to counts per volume of gas. Along with the pressure there is also a variation of background counts per volume of gas that must be considered.

The principal advantages of high-pressure counters are (1) the elimination of the pressure reduction stage and (2) the higher data rate due to the more concentrated aerosol.

The in-line counter HP-LAS represents a special case in that it is separately dependent on the gas velocity at the axis of the sampling cell. The relationship between pipeline flowrate and gas velocity can be established by a calibration[13] as shown for example in Figure 14 for a one inch cell. Measurements in our laboratory[13] have shown, however, that even after accounting for pressure and flow velocity, the HP-LAS is nonlinear in its response to changing flow. For this reason, the family of curves in Figure 15 is not parallel and straight.

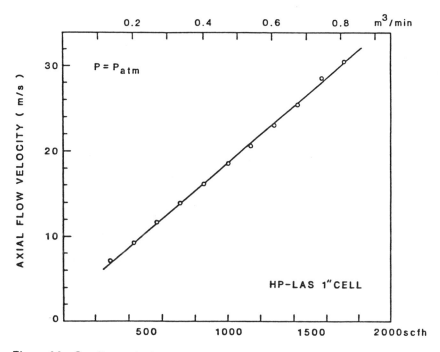

Figure 14: Gas flow velocity at axis of sensing volume of HP-LAS as a function of volumetric flow rate through the pipe line.

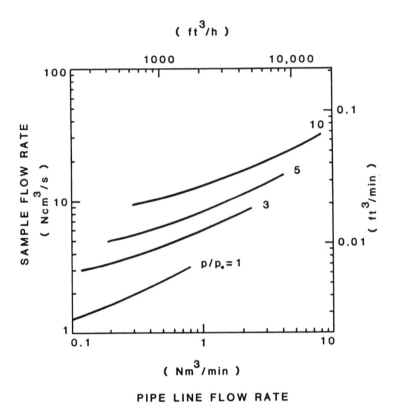

Figure 15: Sampling flow rate of HP-LAS (1" cell) as a function of pipe line flow rate and pressure p/p_o (p_o = 1 atm).

CONDENSATION NUCLEI COUNTERS

CNCs have only recently entered the market for counters in high-purity gas applications. Yet they are very attractive as monitoring instruments due to the wide size range covered, as well as due to the virtual absence of background noise problems.

The most obvious drawback is their inherent inability to classify particles by size. (Some research instruments can do this to a certain extent.) The use of organic liquids in CNCs also limits their applicability to reactive gases such as O_2 or H_2 quite severely.

While the limitation of CNCs to inert gases is a given for the time being (other solutions are still on the drawing board), the lack of size discrimination can practically be circumvented. This matter will be taken up in the next section.

The Operating Principle

The operating principle of Condensation Nuclei Counters is similar to that of the Wilson Cloud Chamber. A momentarily supersaturated vapor of water or alcohol rapidly condenses onto existing particles ("condensation nuclei") to relieve its thermodynamically instable state. Thereby, large droplets of typically 10 μm are formed which can be easily detected even if CNCs use simple insensitive light-scattering counters as droplet detectors. Thus, CNCs are actually hybrids of a particle enlarger by condensation and an optical particle counter. The idea of a CNC goes back to Aitken (1888) who used it for meterological purposes. After some development in the 1930s, the method received a boost during World War II as a detector for diesel exhaust plumes from snorkeling submarines.

Numerous experimental systems have been developed since, of which we shall only mention the continuous flow types of Sinclair and colleagues,[15] Bricard et al.[16] and Kousaka et al.[17] due to their direct impact on commercial equipment suitable for high-purity gases. For further reference see Lundgren et al.[18]

Commercially Available CNCs

This section is currently very short, since only so called continuous flow counters will be considered here as suitable for detecting particles in ultraclean gases. (The nomenclature relates to another family of condensation nuclei counters including most early instruments and their descendants, which operate intermittently with alternating cycles of sampling flow and droplet growth.)

The only devices currently on the North American market are the Model 3020 of TSI, Inc. with a flow rate of 5 cm^3/s (0.01 ft^3/min), and the Model 3760 with a flow rate of 25 cm^3/s.

In Japan, Kanomax is advertising its Model 3851 with a flow rate of 50 cm^3/s (0.1 ft^3/min).

Size Sensitivity and Lower Detection Limit. CNCs of the type discussed here are inherently unsuited for particle size measurements since no direct relationship exists between the size of the original "nucleus" and the size of the droplet condensing around it. (Size discrimination is possible only with experimental devices allowing precisely variable supersaturations.) However, the onset of droplet formation and thus the detectability or nondetectability of a given particle is a function of its size in that nuclei below a certain diameter are not activated to form a droplet. As usual, the threshold is not sharp and may further be influenced somewhat by the chemistry of the condensation nucleus.[19] Recent observations by Wagner et al.[20] seem to indicate that solubility and wettability contribute to a smaller detectable particle size at a given operating mode of the CNC.

The gradual decrease of detection probability of CNCs with particle size is readily measurable with the appropriate equipment. Figure 16 includes so far unpublished results of Niida et al. (referenced in 13) for the CNC Model 3020 manufactured by TSI, Inc. The results show no significant influence of particle composition and a 50% point at about 0.015 μm. This agrees well with earlier measurements by Aggarwal and Sem.[22] Since there is some sensitivity to the operating conditions and environment of a counter, caution should be exercised in comparing concentration data for aerosols with a fraction of ultrafine particles.

Upper Detection Limit. The detection probability of a CNC in the limit of very *large* particles is as important as its *lower* detection limit. It is influenced mainly by losses of particles inside the device and can be derived from counting efficiency measurements such as those shown in Figure 16 for the TSI counter. These recent data by Wen and Kasper[13] indicate an upper 50% detection efficiency at about 3 μm. Again, the cut off is very gradual as shown in Figure 16.

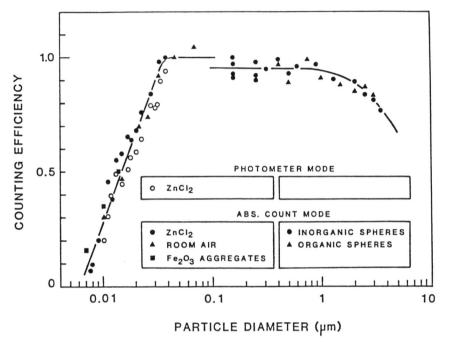

PARTICLE DIAMETER (μm)

Figure 16: Counting efficiency of CNC 3020 combining the data of Reference 13 and unpublished measurements by Niida, Reischl and Kasper (1982) for the size range below 0.1 μm.

Internally Generated Counts. "Background noise" as it was defined earlier does not seem to be a problem of condensation nuclei counters. Measurements in our laboratory for the TSI counters Model 3020 and Model 3760 indicate levels of less than one event per day which converts to less than 0.1 count/ft^3. For the Kanomax device no data are available at present.

A plausible explanation for the virtually zero background level of CNCs is the fact that the droplets to be counted by these devices are so large that their optical detection systems are not as critically sensitive as high performance laser counters to electric interference, cosmic radiation, etc.

ON-LINE MEASUREMENT OF PARTICLE SIZES IN VERY CLEAN GASES—A SYNOPSIS

The aim is to measure particle size distributions over the size range of at least 0.02 to 2 μm. What we have discussed so far are the basic properties and operating characteristics of various particle analyzers, some of which are only capable of measuring total concentration (e.g. CNCs) while others permit the simultaneous determination of size and concentration. Unfortunately no single piece of equipment is available to cover the entire range of interest completely in both size and concentration.

Laser counters, as discussed in great length and detail, are limited to sizes above 0.1 or even 0.2 μm because of rising background levels and decreasing flow rates which make the most sensitive counters less suitable for very clean gases. On the other hand, CNCs would be capable of handling the required operating range but cannot distinguish particle sizes.

We will now evaluate several possible solutions to our problem and discuss their suitability for high-purity gases and clean rooms.

Size Classification With LPC-CNC System

Figure 17 shows the superimposed operating ranges of a typical CNC and a typical LPC. The figure illustrates that the operating ranges of the two types of instruments overlap in such a way that the CNC covers practically the entire detection range of the LPC. This leads us to a simple but very effective method.

The total particle concentration between the lower detection limits of the two instruments (the shaded area in Figure 17) can be estimated quite accurately by subtracting total LPC concentration from total CNC concentration. By combining the two types of equipment, one obtains a fairly detailed picture of the size distribution above ca 0.1 μm (from the LPC) while the size range below is treated essentially as one size class.

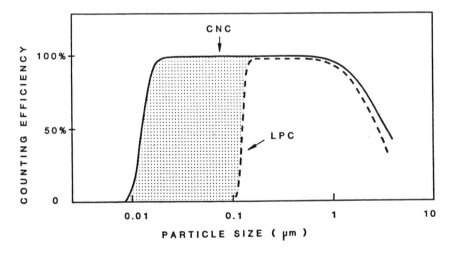

Figure 17: Characteristic counting efficiency curves of a condensation nuclei counter (CNC) and a light scattering particle counter (LPC). The shaded area represents the range of particle sizes covered only by the CNC.

Figure 18(a) shows a size distribution of particles in ultrapure nitrogen gas from an industrial pipeline which was obtained by this method. The data represent a 24 hour average. Figure 18(a) shows nicely that the total count above 0.2 μm was 1/ft³; between 0.17 and 0.20 μm the average was 2 counts/ft³ and below 0.17 μm (in the CNC range) there were an additional 10 counts. Of course we do not know whether these latter 10 particles had a size near 0.1 μm or whether they were much smaller. Figure 18(b) shows the same data presented as a frequency distribution that peaks between about 0.1 and 0.2 μm.

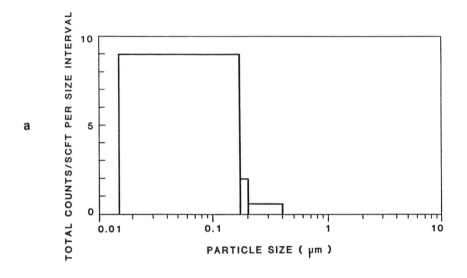

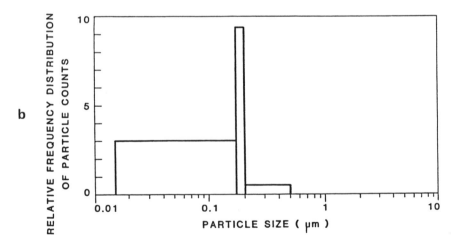

Figure 18: Size distribution of residual particles sampled from ultraclean N₂ originating from a 150,000 ft/³h industrial installation. (a) represents total counts/ft³ in each (unequal) size internal; (b) represents the same data normalized as a relative frequency distribution.

The above described simple technique requires that CNC and LPC both be exposed in parallel to the same aerosol and that their respective sampling lines be properly designed to avoid unequal particle losses. In addition, one has to take into account some of the phenomena at very low particle concentrations, such as background noise of counters. The data shown in Figure 18 carefully eliminate all background counts. Another peculiar problem is the fact that a discrete particle which by chance enters the CNC cannot, of course, at the same time be recorded by the LPC. Even if the probability of entering either instrument were exactly 50%, such measurements are meaningless if they are based on only 2 or 3 particles. For this reason, Figure 18 is an average of 24 hours.

Differential Electrical Mobility Classification

An electrostatic differential mobility classifier (DMC) is a device which does not actually count particles, but "presorts" them on-line into size classes for subsequent detection by a single particle detector such as a CNC. In other words, one has to combine the DMC with another instrument to actually obtain a size distribution.

The DMC sorts particles by their electric mobilities. It works best in the particle size range of roughly 0.01 to 0.2 μm. The ideas behind the DMC have been known for many years. Numerous variations are described in the literature, a few of which have been commercialized. We shall not go into the intricate details and limitations of electric mobility classification in general but refer you to some of the literature.[18] Rather, we will look at the feasibility of this concept for very low particle concentrations.

First of all, a DMC classifies only charged particles of one polarity; the rest of the aerosol is discarded. At the elevated aerosol concentrations for which the instrument was originally designed, this is not a problem because the charging statistics give a very accurate partition into uncharged, single charged, double charged, etc. fractions. At concentrations of $10^{-4}/cm^3$ not much can be said as to whether a given particle will be positive, negative or uncharged. From this point of view, the DMC is not ideal.

A second serious limitation comes from the sequential mode of operation of a DMC: Since the instrument is basically a "band pass filter," it has to go through all size classes from the lowest to the highest in sequential voltage steps. At each step, the DMC must dwell until the particle detector (e.g. the CNC) has counted a statistically valid sample. This may take one hour per step or more. Obviously, there are few particle sources which remain stable enough over many hours to complete a meaningful data run.

Figure 19 illustrates the capabilities of the technique at elevated concentrations as well as the difficulties. On the left side is an actually measured size distribution of ultrafine particles produced by a heated tube. We see that the distribution peaks somewhere near 0.03 μm and that it contains very few particles above ca 0.07 μm, thus representing an aerosol of the evaporation-condensation type. This measurement was made with a DMC Model 3070 (TSI, Inc.) at a total concentration of $10^7/ft^3$. On the right side of Figure 19 we see three consecutive measurements of the same aerosol as on the left, but after dilution to roughly 200 particles/ft^3. None of the three measurements resembles the real distribution even remotely, due to the statistical fluctuation of the particle concentration during the runs.

Practically, the useful limit of a DMC can be placed in the vicinity of 1 particle/cm³ equivalent to 10^4-10^5/ft³ , and above.

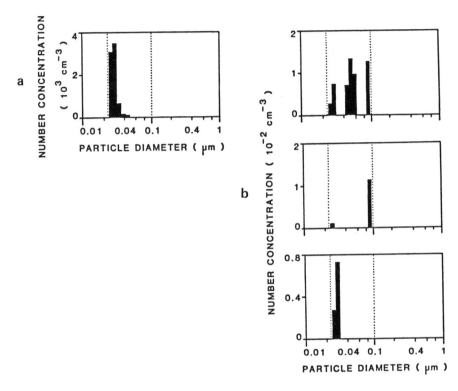

Figure 19: Size distribution data of ultrafine aerosol formed by evaporation-recondensation obtained by a DMC. (a) (at left) is actual distribution obtained at about 10^7 particles/ft³. (b) (at right) is the result of data inversion at very low particle concentration.

Diffusion Batteries

A diffusion battery (DB) performs a function very similar to a DMC, in that it sorts an aerosol into size classes without actually detecting the particles. The detector function is typically left to a CNC.

The DB sorts particles by their diffusion coefficients. It works best below about 0.1 μm, although some measurements have been carried up to almost 1 μm.[23] Numerous variations of the principle have been tried, a few of which have been commercialized. For the basic information we shall again refer to Reference 18 from where further leads can be obtained.

In very clean gases, the DB has similar limitations as the DMC. First, the theory of Brownian diffusion permits statistically valid predictions only at elevated particle concentrations. No statement can be made for a single particle as to where it will be deposited by diffusion.

Second, a DB requires sequential sampling from a series of stages to establish a distribution function. This presents a similar limitation to higher concentrations as for the DMC, unless the particle source is extremely stable.

In principle, it is possible to modify the technique so that actually several DBs of different lengths are operated in parallel. If each DB has a different cut point, one can indeed speed up the measurement of a cumulative size distribution, however, at the price of having to operate a separate CNC with each DB. Even with this arrangement it is necessary to sample for many hours in order to obtain a valid distribution, because individual discrete particles can of course only be detected by one of the parallel stages. This was already pointed out earlier with the CNC-LPC systems approach.

A successful measurement with a DB has been described for a dormant clean room,[24] but further experiments would be desirable to validate this technique.

PARTICLE SAMPLING

Sampling particles for the purpose of on-line counting is a task which requires just as much attention as the performance of the particle analysis techniques which we have discussed in the preceding chapters. In our case, sampling involves taking a portion of the main gas flow from a pipeline and bringing this side stream to the inlet of the counter without distorting the information contained in the gas with respect to either concentration or size distribution, either by losses or by addition of particles.

Much has been said and published on proper methods of sampling particles from flowing gas as well as on transporting an aerosol stream through pipes and tubing without undue losses.[3] The subject is in no way specific to high-purity gases and far too broad to be covered by this article. Rather than expand on the classical problem of sampling, we shall focus on two points which are of specific interest to high-purity gases in general, and to compressed gases in particular, namely on pressure reduction and contribution of particles by the sampling lines.

Particle Generation by the Sampling Line

When particle sampling is discussed in general, one rarely considers the possibility of adding particles to the gas stream by reentrainment from the walls of the sampling tubes. Yet for very clean gases this can be a much more important effect than the various loss mechanisms.

Reentrainment means that particles are dislodged from the surfaces of a duct, valve, etc. and carried along with the gas stream, thereby changing the original particle concentration. Mostly, this undesirable contribution from the sampling system does not amount to a very large number of particles and can therefore be ignored in the context of conventional particle technology. However, at concentrations of only a few particles/ft^3 in a very clean gas, the sampling line contribution often surpasses the original particle concentration by far, thus calling for more attention than the loss mechanisms.

Particles are dislodged and entrained primarily by shear action of the gas as well as by mechanical shock. This point was already discussed in the section on particle sources and formation mechanisms. It is useful to recall the fact that very minute amounts of localized particulate contamination can serve as very substantial sources (Figure 8). This is especially important for the sampling line of a gas distribution system because such a source can totally falsify the results of particle analysis.

The cleanliness of a sampling system must therefore be ascertained by a separate measurement. An arrangement as shown in Figure 20 can be used to determine the sampling line contribution to the total particle concentration under prevailing flow conditions, provided the filter itself has previously been tested.[8]

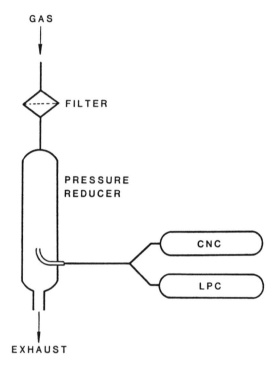

Figure 20: Flow diagram for purging a particle sampling system. The filter shown is connected only for the purpose of delivering a "zero" gas to determine the sampling system contribution to the background particle count.

Pressure Reduction

Pressure reduction is a necessary step in sampling particles from compressed gases because the majority of particle counters operate only at or very near atmospheric pressure.

Pressure reduction is usually accomplished by expanding the gas through a capillary tube or a critical orifice [Figure 21(a) and (b)]. Both solutions have been commercialized. Both devices are built on sound hydrodynamic principles. In either case the expanded aerosol stream is slowed down and then sampled coaxially so as to avoid sampling errors at this point. We will not discuss this part of pressure reduction any further. Rather, from our point of view the task presents itself as one of bringing down the gas pressure without undue *losses* or *generation* of particles.

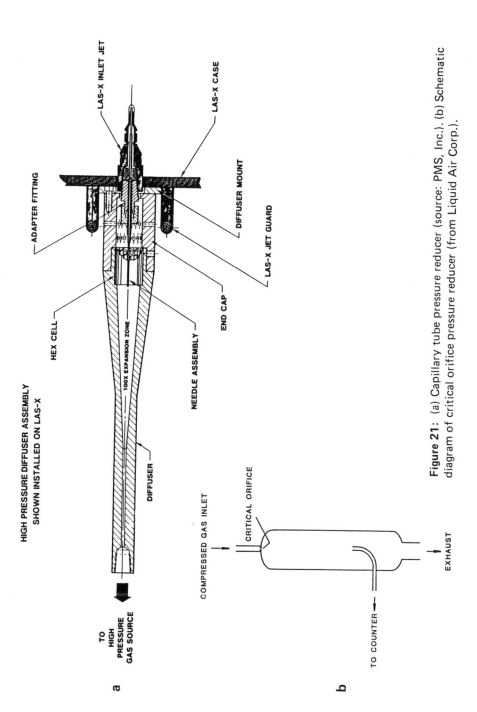

Figure 21: (a) Capillary tube pressure reducer (source: PMS, Inc.). (b) Schematic diagram of critical orifice pressure reducer (from Liquid Air Corp.).

We will now evaluate the latter device from this aspect with special attention given to the critical orifice itself.

Particle losses tend to occur at the up stream side of the orifice plate, where the gas stream narrows down (Figure 22) from the original sampling tube diameter A to the orifice diameter B. Losses at this point occur primarily by inertial impact of particles. (Interception is negligible because the particles are much smaller than the orifice, namely about 1:500.)

Impaction losses in such a configuration can be estimated either by solving the Stokes-Navier equations and then the equation of motion for the particles (which is possible but cumbersome[25]) or by using an analytical approximation given by Pich.[26] The latter method estimates losses on the high side[23] and thus represents a worst case estimate.

IMPACTION LOSSES ON AN ORIFICE PLATE

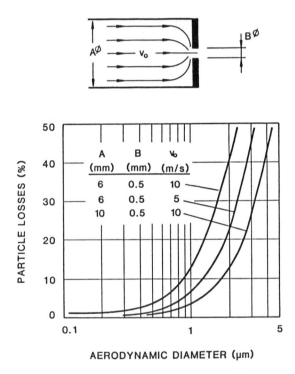

Figure 22: Top: Schematic of gas stream lines converging on a critical orifice. Bottom: Particle loss calculations for various flow conditions and geometries.

Figure 22 shows particle losses by impaction on an orifice plate for three combinations of geometry and flow velocity using Pich's method. The important conclusion is that losses for the particle size range of interest (0.02 to 2 μm) can be held at a relatively insignificant level. The curves show also that inertia effects increase drastically as the particle size increases beyond about 1 μm. The "aerodynamic diameter" of a sphere of diameter D is given by the approximate formula:

$$D_{AE} = D\left(\frac{\rho}{\rho_0}\right)^{1/2}\left(1 + \frac{0.16}{D} \cdot \frac{p_0}{p}\right)$$

where ρ/ρ_0 is its dimensionless mass density and p_0 is the atmospheric pressure. For example an iron oxide sphere with $D = 0.5$ μm and $\rho = 5.24$ g/cm^3 would have a D_{AE} of roughly 1.32 μm.

After establishing that losses on the orifice plate can be held acceptably low, we have to deal now with the issue of particle generation. From what was said earlier about the relationship between shear flow and particle reentrainment, it is obvious that the critical flow at the orifice will cause a very undesirable gas contamination unless the orifice is superbly cleaned. This point will be made by comparing particle count data from two orifices of same geometry. One was "carefully cleaned;" the other was cleaned by procedures specifically developed for high-purity gases. Both orifices were purged with clean compressed nitrogen gas (130 psi) containing less than 1 particle/ft^3. In either case, particle concentrations were monitored with CNCs and optical counters, starting immediately from the moment when the orifices were installed (Table 5). We see that the conventional orifice does not clean out very well (which makes it unsuitable for clean gases) while the specially treated plate can be used almost immediately.

Table 5: Particle Generation by Critical Orifices

```
hourly counts for CNC (0.01 um)
two-hourly counts for LPC (0.2 um)
Pressure drop 120 psi
Orifice A cleaned by conventional procedures;
Orifices B and C clean by special method.
```

	ORIFICE A		ORIFICE B		ORIFICE C	
TIME [hr]	LPC [ft^{-3}]	CNC [ft^{-3}]	LPC [ft^{-3}]	CNC [ft^{-3}]	LPC [ft^{-3}]	CNC [ft^{-3}]
1	0	8	4	6	0	2
2		6		2		0
3	2	5	8	3	3	0
4		2		3		0
5	4	3	2	3	1	0
6		0		2		0
7	1	10	0	0	0	0
8		6		0		0
9	10	6	0	0	0	0
10		21		0		0
11	6	14	2	0	0	0
12		24		0		0
13	4	13	1	0	0	0
14		195		0		0
15	110	32	1	0	0	0
16		78		0		0
17	55	8	1	2	0	0
18		96		0		0
19	97	88	0	0	0	0
20		5		0		0
21	116	178	0	0	0	0
22		133		0		0
23	139	90	0	0	0	0
24		107		0		0

These very typical results indicate that it is necessary to test any sampling and pressure reduction system before installation. They also show how sensitive critical orifices or capillary tubes are to particle generation by shear flow. In general it is very difficult to clean capillary tubes.

A comment is also due here on the use of valves for pressure reduction. In such an application, the valve will function as a critical orifice, directing the jet of gas of sonic or supersonic velocity against a close-by surface of the valve body. From the foregoing discussion one can imagine that most of the original aerosol particles will be lost by impaction and are probably replaced by other particles generated inside the cracked valve by the high shear. Obviously such a choice of pressure reduction defeats the purpose.

PARTICLE SAMPLING FOR OFF-LINE ANALYSIS

This section is dedicated to a brief overview of particle collection techniques for "off-line" particle identification by physical or chemical methods such as microscopy, x-ray analysis, etc. For this purpose, particles have traditionally been collected on filters which are available in a variety of chemical compositions and a wide range of pore sizes to suit most sampling needs.

Filter sampling works quite well as long as there is an abundance of particles in the gas or, if the particles are large enough to be detected comfortably on the filter surface. (Many older ASTM standards are written for particles above 5 μm.) However, at the very low particle concentrations required for semiconductor gases, filter sampling has some disadvantages. Since the few particles contained in the gas are scattered randomly over the filter, it becomes rather tedious to search for and analyze them. Furthermore, these particles can be quite small (recall Figure 2) and therefore become entrapped in the interstices of the filter where they are virtually undetectable (Figure 23). These difficulties can be circumvented by using *impactors* which concentrate the deposit in tiny spots on the surface of virtually any smooth, clean substrate material.

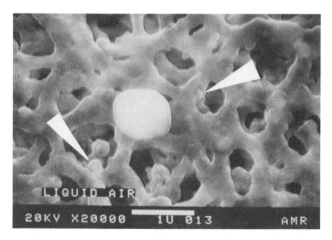

Figure 23: SEM photo of 0.05 μm membrane filter surface with deposited NaCl particles of various sizes marked by arrows. (AMRAY 1600.)

Inertial Impaction

In its simplest embodiment, an impactor is a jet of gas of appropriate velocity directed against a solid flat surface ("impaction plate," "sampling substrate") oriented normal to the jet axis. As the gas flow changes direction, all particles above a certain aerodynamic diameter (see definition in previous chapter) will impact due to their inertia, while smaller particles than the cut-off diameter remain in the gas stream and are not deposited.

The theory of inertial separation is today quite well developed and permits accurate calculations of cut diameters as a function of nozzle geometry and flow conditions. Without going into any of its details, it is useful to remember that the smaller a particle to be impacted, the higher the jet velocity and the smaller the nozzle diameter generally have to be. In this way, cut points of 0.015 μm have been reached.[27]

Most impactors today are *cascade impactors,* consisting of a series of stages with progressively smaller cut points. This permits a size fractionated deposition of the particles and, in addition, reduces the problem of particle bounce which will be discussed below. Since the first reported design by May (1945), numerous cascade impactors have been designed and tailored to specific needs. The latest in the series of applications are impactors for clean room and high-purity gas sampling.

It is tempting to design an impactor for high-purity gas applications as a single stage covering as wide a size range as possible, to reduce the sampling time. The problem incurred hereby is particle bounce. When solid particles impact with excessive velocity on a solid surface, they tend to rebound, return into the gas steam and get carried away. This phenomenon always exists to a certain extent, unless the sampling substrate is coated with a greasy layer to make particles stick. (This is generally not a very practical solution for very clean gases.) However, large particles tend to rebound at much lower velocities than smaller particles. In a single stage impactor, near sonic jet velocities are required to deposit the very small particles and this compounds the bounce problem for the large particles. To minimize particle bounce, a cascade impactor should have as many stages with closely spaced cut-off diameters as possible. Obviously, an impactor that is optimized for high-purity gases will be weighted somewhat in its design toward a wider stage spacing.

A 4-stage cascade impactor tailored specifically toward high-purity gases (and also clean rooms) was recently described.[28] The device is available from Liquid Air's Chicago Research Center.

This impactor covers the aerodynamic size range from 0.03 μm upward to about 3 μm (Table 6). The device was optimized to limit the problem of bounce while permitting size fractionated particle collection with moderate sampling times. Acceptable deposit densities have been obtained in 3 hours at a sample intake rate of 5 liters/min from pipeline gases containing on the order of 200 particles/ft³ total concentration.[29]

Figure 24 shows a photograph of the device and Figure 25 gives a schematic drawing of an earlier version[28] with 3 stages. The device is very small to be suited for direct insertion into a pipe line.

Table 6: Impactor Covering Aerodynamic Size Starting From 0.03 μm

IMPACTOR CHARACTERISTICS

STAGE	AERODYN. CUT POINT [μm]	ORIFICE DIAM. [mm]	# OF ORIFICES
4 (top)	1.0	0.40	68
3	0.32	0.30	24
2	0.10	0.25	15
1	0.03	0.15	56

nominal inlet flow at p_{atm} = 5 l/min (ca. 0.2 scfm)

The Liquid Air Cascade Impactor permits the use of a multitude of collection substrates, including polished silicon wafer surfaces, to suit the analytical technique to be applied to the particles. Figure 26(a) shows an aluminum foil substrate. Figure 26(b) shows one enlarged deposit spot from stage 2 (0.1 to 0.3 μm) obtained by sampling N_2 pipeline gas for 3 hours.

There is, in principle, another solution to particle collection, the so-called *impinger*. It works like a single stage impactor whose jet is directed against a liquid surface.[21] The device is sometimes used to sample particles for analysis by atomic absorption or similar techniques. While the idea seems elegant, little is known today about its performance. Specifically it is not clear what the deformation of the liquid surface under the impact of the jet does to the deposition efficiency.

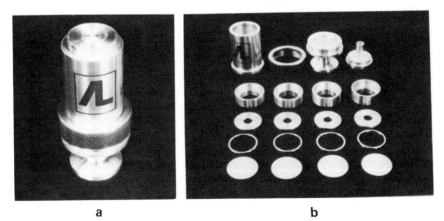

a   b

Figure 24: Photograph of 4-stage impactor for high-purity gas sampling. (a) Assembled. (b) Disassembled.

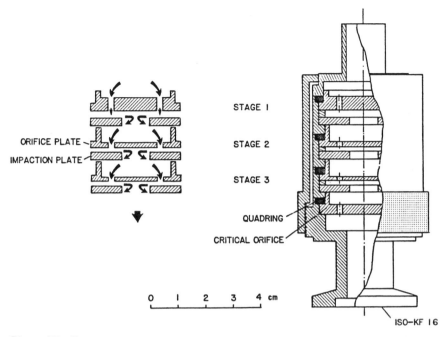

ORIFICE PLATE

IMPACTION PLATE

STAGE 1

STAGE 2

STAGE 3

QUADRING

CRITICAL ORIFICE

0   1   2   3   4 cm

ISO–KF 16

Figure 25: Engineering drawing and flow diagram of 3-stage version of impactor.

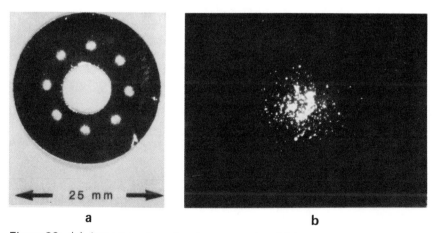

a   b

Figure 26: (a) Impactor deposit substrate made of 25 mm diameter aluminum foil with particle deposits shown. (b) Photomicrograph of a single deposit spot.

REMOVAL OF PARTICLES FROM COMPRESSED GASES BY FILTRATION

So far, this entire chapter has dealt mainly with issues of particle analysis and sources of particulate contamination in very pure semiconductor gases. The chapter would be incomplete, however, if we could not also demonstrate the ability to generate virtually particle-free gases.

The subject of filtration of compressed gases is extensive and we will again only touch specific points pertaining to very clean gases. From the perspective of filter media we will bypass all fibrous materials which are mostly installed as protective prefilters and hence are not essential to achieve ultraclean gases. We will deal only with membrane filter media such as PVDF, Polysulfone, Nylon, PTFE, PFA, etc. of relatively tight pore size ratings ("0.2 µm") and show in one example that the level of less than 1 particle/ft^3 >0.01 µm is attainable *as far as the filter medium is concerned.* We will also touch upon some of the limitations to this very good result that we must expect in industrial installations.

Elementary Filtration Kinetics in Gases

Commercially available filters are usually rated in terms of their ability to retain particles by an effective pore or particle diameter. Electronic grade membranes are usually rated at 0.22 µm. These ratings almost always relate to liquid challenges where certain types of bacteria are used in standardized procedures to determine filter retention. For example, if the filter holds back *Pseudomonas diminuta,* an oval bacterium of about 0.4 x 0.7 µm, it is rated "absolute" at 0.22 µm. This historical terminology is useful for pharmaceutical and food processing work when properly understood, but for gases its significance is nil beyond the fact that coarser liquid ratings can also mean somewhat poorer performance in gases.

There is a striking difference between the filtration kinetics of gases and liquids. A filter in liquid service is believed to retain particles essentially by its sieving action (zeta potential effects being ignored for the present). Thus, particles above a certain size get stuck in the membrane channels, while below this size they mostly pass. In this context the word "rating" refers to a "cut-off size."

In gas filtration, the primary removal mechanisms are diffusion, impaction and interception. While detailed mathematical models have been developed for each of these processes, we shall only concern ourselves with the most basic relationship outlined in Figure 27.

According to this figure, particle retention in the filter by diffusion is most effective at very low particle sizes and becomes less important as sizes increase. Diffusion is the result of Brownian motion causing very small particles to undergo a random walk around their path of flight.

Conversely, impaction and interception gain with increasing particle diameter, impaction being due to inertial "impact" on the filter surface as a result of particle mass, while interception is caused by particles touching the pore surfaces due to their physical extent.

The sum of all three curves is the filter efficiency curve as shown schematically in Figure 27. It is typically close to 100% everywhere except in the vicinity of the "most penetrating particle size" where the curve shows a dip. The existence of this filter minimum was first demonstrated by experiments of Spurny and Pich (1963)[30] with membrane filters and has since been reconfirmed in numerous experiments for virtually every conceivable filter geometry (e.g. References 31–33).

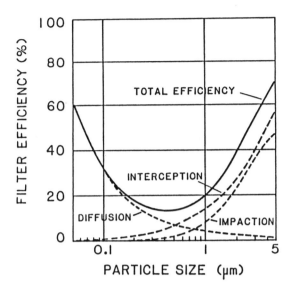

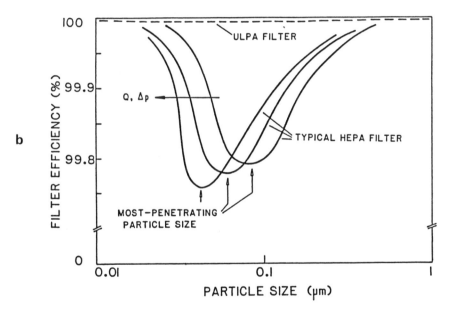

Figure 27: (a) The joint action of particle retention mechanisms shapes the total efficiency curve of a filter. Efficiency curves here are calculated for a disc with cylindrical pores of 15 μm diameter at a face velocity of 10 cm/s using the Spurny-Smutek-Pich Model (source: G. Kasper, 1977). (b) Typical efficiency curves of gas filters. The lower group is for HEPA type filters, such as high quality fiber filters or coarse membrane filters. ULPA designates efficiencies beyond current measuring capabilities on the order of 99.999999%.[34] Q and Δp are flow rate and pressure drop across filter increasing in direction of arrow.

The dip in filter efficiency can range anywhere from 90% for a fibrous filter with low pressure drop to values above 99.9999997% for electronic grade membrane filters rated at 0.45 and 0.2 μm.[34] This is illustrated in Figure 27 for an idealized example.

It should also be pointed out that the position of the dip in the efficiency curve will shift left or right with respect to particle size as a function of flow conditions. Figure 27 illustrates this point by showing an arrow in the direction of increasing flow rate and pressure drop, Q and Δp.

In summary, gas filters exhibit their best performance at very large and very small particles; in between, around 0.1 μm, they show their relatively poorest efficiency. The word "rating" in the sense of "cut-off" has obviously lost its meaning.

While the minimum efficiency traditionally determines the quality of a filter in gas applications, this term looses much of its practical usefulness when we deal with the extreme efficiencies of electronic grade membrane filters. The efficiencies of such filters are generally too high to be measured (hence their qualification "efficiency greater than. . ."). Furthermore, electronic grade gases are seldom dirty enough to require the full capability of a filter. Perhaps it will be possible in the future to agree on other properties such as purge-down times, etc. to provide additional selection criteria.[8]

Achievable Low Levels of Particle Concentration in Process Gases by Filtration

Electronic grade gas filter elements are offered in a variety of materials such as PTFE, PVDF, Nylon, Polysulfone, etc. These media are all similarly rated at 0.2 or 0.1 μm. (Some manufacturers have added a separate rating for gas service, e.g. "0.05 μm." The difficulties in rating high performance filters by a "most penetrating particle size"[32] have been mentioned above.)

The family of 0.2 μm rated filters is generally very performant in terms of particle retention efficiency. For example, the Millipore Wafergard brand has a penetration of less than 10^{-9}.[34] In most cases this exceeds by far the level required to create a "zero gas." Nevertheless, it seems to be common knowledge that particles are indeed detected in filtered process gases even though at very low levels of perhaps as few as 10/ft^3. This conflicting information tends to cast a shadow of doubt on the meaning of filtration theory. Sometimes, the opinion is offered that filtration at ultralow concentrations follows different rules.

A careful study[8] has shown however, that membrane filters are indeed capable of producing gases containing less than 0.1 particle/ft^3 total concentration above 0.01 μm over long periods. The study seems to indicate further, that secondary contamination sources such as shedding of particles from the filter itself or from surfaces of the gas ducts is responsible for the residual particle concentration in the gas.

The study was conducted with standard 10" PVDF membrane filter cartridges using the same filter medium as was used in the study by Accomazzo et al.[34] It was therefore known that such a filter would have a penetration of less than 3×10^{-9}.

The 10" cartridges were challenged with a steady flow of nitrogen containing on the order of 10^6 particles/ft^3 above 0.01 μm with a peak near 0.08 μm.

Downstream concentrations were monitored with a CNC and a LPC (Model LAS 250-X); all measurements included careful studies of internally generated counts as well as contributions from the sampling line. The sum of both of these noise sources was approximately 2 counts/ft^3 for the LAS 250-X and about 0.2/ft^3 for the CNC. These values must be subtracted from all the following data before making comparisons.

Table 7 shows test data for Run #1 with the newly installed cartridge. Both instruments independently record a virtually zero particle concentration.

Table 7: Results of Particle Counting Downstream of a 10" Membrane Filter Cartridge (Millipore Durapore) Using a CNC and a LPC

Run #1:   newly installed filter
          275 ft^3/h at atmospheric pressure
          Δ p = 0.26 psid

| HOURS | CNC | | LPC | | |
	COUNTS / HOUR	COUNTS / SCFT	COUNTS / HOUR 0.2-0.3 μm	> 0.3 μm	TOTAL SCFT
1	0	0	15	1	2.5
2	0	0	15	1	2.5
3	0	0	13	0	2.1
4	0	0	17	0	2.7
5	0	0	15	0	2.4
6	0	0	14	1	2.4
7	0	0	8	1	1.4
8	0	0	20	0	3.2
9	0	0	19	0	3.0
10	0	0	14	1	2.4
11	0	0	16	0	2.5
12	0	0	16	0	2.5
13	0	0	15	0	2.4
14	0	0	17	0	2.7
15	0	0	18	0	2.9

including internal counts

Table 8 shows the averages for the first four runs at flow rates up to 1900 Nft^3/hr. We see a slow but systematic increase in concentrations recorded by both counters, indicating that particles in the size range above 0.2 μm are released. There is a very high probability that these particles originate from surfaces downstream of the filter itself because after precision cleaning of these surfaces (tubing, etc.) the concentration at 1900 N/ft^3/hr dropped to almost zero (Run #5 in Table 8). Consecutive experiments at similar flow rates with other cartridges showed similar results (Runs 7, 8, 9).

These results demonstrate the capabilities and achievable levels of gas cleanliness for membrane filters. However, they do point to the limitations for industrial installations. Since it is virtually impossible to handle materials during construction without some minor contact with ambient particles, there is always a possibility of small but measurable particle concentrations downstream of any absolute filter.

**Table 8: Average Levels of Particle Concentration Measured After 10"
Filter Cartridges at Various Operating Conditions**

	Filter Operating Conditions $[ft^3/h]$	[psid]	Average counts/ft^3 *) CNC	LPC
Run #1	275	0.26	0	2.5 ± 0.4
Run #2	740	0.63	0.2	2.3 ± 0.7
Run #3	1200	1.0	0.9 ± 1	5.5 ± 2
Run #4	1900	1.6	2.5 ± 2	7.1 ± 3
after cleaning:				
Run #5	1900		0.1	2.3 ± 0.7
Run #6	1000		0	2.0 ± 0.8
new filter cartridge:				
Run #7	1900		0.1	2.2 ± 0.6
Run #8	1000		0	1.9 ± 0.6
Run #9	2600		0.5	2.9 ± 0.7

*) incl. internally generated counts and sampling line contribution.

REFERENCES

1. Kasper, G.K. and Wen, H.Y., "Particle technology for high-purity gases," *Proc. 31st Annual Conf. Inst. Environmental Sci.,* Las Vegas (1985).
2. Friedlander, S.K., *Smoke, Dust and Haze,* John Wiley & Sons, New York, NY (1977).
3. Hidy, G.M., *Aerosols–An Industrial and Environmental Science,* Academic Press, Inc., San Diego, CA (1984).
4. Kerker, M., *Adv. Colloid Interfact Sci.,* Vol. 5, pp. 105-172 (1975).
5. Wen, H.Y., Kasper, G. and Chesters, S., "A new method to fast and accurate measurements of total non-volatile impurity concentrations in pure liquids at sub-ppm levels," *Microcontamination J.,* Vol. 4, No. 3, p. 33 (1986).
6. Whitby, K.W., *Proceedings Annual Mtg. of Gesellschaft f. Aerosolforschung,* Bad Soden, W. Germany (1974).
7. *Handbook of Radiation Measurements Procedures* Second Ed., Nat. Council Rad. Protec. and Meas., Bethesda, MD (1985).
8. Wen, H.Y. and Kasper, G.K., "A gas filtration system for concentrations of 10^{-5} particles/cm^3," *Aerosol Sci. Technology,* Vol. 5, p. 167 (1986).
9. Kerker, M., *The Scattering of Light and Other Electromagnetic Radiation,* Academic Press, New York (1969).
10. Hulst, H.C. van de, *Light Scattering by Small Particles,* Dover, New York (reprinted from 1957 ed. by Wiley) (1981).
11. Roth, C., Gebhart, J. and Heigwer, G., *J. Colloid Interface Sci.,* Vol. 54, p. 265 (1976).
12. Knollenberg, R.G. and Luehr, R. (1976) "Open cavity laser 'active' scattering particle spectrometry from 0.05 to 5 microns," in *Fine Particles* (B.Y.H. Liu ed.), Academic Press, New York (1976).

13. Wen, H.Y. and Kasper, G., Counting efficiencies of 6 commercial particle counters, *J. Aerosol Sci.,* Vol. 17, p. 947 (1986).
14. Gebhart, J., Blankenberg, P., Borman, S. and Roth, C., *Staub,* Vol. 43, p. 439 (1983).
15. Sinclair, D. and Hoopes, G.S., *J. Aerosol Sci.,* Vol. 6, p. 1 (1975).
16. Bricard, J., Delattre, P., Madelaine, G. and Pourprix, M., "Detection of ultrafine particles by means of a continuous flux condensation nuclei counter." in *Fine Particles* (B.Y.H. Liu ed.), Academic Press, NY (1975).
17. Yoshida, T., Kousaka, Y. and Okuyama, K., *Industr. and Engr. Chem. Fundam.,* Vol. 15, p. 37 (1976).
18. Lundgren, D.A., Harris, F.S. Jr., Marlow, W.H. et al., *Aerosol Measurement,* University Press of Florida, Gainesville (1979).
19. Madelaine, G. and Metayer, Y. Technical note. *J. Aerosol Sci.,* Vol. 11, p. 358 (1980).
20. Porstendorfer, J., Scheibel, H.G., Pohl, F.G., Preining, O., Reischl, G. and Wagner, P.E., *Aerosol Sci. Technology,* Vol. 4, p. 65 (1985).
21. Mercer, T.T., *Aerosol Technology in Hazard Evaluation,* Academic Press, New York, NY (1973).
22. Agarwal, J.J. and Sem, G.J., *J. Aerosol Sci.,* Vol. 11, p. 343 (1980).
23. Kasper, G., Preining, O. and Matteson, M.J., *J. Aerosol Sci.,* Vol. 9, p. 331 (1978).
24. Locke, B.R., Donovan, R.P., Ensor, D.S. and Caviness, A.L., *Proc. 31st Annual Conf. Inst. Environmental Sci.,* Las Vegas (1985).
25. Smith, T.N. and Phillips, C.R., *Environmental Sci. Technol.,* Vol. 9, p. 564 (1975).
26. Pich, J., *Staub,* Vol. 24, p. 60 (1964).
27. Berner, A., "Design principles of AERAS low pressure impactor," *Proc. First Int. Aerosol Conf.* (B.Y.H. Liu et al. eds.), Minneapolis, MN (1984).
28. Kasper, G., Wen, H.Y. and Berner, A., *Proc. 31st Ann. Mtg. Inst. Environmental Sci.,* Las Vegas (1985).
29. Kasper, G., Wen, H.Y. and Chesters, S., *Proc. 32nd Ann. Mtg. Inst. Environmental Sci.,* Dallas, TX (1986).
30. Spurny, K.R. and Pich, J., *Coll. Czechoslov. Chem. Commun.* (Engl. Ed.), Vol. 28, p. 2886 (1963).
31. Spurny, K.R. and Lodge, J.P. Jr., *Staub,* Vol. 28, p. 179 (1968).
32. Lee, K.W. and Liu, B.Y.H., *J. Air Poll. Control Assoc.,* Vol. 30, p. 377 (1980).
33. Rubow, K.L., Ph.D. Dissertation, Univ. of Minnesota, Dept. Mech. Engr. (1981).
34. Accomazzo, M.A., Rubow, K.L. and Liu, B.Y.H., *Solid State Technol.,* Vol. 27, pp. 141-146 (1984).
35. Whitby, K.T. and Liu, B.Y.H., *Atmos. Envir.,* Vol. 2, p. 103 (1968).

10

Contamination Control and Concerns in VLSI Lithography

Mary L. Long

INTRODUCTION

The masking or photoresist process has traditionally been the scape-goat portion of semiconductor manufacturing. When not understood, the imaging process took on the quality of black magic, and many yield limiting trouble spots were noted. The technology has become more sophisticated, defect sensitivity of imaging processes has increased, and as fast as one set of problems is solved, another is discovered.

This discussion of contamination concerns in VLSI lithography will address the nature of resist materials, the expectations of resist performance, and specific contamination problems in lithographic processes. Because most imaging processes are still optical, the main discussion will center around optical imaging technologies, although many of the advanced, nonoptical technologies have similar contamination problems.

PHOTORESIST: WHAT IS IT?

Photoresist is a polymeric material, classified as negative or positive depending on the image produced; negative resist remains where the mask is clear, and positive resist remains where the mask is opaque. In both cases, the exposure of the resist to ultraviolet-visible light (typically 350 to 450 nanometers) produces a change in the polymeric material, and that change causes a difference in solubility.

Chemistry of Negative Photoresist

A number of different negative-acting resist materials have been used in the lithographic industries (newspaper printing, magazine and color print replication) in the last half century, but the primary resist systems selected for adaptation to semiconductor industry requirements were the natural rubber or synthetic isoprene based polymers. These materials exhibited the good adhesion and etch resistance needed for the semiconductor processing of the early and mid-1960s. Numerous versions of these materials have been developed and marketed, but the basic structure and chemistry is the same. Figure 1 shows the basic structure of cis-polyisoprene.

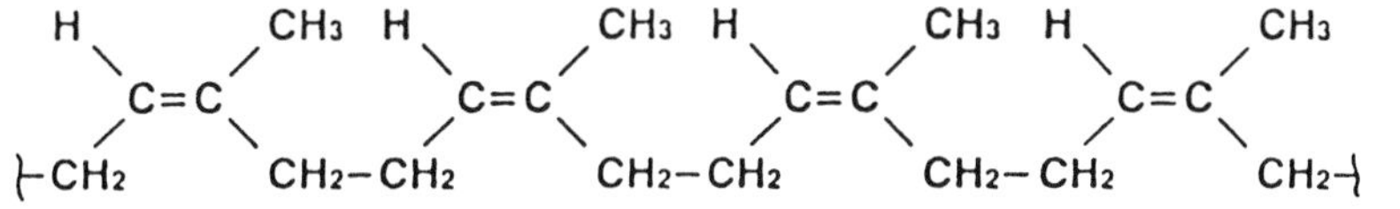

Figure 1: Polyisoprene.

The molecular weight of the polyisoprene is carefully controlled, especially in recent formulations, and the polyisoprene is cyclized to give it the characteristics needed for processing. The cyclical structure, shown in Figure 2, is carefully controlled to enhance adhesion and improve etch resistance.

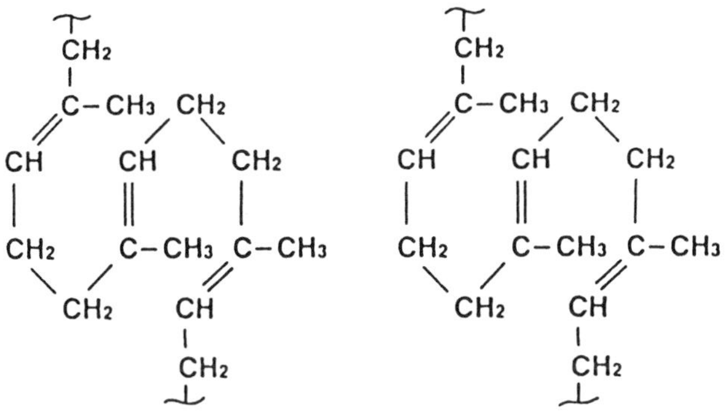

Figure 2: Cyclized polyisoprene.

Because polyisoprene has minimal sensitivity in the ultraviolet-visible energy range, a photoinitiator must be added to provide reasonable photosensitivity. The typical photoinitiator is a diaryl diazide with a structure as shown in Figure 3.

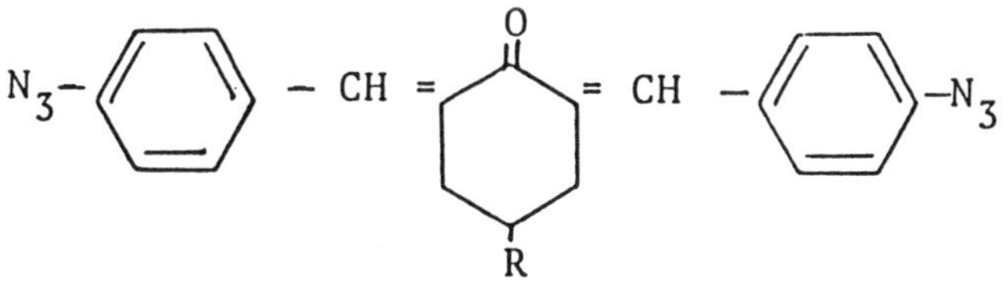

Figure 3: Diaryl diazide.

Spectral sensitivity for negative resists extends from 300 to 436 nanometers, with peak sensitivity occurring around the wavelength (mercury vapor source) of 365 nanometers. The triple nitrogen (azide) is sensitive to the ultraviolet-visible energy, and a series of reactions results in the breaking of double bonds in the isoprene polymer and the linking of two or more polymer chains together to form a new polymer chain of significantly higher molecular weight, as shown in Figure 4.

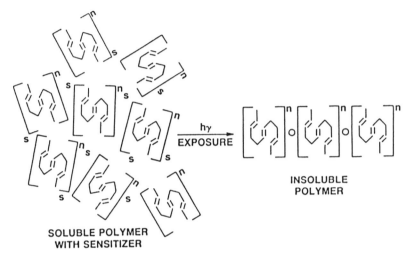

SOLUBLE POLYMER WITH SENSITIZER

INSOLUBLE POLYMER

Figure 4: Exposure of polyisoprene.

The change in the molecular weight also changes the solubility character-istics of the polymer. This change is not abrupt, because the molecular weight is not exact, but is a molecular weight *range*, as shown in Figure 5. The effect of exposure, then, is to shift the molecular weight range of the exposed area into a range with different solubility characteristics.

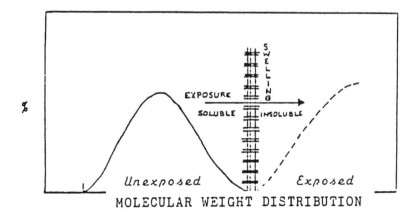

MOLECULAR WEIGHT DISTRIBUTION

Figure 5: Molecular weight vs. exposure.

It is obvious that the molecular weight range of the photoresist is an important factor in the resolution potential of the resist. If the initial range is broad, a heavy exposure dose is needed to insure a clear differential between the exposed and unexposed ranges. On the other hand, a narrow molecular weight range improves the precision of the exposure and gives a clear differential between the exposed and unexposed areas.

Negative Photoresist Developers. Solvents and developers are an important part of any resist formulation and processing package. For negative resists, organic solvents are used exclusively. The organic solvents of interest can be classified by solvent type as shown in Table 1.

Table 1: Polyisoprene Solvents

ALCOHOLS – Non-solvent	
ACETATES	Weak Solvent
ALIPHATICS	Mild Solvent
AROMATICS	Strong Solvent
CHLORINATED HYDROCARBONS	Harsh Solvent

For most applications, aliphatic solvents are preferred because they are mild solvents for the unexposed resist. The two most commonly used aliphatic solvents are decane and Stoddard Solvent. In early processes and with early resist formulations, aromatic solvents, primarily xylene, were used, but these solvents cause swelling and are unacceptable as developers for high resolution applications. Xylene is, however, the primary solvent in the resist formulation. As a class of solvents, the alcohols were often used for rinsing, but they are nonsolvents for isoprenes and can cause precipitation of polymer from the developer solution. Many alcohols cause hardening of the resist, and the resulting brittleness is detrimental to resist performance, especially where the resist is distorted from swelling caused from too strong a developer. Acetates, such as butyl acetate, have become the preferred rinses because, as weak solvents, they do not continue the developing action, but rinse without precipitating polymer from the developer and without damaging adhesion.

The use of negative resist was predominant in the semiconductor industry until the early 1980s. Environmental, as well as performance, factors have contributed to the shift from negative to positive resist. Xylene, as a photochemical pollutant, has been listed in emission control regulations. From a performance point of view, negative resist can handle imaging tasks into the 2.5 to 3 micron range with little difficulty, and with adequate quality control of developer and rinse and reasonable process control monitoring, critical dimensions can be maintained.

The typical characteristics of negative photoresist may be summarized as follows:

- A natural or synthetic isoprene polymer of controlled molecular weight and a diaryl diazide photoinitiator dissolved in xylene (or similar organic solvent).

- Adheres well to a variety of substrate surfaces.

- Elastic and flexible, even when cured.

- Photoreaction involves a shift of molecular weight causing differential solubility between exposed and unexposed resist.

- Spectral sensitivity peaks at 365 nanometers (mercury vapor source).

Chemistry of Positive Photoresist

Positive photoresist, unlike negative resist, experiences no change of molecular weight during exposure. The solubility differential is provided by a functional group change. This is true of virtually all the optical positive resists. Spectral sensitivity for most positive resists peaks around a wavelength of 405 nanometers, but some newer materials are targeted for shorter wavelengths, including wavelengths in the deep ultraviolet range.

The polymeric base of positive resist is the novolak-type resin. This novolak resin, shown in Figure 6, forms the polymeric binder which can be mixed or reacted with the photoactive compound, a diazoketone, shown in Figure 7. In most early resist formulations, the photoactive compound was mixed in solution with the binder resin, but in later formulations, the photoactive compound has been bonded to the resin, forming a sensitized resin.

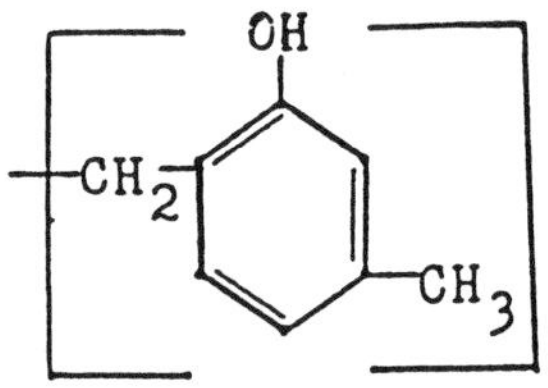

Figure 6: Novolak resin.

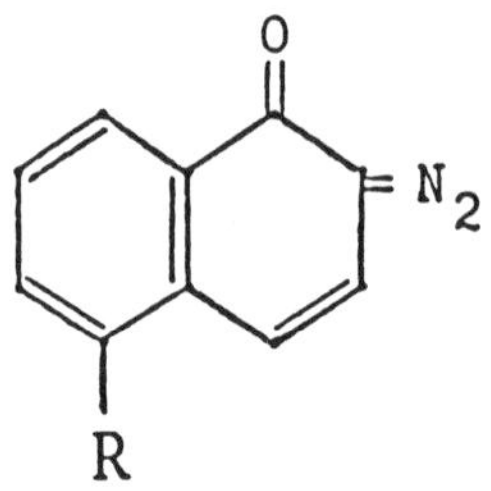

Figure 7: Photoactive compound.

Figure 8 shows that the photoreaction which takes place upon exposure to ultraviolet-visible illumination changes only the functional group of the photoactive compound, from a diazoketone, a dissolution inhibitor, to a carboxylic acid, a dissolution enhancer. The polymeric binder is not changed. The solubility characteristics of the resist film are controlled by the photoactive compound.

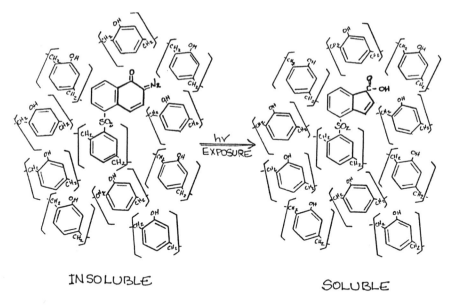

INSOLUBLE   SOLUBLE

Figure 8: Photochemical reaction of diazoketone group to a carboxylic acid.

Figure 9 shows the photoreaction of the diazoketone. Note that during the reaction, significant quantities of nitrogen are given off, and for the reaction to proceed to completion, trace quantities of water are necessary.

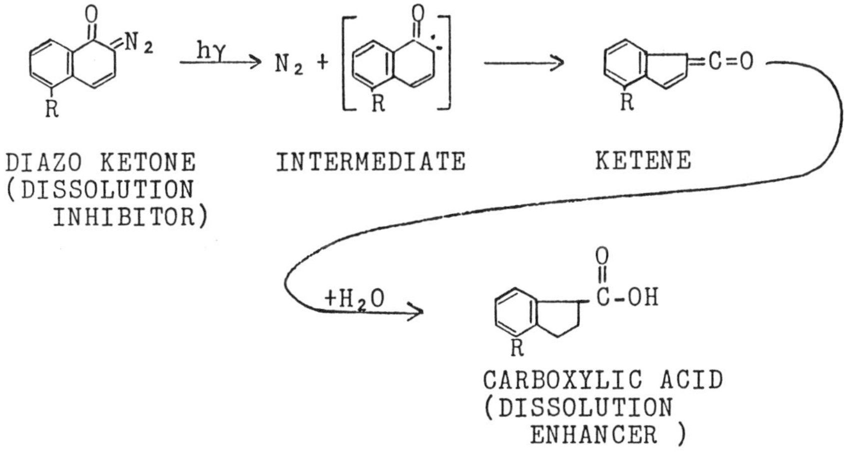

DIAZO KETONE   INTERMEDIATE   KETENE
(DISSOLUTION
   INHIBITOR)

CARBOXYLIC ACID
(DISSOLUTION
   ENHANCER)

Figure 9: Reaction of the photoactive compound (PAC).

The nitrogen release can cause contact problems in contact printing if high intensity illumination is used, because the burst of released nitrogen can force the mask and wafer (or copy plate) apart. The nitrogen released during exposure has been estimated to occupy a volume up to 10 or 12 times the volume of the resist it is released from, at standard temperatures and pressures. It is obvious that the rapid release of this nitrogen, especially with a predominantly clear-field mask, represents a significant force. If the adhesion of the resist to the substrate is marginal, high intensity exposure may cause the resist to pop off or lift from the surface during exposure.

The importance of trace water content in the resist must also be appreciated. If, due to low relative humidity, the water content is too low, photospeed is affected. If no moisture is present and a high intensity exposure source is used, the carboxylic acid cannot be formed, and a different reaction, acylation, takes place. With acylation, as shown in Figure 10, the ketene intermediate reacts with the OH groups of the resin, and crosslinking takes place.

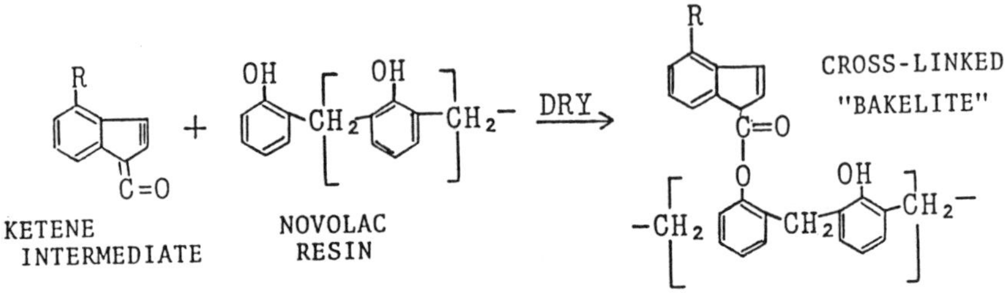

Figure 10: Acylation reaction.

The acylation reaction is desirable when it occurs as a part of post baking, because it increases thermal stability. At other times, it may compete with the formation of carboxylic acid if intensity is too high and humidity is low. Even after exposure and development, the acylation reaction must be expected in subsequent operations. In a vacuum with high energy, such as a plasma etcher or ion implanter, acylation may cause heavy crosslinking, resulting in resist removal problems.

Positive Photoresist Developers. Positive photoresist developers are aqueous solutions of organic or inorganic bases. Early developers used sodium hydroxide, but more selective solutions have been developed. In response to MOS surface sensitivity, nonionic developers were formulated from organic bases such as the ethanolamines, tetramethylammonium hydroxide and most recently, choline (trimethyl 2-hydroxymethyl ammonium hydroxide). Surfactants and buffers are included in many formulations. Developers may be provided in concentrated form, but premixed developers can minimize variations in developer strength due to the inaccuracies of on-line, small batch mixing.

The dissolution reaction which takes place during development is quite different from that of negative resist. The reaction of the carboxylic acid with the basic developer leads to the formation of a salt which appears to enhance the dissolution of the exposed resist. However, the resin and salt do not both go into solution immediately, and some part of the exposed resin/sensitizer area appears to slough off and fall to the bottom of the developer bath. In a recirculating bath, the precipitate of developed resist can be removed by filtration, extending the bath life. Spray development of positive resist is possible if the spin speed is slow enough to maintain a sheet of developer on the wafer surface, permitting the reaction of acid and base to take place.

The selectivity of a positive developer is controlled by formulation components and concentration of the base. With high selectivity, nearly vertical sidewalls can be seen, and minimal resist loss is noted in the unexposed regions. For VLSI applications, this translates into improved control of critical dimensions.

Positive photoresists have predominated in circuit board technology and in hard surface photomask fabrication for many years. Only when linewidths moved below 4 microns and critical dimension requirements became difficult to meet with negative resist did positive resist use become significant in wafer fabrication. The higher cost of positive photoresist was usually offset by yield improvements, and the aqueous based developers were more acceptable from an environmental perspective. Newer formulations have replaced the traditional solvent, primarily ethylene glycol monoethyl ether acetate (EGMEA) with the propyl version of the solvent, the latter being considered a 'safer' solvent for personnel working with the resists.

Positive photoresist can be characterized as:

- Novolak-type resin and diazoketone photoactive compound in solvent media.

- Adhesion depends on appropriate surface preparation.

- Resin is brittle and stress prone.

- Solubility differential is caused by a functional group change during exposure.

- Spectral sensitivity peaks at 405 nanometers (mercury vapor source).

Nonoptical Resists

Numerous exposure techniques have been proposed to push microlithography into the submicron range, including deep UV, E-beam, X-ray and focused ion beam. Resists for these new technologies present special challenges because of the high resolution and sensitivity needed for imaging, and the requirements for these resists to withstand high energy plasma etching. In many cases, compromises must be made between resolution and etch resistance, and this limits the use of the beam technologies for wafer fabrication. For maskmaking, electron resists have been developed with acceptable resolution and sensitivity, and they permit patterning of chrome masks because the chrome substrate is thin, flat and conductive. For most chrome masks, the film thickness to be etched is less than 1000 Angstroms, therefore, wet etching can be accomplished without significant image degradation from lateral etch. When this technology is transferred to wafer fabrication, several problems emerge. Wafer topology requires use of thicker resist or planarizing techniques. Thicker resist causes resolution problems due to scattering of beam energy and proximity effects within the resist film. To maintain dimensional control, dry etching techniques are preferred, but many electron-sensitive resists are also sensitive to the plasmas used for etching.

Despite the problems encountered with beam technologies, work continues in both resist and equipment development. Typical negative-acting electron or X-ray resists function much like negative photoresists, increasing molecular weight with exposure, causing a differential solubility in exposed and unexposed areas. Electron positive resists, on the other hand, behave differently than positive photoresists, in that they experience chain scission reactions rather than functional group changes. Differential solubility is created between the high molecular weight unexposed areas and the exposed areas where molecular weight is lowered by chain scission.

PMMA (polymethyl methacrylate) responds to a large range of energies with a chain-scission reaction, and therefore it has become the baseline process material for many technologies, including deep UV, X-ray, E-beam and focused ion beam lithography. When properly processed, PMMA gives sharp edges and steep sidewalls. PMMA is developed with organic solvents such as chlorobenzene, methyl ethyl ketone or methylisobutyl ketone blended with nonsolvents to give the desired dissolution differential. PMMA has been especially useful as a planarizing layer for improved functional resolution with optical imaging, and similar techniques are being investigated for use with the beam technologies.

Another resist material which differs from standard optical negative and positive resists is the deep UV negative acting resist, reported by Hitachi, which relies on a functional group change for differential solubility. Use of this resist has been limited, primarily of interest for deep UV projection, contact or proximity exposure systems.

It is virtually impossible to characterize all the nonoptical resists, because there are so many polymer systems being investigated. It is not clear which, if any, of the various resists will best meet the variety of requirements for imaging and processing. Many formulations exist as special purpose resists for a limited range of applications.

PRIMERS

Hexamethyldisilazane (HMDS)

Primers are used in many applications to improve adhesion between dissimilar materials. In semiconductor processing, primers help prepare the hydrophyllic surface to receive a hydrophobic resist coating. The most widely used primer is hexamethyldisilazane (HMDS). HMDS can be applied by spinning or by vapor, and it reacts with both physisorbed and chemisorbed moisture on the wafer surface. Physisorbed moisture, as water, reacts to form hexamethyldisiloxane, a volatile product which is exhausted, and ammonia, also exhausted. Chemisorbed moisture, as a hydroxide of silicon, reacts with HMDS to change the surface to an organosilane and the ammonia by-product, which is again exhausted. HMDS functions as an effective primer for both negative and positive photoresist.

PHOTORESIST FUNCTIONAL REQUIREMENTS

Photoresist was once only required to faithfully reproduce a given image and offer protection from chemical etch baths. Not only have imaging and etch resistance requirements become more demanding, with tighter linewidth specifications and the use of plasma etching techniques, but a series of new tasks have also been added. A production-worthy resist is also expected to withstand and mask against ion implantation, function as a lift-off mask, and possibly even be used in a multilayer system if planarization is required. A survey of resist performance expectations follows.

Image Fidelity

Typical resolution requirements through the mid-1970s were no more demanding than 5 microns. With improved imaging systems, particularly those which prevented contact of mask and wafer with subsequent damage to both, high quality masks became economically practical. The effort and cost of making a 'defect free' mask was incidental in comparison with the yield potential as that mask was used for hundreds and thousands of projection image transfers. By the early 1980s, design rules had moved to 3 microns in high volume production, and research efforts were fast bringing 1.5 micron minimum dimensions into range. By the middle of the decade, 1.5 micron images were being realized in production, primarily with wafer step and repeat systems, and submicron research was in full swing.

Resolution, as a term, can be interpreted several ways. In a simple sense, resolution is the ability to distinguish between two adjacent features. One optical definition of theoretical resolution is 0.61 times the wavelength, divided by the numerical aperture of the imaging system. Neither approach is sufficient for the semiconductor application of microlithography, where images must not only be clearly resolved, but the size of the transferred image must not deviate more than a fraction of a micron from the specified dimension. Unless these critical dimensions are held to very tight tolerances, yield and circuit performance are affected. In the 1970s, with 5 micron minimum geometries, a typical specification might give a range from 4.5 to 5.5 microns (± 0.5 micron), but as minimum linewidths shrank, so did the acceptable range. Current critical dimension specifications are often in the range of ± 0.1 micron.

This, of course, brings up problems of linewidth measurement, measurement techniques, measurement equipment and linewidth standards. For practical purposes, arbitrary standards, which correlate critical dimensions with yield, can provide adequate process control, but these are inadequate when exact dimensions are required for process transfer or mask purchase specifications. Considerable work has been done by the National Bureau of Standards to provide traceable standards for this increasingly demanding application. (Although numerous papers have been presented and articles authored by the NBS group, the SPIE Proceedings, Volume 342, *Integrated Circuits Metrology*, May 4-5, 1982, edited by Diana Nyyssonen, provides a good overview of the work to establish measurement standards.)

Another aspect of critical dimension control is the technique called dimensional shifting. Dimensions may be shifted during the image transfer process by adjusting the exposure and development parameters. Skilled photoresist engineers were often able to compensate for an incorrectly sized mask by such procedures when linewidths were in the 4 to 5 micron range, especially with negative photoresist. It was quickly discovered, when linewidths decreased and positive resist was used, that dimensional shifting became more difficult, and was limited to shifting accomplished by overexposure. Current requirements for VLSI technology often mandate exposure latitude, or the lack of linewidth shift with exposure variation. Such exposure latitude eliminates the option of dimensional shifting for linewidth adjustment.

Etch Protection

Resist films must be continuous, with no voids or pinholes, they must adhere to the substrate surface, and they must provide a barrier against the etchant. Etching processes may be either wet or dry (plasma), and the etching species is chosen for its selectivity of thin film to be etched over the under- lying substrate. For many applications, wet etching is the simple solution, but some substrate materials, especially polysilicon and silicon nitride, are more easily etched with plasma techniques. Smaller dimensions have caused a shift toward plasma etching techniques because of reduced or eliminated lateral etching. High resolution imaging requires equally high resolution etching.

Resist films must still provide protection from a variety of wet and dry etchants for an increasing number of substrate materials. Most wet etchants are isotropic rather than directional in nature; therefore, some degree of lateral etching must be expected. If substrate films are thin and reasonably flat, lateral etch problems are minimal, provided the resist exhibits good adhesion. When the lateral etch of a wet etch process presents a problem, dry etching tech- niques, including barrel and parallel plate plasma, reaction ion etching, sputter- ing and ion milling, may be preferred. In most applications, the resist withstands the etching media with little difficulty for typical substrate thicknesses, but newer substrates and some of the more anisotropic etchants challenge typical resist systems. Stabilization techniques have been developed to add to the etch and thermal resistance of the resist.

Ion Implant Mask

The discovery that photoresist could effectively mask implant ions elimi- nated the oxidation and etching steps previously required and improved the planarity of the resulting circuit. From a baseline, where the resist has been proven to stand up to the implant, implant doses may be increased, requiring thicker resist for masking, and simultaneously, linewidths may be decreased, possibly requiring the use of thinner resist. As implant doses increase, removal of the resist becomes more difficult, due to the acylation reaction which takes place as the resist is irradiated with high energy ions in a vacuum environment. Evidence indicates that the acylation reaction begins at the top surface of the resist and proceeds downward as a function of dose. Increasing the resist thick- ness helps minimize the removal problem, but decreasing linewidths may limit resist thickness.

Lift-Off Mask

Photoresist, especially positive, can be used to define metallization patterns by a lift-off, a technique that can provide higher resolution than traditional etching. The resist pattern is defined prior to metallization, and metal is deposited over the resist image. The resist is then exposed to solvent, and the metal on top of resist is lifted off with the resist, leaving a sharply defined metal pattern in the developed areas. The primary issues in lift-off techniques are the resist profile, thermal stability of the resist, and scum-free development. The resist profile should be steep or negatively sloped to provide gaps in metal coverage and easy access of solvent to the resist to be lifted. The resist must be thermally stable at the temperatures expected in the metallization system. Lateral distortion of the pattern is unacceptable, and loss of edge acuity would hamper the lift-off process. The substrate surface must be clean after development to insure good adhesion of metal to substrate.

The use of a lift-off technique may be required when metallization patterns cannot be chemically etched without attacking the underlying substrate or where etchants are especially toxic or dangerous. High resolution metal patterns with outstanding aspect ratios (height to width) can be achieved with lift-off.

Multilayer Mask

A more recent process involving the use of photoresist is the multilayer or PCM (Portable Comfortable Mask) process, first described by B.J. Lin in the 1979 SPIE Proceedings. This process uses one resist material as a planarizing and transfer layer, and another for high resolution imaging. The most popular combination has been PMMA for planarization and optical positive resist for high resolution. Because the substrate surface has been planarized with the PMMA (1.5 to 3 microns thick), only a thin layer of positive resist is needed for imaging the top surface as compared to standard resist thicknesses required for adequate step coverage. If the image is transferred into the PMMA by exposing it with deep UV, at a wavelength of 220 nanometers, through the developed positive resist layer, 0.2 micron of the latter has been found to be opaque to the deep UV.

Multilayer processes have been defined using the top layer for imaging, and transferring the image through the planarizing layer by plasma or reactive ion etching techniques. Some processes have a deposited layer between the imaging and planarizing layer, and the bottom layer may be any of a number of polymeric materials. These processes have been found to have higher resolution than single layer processes, but less resolution than the bilayer deep UV process.

PHOTORESIST CONTAMINATION CONCERNS

Contamination issues in photolithographic processes include particulate control in resist manufacturing and packaging, resist degradation due to aging, improper storage or transport of materials, and contamination control in actual processing. As a general rule, prevention of contamination is the desired approach, and it is especially true in microlithography. Small particles are very hard to remove, particularly if steps and linewidths are of similar dimensions. Particulate contamination translates directly into yield loss for nearly all process steps, and therefore, significant efforts have been launched in the past few years to assemble an effective contamination control program for microlithography.

Purity

Filtration of photoresist was a topic of great interest in the earliest Kodak Microminiaturization Seminars in the mid-1960s. At that time, resists designed for other purposes were being adapted for use in the new semiconductor industry. The cleanliness requirements of this new application were different than for the standard chemical milling or lithographic applications. Methods of preparing existing resists for semiconductor applications included centrifugation and electrophoresis as well as filtration.

Early filtration efforts were hampered by the gelatinous nature of negative resist contaminants. Catalysts used for cyclization, especially tin, caused nucleation of contamination as the resist aged. Incompletely dissolved polymer could also be found as 'gel slugs.' In addition, most of the early filters were not really compatible with the solvent system of the resist, and they would swell and clog; therefore, many photoresist users relied on nonfiltration methods of resist purification.

The rapidly growing microelectronics industry put pressure on resist manufacturers to improve the purity and performance of their photoresists and on filter manufacturers to provide usable filters. Special photoresists are now readily available for the microelectronics industry, and competition among the resist manufacturers who supply this industry is increasing. New filter membrane materials, including Teflon, propylene and nylon, are unaffected by the solvents and permit effective filtration to submicron levels, and low pressure filtration systems with large filter surface areas are able to remove gelatinous material without premature filter clogging. Filtration to 0.2 micron is now an industry standard, and there is pressure to move to even smaller pore sizes.

Positive resists present a special challenge due to a degradation reaction that causes resist to self-contaminate. Many of the standard positive resist formulations contain a mixture of resin and photoactive compound. As the material ages, the photoactive compound degrades, causing a particulate fall-out and an outgassing of nitrogen. Resists that have been microfiltered may require refiltration after four to six weeks to maintain performance. Newer formulations which have the photoactive compound bonded to the resin exhibit an extended shelf life.

Although the primary focus is on photoresist cleanliness, it is important that all processing chemicals be filtered to the same specification as photoresists. Unfiltered primers, developers, rinses, etchants or strippers can be as potent a source of particulate contamination as the resist.

Packaging

Control of contamination during photoresist manufacturing and submicron filtration techniques are only part of the overall contamination control problem. Packaging is a major issue for resist manufacturers. Resist that has been filtered to 0.2 micron must be put into a particle-free bottle if cleanliness level is to be maintained. Bottle washing techniques are expensive and ineffective for removing small particles due to electrostatic forces attracting particles to bottle sidewalls. A special supply of bottles, manufactured and capped under clean room conditions is preferred. For most applications involving solvents, especially photoresist, glass bottles provide the best level of cleanliness.

Plastic bottles are required for positive resist developers and for solutions containing hydrofluoric acid. Because they are break-resistant, plastic bottles would be preferred for other hazardous materials. Special care must be taken to insure that the plastic has a stable plasticizer and that the bottle does not shed under the stress caused by shipping or aging. The composition and integrity of the cap must also be considered as a part of the whole package.

Particles on the outside of the bottle, from shipping and handling outside of clean room environments, can be a source of contamination in the processing environment. To minimize this problem, bottles are often double bagged, placed in two plastic bags before they leave the clean filling room. The outer bag is removed, taking with it the transit contamination, when the bottle is placed in inventory within a clean storage area. The second bag is removed when the bottle is taken into the processing area.

Bulk packaging of chemicals in 5 to 25 gallon canisters is an effective method of transporting and providing chemicals for the clean room. The canisters are kept outside the clean room, and chemicals are delivered by pipeline to the point of use. Such a system minimizes bottle handling and disposal problems in the clean room. The initial cost of canisters and the piping system is substantial, involving not only the specialized piping, but also a stock of canisters sufficient for inventory and transit as well as those in use. The canisters must have an interlock system to prevent accidental contamination, and an effective method for canister cleaning in case of such contamination. Construction of the canisters, including nonreactive and nonparticulating liners, must meet quality and safety standards for pressurized use in the lab, and various transportation codes must be met. Special connectors, coded to prevent accidental contamination of canister or line by excluding all but the correct fittings are typically required for filling and for using the chemicals. Standardization of fittings would permit use of canisters by several companies, and widespread usage would permit the system to operate economically. From a contamination control aspect, filters would be required on pressurized gas lines feeding in to the canisters as well as on outgoing lines. In addition, a point of use filter is recommended at the dispense point.

The final packaging issue is that of labeling. Labels must clearly give all pertinent information about the product, including lot and batch numbers for material traceability. Information must be clearly visible through protective packaging material. Date codes are recommended for any materials with limited shelf life. Temperature sensitive materials may be shipped with a temperature monitor, dots mounted directly on bottles or a limited time recorder enclosed in the shipment, to protect against resist degradation due to improper shipping or storage conditions. For monitoring and packaging procedures to be effective, receiving and stockroom personnel must be made aware of their significance.

Processing

The greatest leverage in the contamination control strategy is in the processing area. Materials may be provided contamination free, point of use filters may guarantee clean dispensing of all liquids, and Class 10 clean rooms may be provided, but without proper attention to the process and handling, uncontrolled contamination may still cause loss of yield. Defect density is defined as a function of minimum linewidth, but minimum film thicknesses must also be considered. Often, film thickness dimensions are an order of magnitude smaller than linewidths, and therefore, the sensitivity of a 0.1 micron field oxide, for example, must be considered as well as that of a 0.8 micron linewidth.

CONTAMINATION CONTROL IN PHOTORESIST PROCESSING

To effectively address potential sources of contamination in the photoresist processing sequence, each step of the process will be examined for potential contamination problems. Attention to details in processing and effective training of personnel can make a substantial difference in yield and profitability. In fact, personnel working habits can be more important than the most expensive, high technology clean room facility.

Surface Preparation

The purpose of this step is to prepare the wafer for subsequent processing steps. The surface must be prepared for imaging, etching, diffusion or deposition processes. In addition to particulate contamination, organic films or inorganic residues must be removed from the substrate surface, and the surface must be treated for optimum adhesion. Surface preparation may consist of cleaning steps, dehydration, and priming. Wafer cleaning processes must be defined to remove specific contamination, therefore, it is important to know what types of contaminants are present.

Particulate contamination is most easily removed by scrubbing, either with a specially designed brush or with high pressure water. Care must be taken to provide adequate brush rinsing, or contamination may be transferred from one wafer to another, and surfaces may be scratched in the process. Wafers with topography may accumulate particulates on one side of the patterns if spray and brush directions are not correctly controlled. A scrub does not insure complete removal of particles, so an inspection may be advisable.

Acid cleaning with solutions such as sulfuric acid and hydrogen peroxide, mixed approximately 4 parts acid to one part peroxide, can be used to remove organic contamination. In some cases, particulates will also be attacked by this aggressive solution as well. This solution, often called "piranha," is best used immediately after mixing, heated only by the exothermic reaction of mixing. The hot acid reacts with any organic substance on the wafer, and the peroxide converts the resulting carbon to carbon dioxide, which evolves as gas. As the solution cools, the peroxide loses oxygen and becomes water, thereby diluting the remaining sulfuric acid. Attempts to prolong the life of the bath by heating results in an ineffective clean in diluted sulfuric acid. One alternative is to mix the bath at a 10 to 1 ratio and recharge several times with additional peroxide. An active bath will turn dark when organic materials are introduced, but then will become clear as the carbon is oxidized. Another alternative is to use ammonium persulfate in a heated sulfuric bath. Particulate contamination can be a problem in these baths, but few materials will not be attacked by the strong solution.

Water rinsing to high resistivity by overflow, cascading or quick dump methods, follows the acid clean, and a final spin rinse-dry completes the cleaning operation. In the rinsing operations, standing baths are avoided, therefore the most probable contamination point may be the spin rinse-dry operation where electrostatic charges may build up on the wafers and carriers. Ionizers in the nitrogen line and appropriate selection of carrier materials can minimize this problem.

Dehydration follows cleaning procedures. Various systems are in use, from simple ovens to infrared tracks, microwave stations, or hot plate chucks with or without vacuum. The temperature is raised above the boiling point of water for a time sufficient to remove physisorbed water. Removal of chemisorbed water requires heating above 450°C. The dehydration process is quite simple, but contamination may be introduced by wafer transport systems, by ovens or tracks, or by air or nitrogen purges. Routine cleaning and appropriate filtering can address such contamination problems.

Primer is applied after dehydration and resist spin. Traditionally, HMDS has been spun on, diluted to about 30% in a solvent compatible with the resist. Many process flow sheets refer to this step as the HMDS "wash." Such a designation may be misleading because primer application is not intended as a cleaning step and is not effective as such. Primary concerns at this point are that the HMDS solution has been properly mixed and filtered, and that it has not exceeded its shelf life. Solvent compatibility is important, because some solvents can react with resists to form gel particles. Xylene or Stoddard Solvent is the diluent for HMDS used with negative photoresist, and EGMEA is the diluent for positive photoresist. HMDS must be mixed with dry solvent in a container purged with dry nitrogen. Exposure to atmospheric moisture begins a degradation of the resist. Some contaminants, especially chloride, can precipitate upon mixing with a solvent, and, even though filtration of the particles is possible, there is no guarantee that more particles will not be generated with time.

Vapor prime has become very popular in the past few years. The advantage of combining the dehydration and priming operations is obvious, but other advantages may be noted. In a vacuum vapor prime batch system, wafers are dehydrated and primed with much less HMDS. Vapor prime uses undiluted HMDS and is therefore compatible with both negative and positive resists. Also, if HMDS is introduced into an evacuated chamber, particles will not be carried to the wafers but will fall to the bottom of the chamber. Of course, proper maintenance cleaning is necessary, as is filtration of all gases feeding into the system. The condition of the reservoir must be checked in some systems, and special care must be taken to stay within shelf life limitations of the HMDS, especially due to the small quantity being used. Smaller containers, quarts or pints rather than gallons, are recommended.

Spin

Because the integrity of the resist is at stake, contamination is a major concern during the coating process. The purity of the resist, the patterns of air flow, the balance of exhaust and laminar flow air, as well as wafer transport and handling techniques are factors in the cleanliness equation. Air flow patterns are also important for resist film formation, both in terms of uniformity and solvent removal. The wafer is directly in the path of air currents, because solvents must be exhausted and excess resist must be drained. The configuration of the catch bowl is important in directing excess resist away from the spinning wafer, and exhaust provides an air velocity at the wafer edge, between 150 and 250 CFM for best results, to prevent splash-back or cobwebbing of the resist. This flow of air may create a turbulence that pulls contamination across the spinning wafer; therefore, special care must be taken to prevent particulates in the area of the spinner.

The dispense system is another potential problem area. In older systems, the resist is pumped from bottles under the spinner. Several problems may be seen with this approach. The pump bellows may add contamination to the resist, resist may be contaminated by the dip tube that is put into the bottle, and it is difficult to balance the forward and backward stroke of the pump to provide dispense and suck-back (to avoid unwanted dribbles after dispense) with some filters in place. Solutions to these problems include a new line of pumps that can dispense and suck-back with even disposable filter cartridges in place, and new filters that provide large surface areas for filtration within a rigid structure and small package for point of use filtration on older pumps. To address dip tube contamination, a new collapsible pouch with short stainless steel fittings and bottom dispense has been introduced to the market.

It is probable that contamination during the coating operation will be an area of increased concern as VLSI and VHSIC technologies move into Class 10 clean rooms. Wafer transport mechanisms which are acceptable with Class 100 processing may be sources of major contamination under more stringent specifications. Spin coating technologies may also become more difficult as wafer diameters increase to 6 and 8 inches and uniformity requirements are more severe.

Bake

Curing or soft baking the freshly coated photoresist seems to be a simple task, with a combination of time and temperature to achieve removal of residual solvent and densification of the resist. Typical specifications on resist manufacturers' processing instructions suggest $90^{\circ}C$ to $95^{\circ}C$ for 30 minutes. However, in high volume production, where time is money, techniques have evolved to cut the baking time to 4 minutes or less. Track bakes using infrared radiation from the bottom, top or a combination of both, were able to provide effective bake in 3 to 4 minutes. Microwave energy, which heated the wafer through the transparent resist, cut the time to a few seconds. Current systems often use a conduction hot plate bake, effective in 45 to 90 seconds. In most of the in-line bake systems, air flow is carefully controlled and exhausted. Installation of filters is a simple precaution, but balancing the filtered purge air or nitrogen with exhaust may take more effort. Transport mechanisms are still a concern, as all moving parts may generate contamination. A larger concern may be the possibility of outgassing or sublimation of resist components, which must then be cleaned from ovens or track covers on a regular basis to prevent particulate fall-out. The edges of the wafers are also likely to be a source of particles, especially with positive resist which is brittle when dry. Regular cleaning of receiving cassettes and transport containers is recommended to avoid such particulate contamination.

Exposure

During the alignment and exposure steps, two concerns are superimposed as both wafer and mask surfaces are sensitive to particulate contamination. The development of noncontact imaging systems was critical for reducing defect density caused by wafer to mask contact. Even so, particulate contamination is still a concern for all of the advanced imaging systems, albeit in differing ways.

Mask cleaning, storage and handling techniques are very important because mask contamination can affect the yield of a large number of wafers. The investment in a mask is significant, because masks can now be fabricated with high precision and virtually defect free. Maintaining a mask in a defect free condition is difficult, even though cleaning systems and techniques are continually improved. Covering the defect free mask with a pellicle, a thin optically clear membrane suspended outside the focal plane of the mask, offers the best assurance that subsequent contamination will at least be out of focus. The most contamination sensitive optical system is the 1:1 stepper. A small area is imaged repeatedly in an array over the wafer at no change in image magnification. Any contamination on the focal plane of the mask will be imaged on the wafer at each step, maximizing the impact of contamination. The reduction stepper, on the other hand, reduces the size of a particle by a factor of 5 or 10, and the scanning projection printer does not multiply the defect. Pellicles were designed specifically for the 1:1 stepper, but their adaptation to other systems has reduced contamination caused yield loss.

Wafers, too, are subject to contamination from transport systems or the carriers used to move them from the end of the bake oven to the align and expose station. Wafer handlers can be a source of particulates as wafers are loaded into the imaging system. Because the align and expose operation is one of the most labor intensive steps, personnel training is essential. More attention is being focused on controlling contamination within the optical systems, especially as it becomes clear that it is easier to sweep particles from smaller internal areas than to clean entire work areas.

Develop Cycle

Resist may be developed in bath or in-line track systems. Both have advantages and disadvantages. For large numbers of wafers, bath develop may be able to process more wafers per unit volume, and bath temperature may be controlled for repeatability. To prevent contamination, the bath should be set up to overflow and recirculate through a filter. This prevents the accumulation of contamination on the surface of the liquid, and, for positive resist, the filter also removes resin residues that discolor the solution or drift to the bottom of the bath. Without such recirculation and filtration, the bath quality degrades rapidly, and contamination may be transferred to wafers as they are pulled through the surface meniscus. With positive resist, the water rinse is usually overflowing, helping to remove any particulates that might have been picked up in the developer bath. Bath develop for negative resist too often consists of standing baths, and contamination from both the develop and rinse may be a problem. Compounding the problem is the fact that bath develop may take place in an exhausted hood. Care must be taken to balance such hoods to prevent pulling turbulent, contaminated air past personnel and across wet wafer surfaces.

Spray development is often favored because each wafer is processed identically and with fresh developer. For negative resists, spray development was relatively easy to set up, and a wafer could be processed in less than 30 seconds. The most uniform development was possible with a low pressure, course spray nozzle and a slow spin speed. Deviation from this optimum was not catastrophic, and good results could be achieved. With positive resist, develop times were usually longer, and proper results would be seen with too much nozzle pressure or too fast a spin speed. If the developer formed a mist, the water could evaporate leaving particulate deposits on the inside of the developer bowl and cover. Air flow and exhaust balancing are also important here to minimize contamination potential.

Inspection

The post-develop inspection has always been a process control check point. Prior to etch, implant, deposition, etc., wafers can be reworked if the imaging process is not acceptable. If the process is not well controlled, 100% inspection is needed. However, when processes are optimized and microprocessor controlled, this inspection may focus on a much smaller sample size, checking that the correct mask was used and measuring critical dimensions. Once a controlled process is established, over-inspection may be a liability, causing delays, additional handling and inevitable contamination. Automatic inspection stations offer wafer transport without operator handling, and small details, such as the microscope face shield, may help minimize contamination and yield loss at inspection.

Automatic wafer inspection is more difficult than automatic mask inspection, because images are not just clear and opaque, and require that reflected light rather than transmitted light be used, but progress is being made. Inspection for defects caused by mask contamination is often done using a quartz wafer which permits automatic pattern comparison with transmitted light and gives an accurate representation of the quality of the pattern transfer. Once a mask or reticle is in place, it can be checked for contamination before the lot of wafers is imaged. The end result of new wafer inspection strategies is a higher confidence level with reduced handling and fewer delays.

Etch

Wet etches, because they are standing baths, are potential contamination traps. Most etchants, even when filtered to standard submicron specifications prior to shipment, arrive in the lab with high levels of particulates. The use of an overflow bath, recirculating through a filter, leads to cleaner etch solutions and less potential for contamination. A recirculating system is also easily temperature controlled for higher etch precision. Pumps, filters and overflow baths are available that are compatible with most liquid etchants.

Plasma reactive ion etch, and other dry etching systems are most sensitive to contamination in terms of imaging the contamination through the transferred image. A particle, in an anisotropic etch system, may totally protect the underlying substrate from etching, creating a spike in the middle of an etched area. This problem increases as etching processes become more selective and anisotropic. Any contamination that retards etching will cause a problem in meeting critical dimension tolerances. All gases going into the plasma etch chamber must be filtered to 0.2 micron. The plasma reactions must result in reactants that are totally volatile, minimizing the possibility of redeposition, and reducing maintenance and cleaning requirements. Vents that backfill the chamber with ambient air must not be forgotten as filters are fitted. Traps to prevent contamination from vacuum lines in case of vacuum failure during processing are also recommended.

Resist Mask

When a resist mask is used for implant pattern definition, its integrity is important. Particles or contaminants that might react with the implant beam must not be in the resist or in the developed areas, and the integrity of the resist must not be compromised by particles. The energy of the ion beam causes the resist to react, outgas, and contaminate the system. Outgassing can be minimized by vacuum hardbake of the resist prior to implanation, but the nature of the organic material in the implanter is such that some outgassing must be expected, and regular cleaning is necessary.

Resist images used for lift-off processes must have steep slopes and cleanly developed areas for the metal deposition to adhere. The largest contamination issue in lift-off involves the material that is lifted. In a bath where resist is soaked to cause it to dissolve and the metal to be lifted, the unwanted fragments of metal float off the wafer and remain intact in the solvent. These fragments are huge in comparison with the submicron particles addressed in normal filtration, but they cannot be allowed to deposit themselves randomly on the remaining metal pattern. If a bath is used for lift-off, an aggressive recirculated overflow system with filters is necessary. Spray removal of the pattern to be lifted may be possible, but effective lifting often requires soaking to permit the solvent to penetrate beneath the metal covered resist.

If resist is used for a multilayer process, the integrity of the film is important; therefore, it must be particle free. For the multilayer process using deep UV exposure of PMMA through a positive resist mask, better results are obtained by spray developing the PMMA than by bath development techniques, and the resulting image is crisp and clean.

Resist Removal

Resist removal procedures must be defined around the resist to be removed and the underlying substrate limitations. Organic resists are readily removed in the active "piranha" bath, but this is proscribed if metal surfaces are exposed. Positive resists are easily removed with solutions of organic bases, such as the aqueous amines, unless they have been acylated by implantation or high temperature bakes. Even for these tough resists, new formulations have been developed. Negative resists can be removed with acid cleaning if no metal is exposed, but sulfonic acid and phenol solutions have traditionally been used if metal surfaces are present. Again, new chemical formulations have been developed to reduce the problem of disposing of phenols.

Plasma ashing may be effective for resist removal in most situations, but if dyes or other inorganic additives are present in the resist, a standard plasma process may leave a residue. Proper tuning of the plasma system is essential, or organic resist residues may remain at the end of the process. Inorganic residues may require special plasma processing or wet chemical rinses.

SUMMARY AND CONCLUSIONS

The attention that microlithography is receiving in terms of contamination control is well justified. The imaging process is repeated many times in the process of manufacturing a circuit, and even a small reduction in defect density can boost yield several percentage points. Lithography is also more labor intensive than most other steps, and much can be done to train personnel in handling procedures consistent with high resolution and high yield. The following summary lists some of the areas often overlooked in an effective contamination control program, and may serve as a quick check list.

Filtration

- Filter all processing chemicals.
- Use filtered recirculating baths.
- Filter all gases, including ambient air for backfill.
- Keep laminar flow laminar and get full benefits from HEPA filters.

Equipment Maintenance

- Give careful consideration to cleaning procedures.
- Use filtered solvents and lint-free wipes.
- Verify solvent compatibility.
- Be aware of air flow and exhaust, and establish balance.

Personnel Training

- Work with all levels of personnel to develop an understanding of the effects of contamination of the work being done. Most people can be motivated to do a good job and obey rules if they understand why such rules are important.
- Teach proper handling techniques from a contamination point of view.
- Demonstrate the causes and effects of turbulence.
- Involve personnel in developing strategies to control contaminants that affect them, especially things like make-up.
- Develop an awareness of breath-borne contamination potential of smokers, and institute a practice of rinsing mouth (drinking water) prior to returning to the clean room if masks are not required.

Dirt Traps

Be especially aware of the following dirt traps:

- Containers,
- Cassettes,
- Transport systems,
- Standing liquids.

Contamination control is not just a casual addition to imaging technology. A full program to control contamination sources and remove potential contaminants from surfaces, air and liquids is essential to improve yields and be economically competitive in the semiconductor marketplace. Improvements in technology may offer cleaner working environments, to Class 10 or even Class 1, and cleaner chemicals with which to process wafers, but the biggest factor in any contamination control program is people. Even if robotics and special modules for separating people from the contamination-sensitive wafers are introduced into front-end wafer processing, personnel will continue to make the difference. Only when people are effectively trained to be aware of the issues involved in contamination control, can such programs work.

11

Contamination Control in Microelectronic Chemicals

Mike Naggar

DEFINING CONTAMINATION

In order to better understand this chapter about contamination control for semiconductor chemicals, it is critical to adequately define the term "contamination." Contamination as it will be discussed in this chapter could be plainly defined as any undesirable substance that is introduced to a silicon wafer during processing by a chemical used in the fab area. This contamination could either cause immediate die yield loss or eventual device reliability failures. This chapter will address two specific types of contamination that have proved to be harmful to semiconductor devices. These two contaminants are particle and metal ion contamination. Millions of dollars are spent by semiconductor houses to build new state-of-the-art wafer fabs to eliminate these types of contamination. This book addresses various sources of contamination, specifically particles, and suggests ways of reducing it. This chapter will concentrate on semiconductor chemicals used in the photomasking process and their impact on wafer surface contamination.

Particle Contamination

Contamination could be introduced to the process via the environment, equipment, people, or contaminated material used to make devices. The most common materials are chemicals used in the photolithography process. Such chemicals include: photoresist, which is the backbone of the process; acids; solvents; and bases. Detection of large particle contamination from the process can be observed relatively easily during the various inspection steps throughout the process. In most cases, particle contamination only affects a few devices on the wafer and is not always a killer defect; although, with shrinkage of geometries, and, in order to maximize die yield, controlling and/or eliminating such contamination is a high priority for a fab area.

Metal Ion Contamination

Metal ion contamination is probably the least understood of all contamination. Its detection during the process is extremely difficult and its effect on the devices is not always immediate. Ionic contamination is generally viewed as a potential source for loss of device reliability. The most commonly known and harmful elements are: sodium, potassium, iron, chloride, lithium, etc.

MEASURING THE CONTAMINATION

Particle Detection Technology

Particle measuring techniques vary significantly, depending on the medium in question. Optical microscopy and scanning electron microscopy are standard methods for device inspection. Automatic laser surface scanning equipment is normally used for silicon wafer inspection (see Chapter 12). Liquid and airborne particle counters are acceptable means of evaluating the particle counts and size distribution in air, chemicals and gases. There are several liquid particle counters available today, with various principles of operation. Most particle counters utilize light scattering or light blockage techniques with white or laser light. All techniques have limited detection in the submicron range. The most acceptable and commonly used particle detectors for liquids at the 0.5 to 1.0 μm range is a laser based counter. The Hiac/Royco particle counter is one of these instruments. An important feature that is required in a particle counter is reproducibility. Accuracy is very difficult to assess in particle counting. This equipment has come a long way in the last several years, but many questions about the adequacy of such tests still exist. Questions that must be addressed and taken into consideration are detection limit, calibration, particle uniformity within a container, sampling plan, and correlation, etc. It is important when characterizing particles in a given solution to constantly use the same method of measurement, sample preparation, and cleaning of containers. It is also important to talk about particle concentrations, order of magnitude and in trends rather than counts. Correlation between two different pieces of equipment, even the same type of container, is extremely difficult. In this chapter particle count numbers will appear in tables and examples will be given of counts of various chemicals. All counts have been taken by a Hiac/Royco particle counter and do not necessarily indicate an exact count, but imply trends of that chemical.

Metal Ion Analysis

This contamination is more commonly carried by the chemicals used in the process. Measuring equipment includes: atomic absorption, emission, or inductive coupled plasma systems. Physical equipment, such as Auger or x-ray fluorescence, are used for surface analysis. Electrical testing such as CV or IV plots of surfaces usually give an indication that contamination exists. Measuring equipment for such contamination are more exact when compared to measuring particle contamination. Either flame or flameless atomic absorption spectroscopy is most common for chemical analysis. Inductive Coupled Plasma

(ICP) systems are becoming more common, due to speed and low interference. Emission spectroscopy, especially arc emission spectroscopy, lack sensitivity for today's needs. More sophisticated equipment, such as x-ray and NMR, are less widely utilized for routine analysis, due to their high cost and complexity. Specific ion chromatography is becoming more popular for very specific ionic detection with tremendous improvement in equipment reliability and cost.

Equipment Choices

For analysis at KTI Chemicals, Inc., we use the Hiac/Royco laser particle counter for particle measurement and size distribution, an atomic absorption flameless system for metal ion concentration detection in chemicals and, finally, an ion chromatograph for specific ion detection. There are various types of atomic absorption and ion chromatographs. It is important to choose the correct one for a specific application. Even though flame atomic absorption might be sufficient, it is suggested that flameless capability also be obtained for better detection. With ion chromatography, it is important to choose a system with a good computer and data reduction capability. The cost of this equipment could range between $10,000 and $25,000; particle counters are about $25,000; atomic absorption $15,000; and ion chromatograph $25,000.

Sizing up the Contamination

Metal ion concentration and particle size will dictate the degree of potential damage to the device. Other factors, such as location of the contamination, masking level and type of contamination, are equally critical. Metal ion concentration is measured in parts per million (ppm). In most cases less than one part per million is acceptable. Particles are measured in microns. Generally, two microns and above are easily detected and controlled. With smaller geometries, better detection and control is mandatory. For today's technology, particle sizes greater than 0.5 micron are capable of providing catastrophic failure of the device. At this level, detection is difficult and control is not easily achieved.

A current challenge facing quality control engineers, fab engineers, and chemical suppliers, is to find a direct correlation between contamination in chemicals and surface contamination on the wafer. Due to difficulties in wafer surface analysis and the complex nature of wafer processing, it is impossible to assess the exact impact of a certain level of contamination on yield and device reliability. The following pages contain tables which outline contamination levels of standard semiconductor grade chemicals (SEMI spec) and so-called high purity chemicals. Most of the data is based on a great deal of trial and error, evaluation, and various studies of chemical purity versus wafer surface purity. The importance of the data is to note trends. Metal ion contamination levels are standard in the industry and agreed upon by suppliers and users. Particle counts is a new field that chemical suppliers are now introducing. Correlation and verification of particle specifications is still a major issue. Equipment repeatability, compatibility, accuracy, and detection limits are still being resolved.

Table 1: Particle Content of Semiconductor Grade Chemicals Circa: 1980

Particle Size (microns)	. . . . . . .Particles per Liter			
	1	2	5	10
Sulfuric acid	6×10^7	4×10^7	6×10^6	1×10^5
Hydrogen peroxide	1×10^7	6×10^6	8×10^5	4×10^4
Hydrofluoric acid	1×10^6	5×10^5	1×10^5	2×10^3
Phosphoric acid	2×10^7	2×10^6	3×10^5	5×10^4
Buffered oxide etch (6:1)	4×10^5	3×10^4	5×10^4	2×10^3
Isopropyl alcohol	5×10^5	3×10^5	3×10^4	3×10^3
Xylenes	5×10^5	1×10^5	6×10^4	3×10^3
Trichloroethylene	5×10^5	2×10^5	3×10^4	1×10^3

Comment: In 1980, particle specifications in chemicals was not typical, and control of the particle concentration was very limited.

Table 2: Semiconductor Process Fluid Cleaniness vs Class 100 Air Cleanliness Source: Millipore

Process Fluid	Typical Relative Particle Levels*
DI water (central)	300
DI water (P.O.U.)	10,000
Process gases	30–30,000
Photoresist	300
NH_4OH	300,000
HCl	1,000,000
HF	2,000,000
H_2O_2	2,500,000
H_2O_4	13,000,000

*Class 100 Air–3.5 particles/liter.

Comment: Obviously, wide variations exist in these fluids so dollar investments in clear space and in process fluid filtration systems are also worlds apart. For example, $50 million may be spent on a new fab line to produce the perfect chip. However, the cost of protecting that chip from fluid generated particles will vary widely. Chapter 8 covers the system aspects of particle filtration systems.

Table 3: Low Particle Grade Chemicals

	. . . . .Particle Count per Liter		
	$\geqslant 0.5\,\mu$	$\geqslant 1.0\,\mu$	$\geqslant 5.0\,\mu$
Allied Particu-Lo	—	<10,000	—
J.T. Baker VLSI	—	<10,000	—
KTI Low Particle	<100,000	<10,000	<1,000
Ashland Clean Room L.P.	—	10,000	—
Mallinckrodt	Unknown	—	—

Comment: Published specification for most acids and solvents.

Table 4: 1985 ASTM Proposed Specifications Contaminant Classes*

. . . . .0.5 Micron–5.0 Micron (per liter). . . . .

(1) 10–0	(10) 100,000–5,000	(18) 1,000–250
(2) 100–0	(11) 10–1	(19) 10,000–2,500
(3) 1,000–0	(12) 100–0	(20) 100,000–25,000
(4) 10,000–0	(13) 1,000–100	(21) 10–5
(5) 100,000–0	(14) 10,000–1,000	(22) 100–50
(6) 10–0.5	(15) 100,000–10,000	(23) 1,000–500
(7) 100–5	(16) 10–2.5	(24) 10,000–5,000
(8) 1,000–50	(17) 100–25	(25) 100,000–50,000
(9) 10,000–500		

*In this table, each two-segment number represents one particle con-
tamination class. The number to the left of the dash indicates the
maximum concentration of particles equal to and larger than 0.5
micron per liter. The number to the right of the dash indicates the
maximum concentration of particles equal to and larger than 5
microns per liter.

Comment: Note that the largest number of particles per liter does not
exceed 100,000 for 0.5 micron sized particles.

Table 5: Metal Ion Specification for Selected Acids and Solvents (ppm)
1984 SEMI Spec

	Acetic Acid	Ammonium Fluoride	HF	Hydro-Chloric Acid	Nitric Acid	Acetone	n-Butyl Acetate	Xylene
Chloride	1.0	4.0	TPT	5.6	0.08	0.2	—	0.2
Nitrate	—	10.0	—	3.0	0.2	—	—	—
Phosphate	1.0	1.0	0.05	1.0	—	0.1	1.0	0.1
Sulfate	0.5	2.0	0.5	5.0	0.5	—	0.1	0.2
Heavy Metal (as Lead)	0.5	1.0	0.1	0.1	0.1	0.2	0.05	0.01
Arsenic (AS)	0.005	0.03	0.005	0.03	0.005	0.01	1.0	1.0
Aluminum (Al)	1.0	0.5	1.0	0.05	0.5	1.0	1.0	1.0
Barium (Ba)	1.0	1.0	1.0	0.5	1.0	1.0	0.1	0.2
Boron (B)	0.5	0.5	0.1	0.05	0.1	0.2	1.0	1.0
Cadmium (Cd)	1.0	0.5	1.0	1.0	0.5	1.0	1.0	1.0
Calcium (Ca)	.10	0.5	1.0	1.0	0.5	1.0	1.0	0.5
Chromium (Cr)	0.1	0.1	0.5	0.01	0.1	0.5	0.5	0.1
Cobalt (Co)	0.1	0.5	0.1	0.5	0.5	0.1	0.1	0.1
Copper (Cu)	0.1	0.1	0.1	0.05	0.05	0.1	0.1	0.5
Gallium (Ga)	0.5	0.5	0.5	0.05	0.1	0.5	0.5	1.0
Germanium (Ge)	1.0	0.5	1.0	1.0	0.1	1.0	1.0	0.5
Gold (Au)	0.5	0.1	0.5	0.5	0.5	0.5	0.5	0.1
Iron (Fe)	0.2	1.0	0.2	0.5	0.2	0.1	0.1	1.0
Lithium (Li)	1.0	1.0	1.0	1.0	1.0	1.0	1.0	1.0
Magnesium (Mg)	1.0	0.5	1.0	0.5	1.0	1.0	1.0	1.0
Manganese (Mn)	1.0	0.5	1.0	0.5	0.5	1.0	1.0	1.0
Nickel (Ni)	0.1	0.3	0.1	0.1	0.05	0.1	0.1	0.1
Potassium (K)	1.0	1.0	1.0	1.0	1.0	1.0	1.0	1.0
Silver (Ag)	0.5	0.5	0.1	0.1	0.5	0.5	0.5	0.5
Sodium (Na)	1.0	1.0	1.0	1.0	1.0	1.0	1.0	1.0

Comment: These specifications are not currently controversial. The maximum allowable
concentrations are expected to continue to be driven down as device sensitivity continues to increase.

CONTROLLING THE CONTAMINATION IN CHEMICALS

Materials Control as Supplied

Controlling metal ion and particle contamination in the material is the responsibility of the chemical suppliers. State-of-the-art chemical manufacturing and packaging today typically has not kept pace with semiconductor manufacturing. Still, some significant improvements have been made to allow production of the advanced grades of chemicals needed for today's technology.

Metal ion contamination is usually caused by the purity level of the raw material used during the manufacturing of the chemicals. Special care in the purification process of acids, solvents, or photoresist and the selection of the grade of raw material will dictate the end results. There are some specialized processes, though costly and time consuming, to purify the end products; special distillation and extraction techniques are commonly used. Usually, purification processes will result in an order of magnitude lower than concentration of selected elements, but the process causes a 2 to 3 times increase in the cost of the chemical.

Typically, the metal ion contamination is stable in the material and does not change with time as long as the material is stored in a compatible container. If the container is not totally compatible then the level of certain elements will rise as a result of leaching of chemicals with time. For example, if storing a strong acid or a strong base in a metal container, the level of some heavy metals such as lead, iron, and zinc will increase.

Filtration of Chemicals

For particle control, filtration technique, compatibility of the filter, and, of course, the cleanliness of the container, determine the particle distribution. Generally, undiluted chemicals with low viscosity are relatively easy to filter and achieve a low particle level with a high filter efficiency. High particle raw material with high viscosity will be more difficult to filter to a low particle level. Maximum efficiency of the filter is achieved by an adequate flow rate through that filter. The flow rate of the filter is measured by the amount of liquid passing through the filter as compared to the area and pressure drop of that filter. Typically, for cartridge filters and chemicals, a flush cycle is required prior to filling a container. The cycle depends on the area of the filter media. In most cases, a ten gallon flush is required for a ten inch long cartridge filter, twenty gallons for a twenty inch, thirty gallons for thirty inch, etc., is sufficient. Recirculating the chemical through the flushed filter, prior to filling, will achieve the same result. Recirculating time also corresponds to filter size. I suggest a ten minute circulation for a ten inch filter, twenty minutes for a twenty inch filter, and thirty minutes for a thirty inch filter, etc. Various chemicals and various types of filters will require fine tuning of the flush cycles. These cycles will require particle testing of the material with various flush and/or recirculation sequences to determine the best possible cycle.

Cleaning of Containers

Even though the technique of filtration of chemicals is critical to purity, filtering a chemical is one of the easiest steps in producing low particle chemicals. Achieving an ultra pure container and maintaining that purity level over the life of the chemical in that container are the more difficult steps in high purity chemical production. The most common containers used by the semiconductor industry are:

(1)   one gallon glass or poly bottles

(2)   pouches or plastic bags

(3)   five gallon cans or canisters

(4)   55 gallon drums

Other sizes and/or bulk shipments such as tanker car are not as common. In the case of bulk chemicals, low particle levels are difficult to achieve. Bulk chemicals require point of use filtration.

There are three basic types of container cleans that are commonly used by the chemical supplier. They are:

(1)   high pressure nitrogen and air blow

(2)   hot water rinse followed by hot nitrogen dry

(3)   compatible chemical clean and rinse

Each of these methods are effective for a particular particle size and container type. Due to the critical nature of this step, as a key in achieving low particle levels, each method is discussed in detail.

High Pressure Nitrogen and/or Air Blow: This method can be done manually with a filtered nitrogen or air gun or with automatic high volume dispense equipment. It involves inserting a metal rod inside the container and dispensing high pressure nitrogen from the open end of the rod while swirling the container around the rod or turning the rod inside the container for a specific period. A second recommendation is for the container to be inverted and a vacuum applied at the opening of the container to act as a suction for the loose particles. This method is effective for large size particles specifically free floating particles such as dust, fibers, insects, dandruff, etc. Conversely, it has been determined that this method is not effective for sub-micron particles or any particles that have a strong adherence to the surface of the container.

Hot DI Water and Hot Nitrogen Clean: This method involves a high pressure hot dionized water spray of the container while inverted, followed by a period of gravity draining. The container is dried by a filtered hot nitrogen stream. This method is effective for large range at particle sizes independent of the type of particle. Limitations of this method are that particles with strong adhesive forces to the inner surface of the container will not be removed. Second, the moisture level in the container will be relatively high. This method is recommended for water base chemicals where moisture levels will not be a factor and where the chemical will not eventually remove particles adhering to the surface of the container.

Compatible Chemical Clean: This method involves either pickling or spraying the container with a chemical that is compatible with the specific chemical that eventually fills the container. The pickling or spraying time will depend on the size of the container and the effectiveness of the chemical for particle removal. This method is the most effective but also the most expensive and time consuming.

The table below contains particle count and size distribution for isopropyl alcohol in various one gallon containers that obtained one of the three above-mentioned cleaning methods. Note that particles are stated as particles/ml.

Table 6: Particle Counts per ml of Isopropyl Alcohol (IPA)
Hiac/Royco Laser Counter (0.5 micron and larger)

	1 Gallon Glass Bottle			1 Gallon Plastic Bottle		
	≥0.5	≥1.0	≥5.0	≥0.5	≥1.0	≥5.0
No clean	887	64	4	1,774	178	15
High Pressure N_2	186	17	2	362	43	4
Hot DI water	59	4	0	87	9	0
Solvent clean	17	1	0	29	2	0

Source: KTI Chemicals, Inc.

An obvious conclusion drawn from Table 6 is that package cleaning, prior to filling, is absolutely mandatory. A second requirement is to ensure that, when chemical manufacturers initially receive these containers, they are as clean as possible. Clean containers require chemical manufacturers to work with the various container manufacturers to ensure that their suppliers take steps in their process to achieve minimum particle contamination. The following steps should be included:

(1) Capping at point of manufacture;

(2) Improving the environmental control in the manufacturing site, such as laminar flow hoods;

(3) Lining the interior of the containers; and

(4) Using one of the previously mentioned methods to clean the containers prior to shipping to the chemical manufacturer.

A brief description of the filtration techniques and container cleaning procedures to achieve a low particulate level in packaged chemicals has been discussed. The key to ultimate success is to keep the particle level constant with the life of the material. There are very few compatible materials for long-term packaging of semiconductor chemicals, Teflon and pure stainless steel are the most compatible for the majority of the products. These materials are extremely expensive and not always practical for packaging. Figure 1 shows with time the increase in particle level for selected critical chemicals in selected common containers.

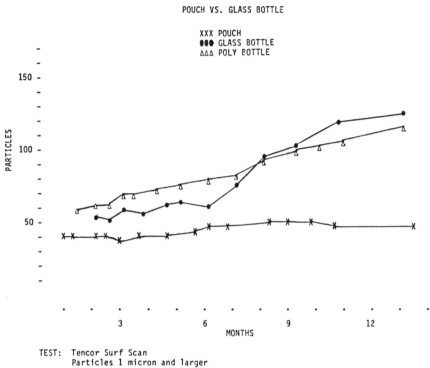

TEST: Tencor Surf Scan
       Particles 1 micron and larger
       Stored at Room Temperature

NOTE: <50 considered acceptable for fab operation

Figure 1: Photoresist shelf life in various one gallon containers.

Chemical Control in the Process

Contamination contributed to the silicon wafers by the chemicals during the process will most likely come during the following steps.

(1) General wafer cleaning, i.e., pre-diffusion clean, pre-metal clean, pre-epi clean, or pre-spin clean;

(2) Resist spin;

(3) Resist develop;

(4) Pattern etching; and

(5) Resist strip.

Assuming that chemical manufacturers supply low contamination materials with an acceptable quality level for processing, there are many places during the process itself where materials can be contaminated in use. In this section a discussion of the areas where contamination can occur will be covered and ideas on how to minimize or eliminate such contribution will be mentioned.

Listed below is a laundry list of some of the possible causes for contamination to the chemical by the end users:

(1) Inadequate storage conditions

(2) Using relatively old chemicals or materials

(3) Incompatible plumbing material and equipment

(4) Poor housekeeping procedures, such as equipment cleaning, changing lines, preventive maintenance, etc.

(5) Insufficient process control and monitoring

(6) Excessive use of one batch of a chemical.

Details of the critical points in each of these items are addressed below.

Storage and Shelf Life

As a minimum, the end user should clearly be aware of the supplier recommendation for storage conditions and shelf life. Each chemical should be clearly marked with either a maximum shelf life or an expiration date with each batch. A temperature range should be specified on each label requiring the end user to provide a safe storage environment. Photoresist as a polymer is by far the most critical chemical requiring specified storage conditions and shelf life. Care must be taken in storage of photoresist to ensure zero gel reformation. Photoresist should also be used as fresh as possible to guarantee the highest purity level. Listed in Table 7 are suggested storage conditions, optimum and maximum shelf life for the various selected packaged chemicals required to ensure the highest quality and lowest contamination level.

Table 7

Chemical	Optimum Shelf Life	Maximum Shelf Life	Storage Temperature Safe Range ($^\circ$F)
Positive photoresist	6 weeks	6 months	40-60
Negative photoresist	3 months	1 year	55-70
Positive developers	6 months	1 year	60-80
HMDS	2 months	6 months	50-80
Acids	3 months	1 year	65-80
Solvents	6 months	1 year	50-70

Generally, the chemical supplier could easily reprocess most material for refreshment and extension of usage life. Although it is important to note that this type of refreshment can only be done once or twice for a particular batch of chemicals after which that particular batch is considered unusable. I should also note that the shelf life or usable life after each refreshment is not as long as the original shelf life. That is true at least in the case of photoresist.

Processing Compatibility of the Equipment

Compatibility with equipment plumbing material is essential. A chemical compatibility chart must be referred to when selecting these materials. The cleaning of this equipment and routine changing of transfer tubing is required to ensure their cleanliness. The cleaning itself must be done with compatible materials. End users should confer with chemical suppliers to identify compatible chemicals with the material in use. Resist spinners are common contributors to contamination in photoresist. The spinner is the most demanding piece of equipment in the photomasking for routine cleaning and maintenance. Spinner cups are cleaned once a day, with line flushed once a week. It is common for spinners to be cleaned without using the most compatible solvent. Acetone is often used to clean positive resist spinners. It is clear that the carrier solvent for positive photoresist is Cellosolve acetate. It should be obvious that it would be the most compatible for cleaning the resist. To say that acetone is not totally compatible with positive resist is not intended, it is simply not the solvent of choice for safest results. For resist spinners, it is common for dispense tubings not to be changed, but simply flushed once in a while. Due to the polymeric nature of photoresist, routine flushing of lines is not sufficient. It is crucial for spinner dispense tubing to be changed on a quarterly basis. A singular focus on the spin process or spinners in general is not intended, the spinner is, however, an excellent example where incompatibility, insufficient maintenance, and general housekeeping can be a major contributor of contamination to the process.

Spraying and Immersion Processes

During the photolithography process, developing chemicals are applied to the wafer, either by a vaporizing spray or liquid immersion. Safeguards can be clearly added when using spray or immersion processes.

Spray Processes: It should be obvious that, when a chemical is sprayed onto a wafer, point of use filtration as close as possible to the spray nozzle would be an effective safeguard against the introduction of particle contamination to the wafer by the chemical. This is true whether the contamination is related to the chemical or to the plumbing. Due to the pulsating action of most pumps used to dispense chemicals and photoresist, some filters will actually cause more damage than if not used at all. The author suggests that, in line, one or two micron rated depth filters should be sufficient to eliminate gross contamination caused by the chemical or the lines, while minimizing any chance of filter breakdown as a result of the pulsating action of the pump.

Immersion Processes: Similar to the case of spraying, point of use filtration is very effective when using an immersion develop process. Immersion processing will be more effective as a result of the recirculation which filtration baths provide. Since recirculation minimizes the effect of pulsating action of a pump, the use of small pore-size filters, such as 0.2 micron, are very effective. Also, the particle reduction factor discussed earlier is better as a result of recirculation. During the immersion process, contamination is introduced into the bath. A period of recirculation and filtration between dips is required to sustain a low particle and contamination level throughout the life of the bath.

In both spray and immersion processes, fluid filters should be routinely changed. A pressure drop check, if available, should be sufficient to check the status of the filter. Normally, filter manufacturers will suggest such a test.

Process Monitoring

The subjects which are discussed here require a great deal of process monitoring by the process engineers. Basically, the author suggests monitoring and controlling the particle level of most chemicals in the process and giving them the same type of attention and priority that aerosol particles are routinely given. New wafer fabrication areas are often certified as being a Class 10 operation with older fabrication areas being rated as Class 100. This classification basically indicates the number of particles in one cubic foot of air. Routine monitoring of particle counts in the air is performed on a daily basis in each fab area. What is needed and suggested here is a classification for chemicals for each process with routine daily monitors in the same manner that aerosol particle counting is performed. The classification should be determined upon running correlation tests between liquid particle counts and the wafer surface particle counts for each step and each chemical used in the process. By knowing the chemical particle classification for each step and whether or not the step includes point of use filtration, the efficiency factor of that filtration can be determined and the maximum allowed particle level for the chemical can be documented. Table 8 contains some suggestions on classification. This information is based solely on in-house studies at KTI Chemicals, Inc.

Table 8: Liquid Particle Classification for Photoresist Processing

Chemical	Method of Counting	Particle Size (μ)	Working Clean Chemical Classification	Point of Use Filtration (μ)	Recirculation Time if Applicable (min)	Efficiency Factor	Incoming Chemical Particle Specification
Photo-resists	Surfscan	≥ 1.	Class 8	Yes, (1.0)	N/A	2	$<1.26*$
Etchants	Liquid counter	≥ 0.5	Class 10	Yes, (0.2)	10	1,000	$<16**$
Primers (spray)	Liquid counter	≥ 0.5	Class 10	Yes, (0.2)	N/A	1,000	$<100,00/\ell$
Developers (immersion)	Liquid counter	≥ 0.5	Class 10	Yes, (0.2	10	1,000	$<100,000/\ell$
Developers (spray)	Liquid counter	≥ 0.5	Class 20	Yes, (0.2)	N/A	500	$<100,000/\ell$

*Particles per centimeters.
**Per diameter inch of wafer.

CONCLUSION

Processing technique and attention to detail are the keys to successfully controlling contamination in lithographic materials and processes. Contamination control for VLSI microlithography is achievable, to a large degree, if a total committment exists in the factory. Equipment and material suppliers must join forces with the semiconductor manufacturers to win the particle war. Lack of communication will inevitably result in duplication of efforts, misunderstanding of needs, and confusion about specifications. Process control, correlation studies and repeatability of testing are essential in understanding real requirements. The Japanese electronics industry is ahead of the U.S. in contamination control because they learned long ago to work as a team and share in the burden of defining needs and upgrading capabilities.

Obviously, contamination control will become a key in the continued progress in VLSI and ULSI technologies. A great deal of attention must be placed on this issue which means that we must continue to invest the manpower and capital to achieve our ultimate goals.

12

Surface Particle Detection Technology

Peter Gise

INTRODUCTION

Effects of Particles on Circuit Yield

The majority of defects which arise during semiconductor wafer processing or photomask manufacture are the result of particulate contamination. This foreign matter on the surface of wafers and photomasks adversely affects the yield of the process.

Die yield, for example, is a function of defect density, chip area and the number of critical masking steps. These three variables are sometimes related by the following equation first proposed by J.E. Price of HP Laboratories based on Bose-Einstein statistics:[1]

$$Y = \frac{1}{(1 + A_e \overline{D})^n}$$

where:

Y = yield

A_e = effective (or critical) chip area

$\overline{D}$ = effective defect density

n = number of critical mask steps

As circuits become more complicated, the chip area becomes larger. Even though the circuit geometries are shrinking, the circuit complexity and consequently the chip size continues to increase more rapidly. Both factors work to decrease wafer yields. The process itself is also, simultaneously becoming more complex, requiring more layers and thus more critical masking steps. Greater numbers of critical masking steps also cause yield to decrease. One factor then, over which a process engineer has control, is defect density. For example, a process with six critical masking steps and a chip area of 0.5 cm^2 will have a random yield of 75% if there are 0.1 defect per cm^2, but only 56% if the defect density increases to 0.2 defect per cm^2. Yield is very sensitive to defect density, so there is a very strong motivation on the part of the process engineer to "clean up the line."

The critical size of the particles which degrade the circuit yields is considered to be 1/10 of the circuit geometry. The pattern size for a typical VLSI memory circuit can range from 2.5 microns for a 64K bit memory down to 0.6 micron for modern 4M bit memories. This means that the defect size which degrades circuit yield is 0.25 micron for a 64K bit memory down to 0.06 micron for a 4M bit memory. These fine particles are beyond the range of unaided visual inspection, so other methods such as automatic wafer and mask inspection systems are now required for particle and defect inspection.

Historical Perspective

For years, the mainstay of wafer inspection has been the high intensity light source and the (unaided) human eye. Lack of proper equipment has led manufacturers in the past to use everything from microscope lights to the innards of Prado® and Kodak® slide projectors mounted in crudely made inspection booths painted flat black and draped with a black cloth.

The first commercial attempt at automation was done in 1980 by Captec. The system consisted of a high intensity tungsten light source which illuminated the wafer as it moved on a mylar conveyor belt under a lens. Scattered light from a defect or particle was collected and focused by a lens onto a photodiode detector. The total number of defects on the wafer was accumulated and compared to a preset counter for pass/fail inspection. The method never equalled the sensitivity or speed of a human inspector and was thus never viewed as a viable inspection technique.

Of greater significance than the Captec system is the still-camera technique pioneered by IBM. This system consists of a vacuum chuck which holds the wafer in front of a high intensity tungsten light. The scattered light from particles and defects is collected by a Polaroid® camera (time exposure) for later manual counting and analysis. Both the number of spots (defects) and their intensity (size) can be determined from the photographs. Since the films used typically have an ASA rating of 3000, and since a time-exposure is used to accumulate the light (as in astrophotography), the technique is quite sensitive. Defects in the one micron size range have been detected. This method was used extensively between 1980 and 1983 at which time the Polaroid® camera was replaced by the lens-vidicon (TV camera) collector-detector. The photograph then became the CRT image which was processed for defect number and defect size using image processing techniques.

A completely different approach to the problem was developed by RCA in Zurich using a scanning laser, collecting lens and photodiode detector. Scanning laser systems have been in use since the early 1970s at Bell Labs,[2] but the RCA system was the first such unit to be produced in significant quantities.

INSPECTION METHODS FOR PARTICLES ON PLANAR SURFACES

Wafer Surface Inspection Systems

Based upon the type of illumination used, wafer and photoblank surface inspection systems fall into two categories: those systems which use high intensity collimated light and those which use laser light.

High intensity collimated light inspection systems arose from the early methods of inspection which used slide projectors to illuminate the planar surface. The inspector views the wafer at an angle other than the angle of reflection of the incident beam so as to avoid the dazzling specular beam. In this method, the defects will scatter a very small portion of the incident beam and appear to the observer as bright spots on a dark background (hence the term dark-field inspection).

A simple dark-field inspection system is shown in Figure 1. The system consists of a light source and a detector. The light source can either be a high-intensity collimated light source or a laser (typically helium-neon). The light source illuminates the surface of the substrate which can either be a polished silicon or gallium arsenide wafer or a chrome-coated photoblank. The specular beam is reflected at an angle equal to the angle of incidence and is directed out of the field of view of the detector. If, however, there is a defect or particle on the surface of the substrate, a very small portion of the incident beam will be scattered. If the detector is placed at the proper location, the scattered light can be detected.

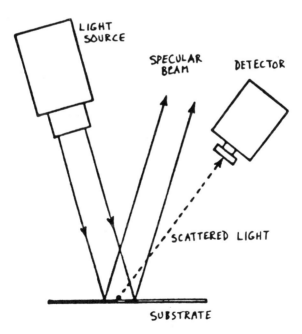

Figure 1: A simple dark-field inspection system.

Light Scattering from Surface Defects

Since surface defects are classified by the measurement of scattered light, it is important to understand the various categories of surface irregularities. The size and spacing of defects on the surface of the substrate relative to the wavelength of the illumination determines the nature of the light scattering which occurs. The three groups which result are: scratches and specular defects which are large relative to the wavelength of the illumination; isolated particles which are comparable to or smaller in size than the wavelength of the illumination; and surface irregularities or fine particles which are much smaller in size than the wavelength of the illumination but which cover large areas of the surface. This latter category is sometimes termed haze. Figure 2 illustrates these categories of defects.

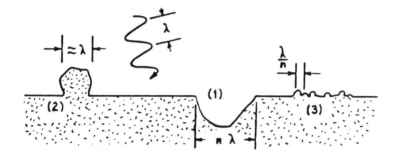

Figure 2: Categories of light-scattering defects.

Defects which are large relative to the wavelength of the illumination are scratches, cracks, epi spikes, slip lines and stacking faults. These defects tend to scatter preferentially and require a light collector with a large collection angle for proper detection. On the other hand, organic particles and silicon "dust" are on the order of the wavelength of the illumination in size, scatter more uniformly than the preferential scatterers and thus may be detected from a multitude of directions. To properly size these defects a large collection angle is required in order to sum up all of the scattered light. Figure 3 illustrates how light is scattered from such preferential scatterers. There may be fine particles, stains or a general surface roughness on the surface of the wafer. These areas will also scatter a portion of the incoming illumination. Such scatter is treated statistically using Kirchhoffs[3] diffraction integral and the condition is termed *haze*. The dependence of Total Integrated Scatter on wavelengths is given by:

$$\frac{\Delta R}{Ro} = 1 - e^{-(4\pi\delta/\lambda)^2} \cong (4\pi\delta/\lambda)^2$$

where:

Ro = the fraction of incident light reflected at all angles, or the total reflected light

ΔR = the fraction of incident light that is diffusely scattered by the haze

δ = RMS height of the surface irregularities

λ = wavelength of the illumination

As an example, if diffuse scattering (Average Haze) equals 1000 parts per million and the fraction of incident light reflected at all angles is known to equal 0.33 (a typical value for polished silicon), then the RMS roughness is 27 angstroms (when measured with a wavelength of 6328 angstroms). For other types of substrates the sample reflectivity must be determined.

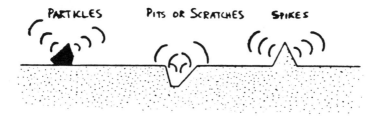

Figure 3: Common sources of preferential scatter.

Basic Light Scattering Processes

Referring again to Figure 1, the vital question becomes how much light is actually scattered by the defect shown on the surface of the substrate. The amount of light scattered has a direct bearing on how "big" a localized defect or particle actually will appear to the detector in the system.

Consider next that the substrate is perfectly reflecting ($R = 1$) and that it is illuminated with plane waves of constant intensity I. Next assume that a particle with an optical cross-sectional area, A, is placed on this surface. Then the amount of light scattered is just I times A. If we divide the scattered light by the illumination intensity, then the result is a scattering cross-section with units of area. (The small size of the detectable defect makes square micrometers an appropriate unit.) Just how small this percentage of scattered light really is can be calculated for a 150 mm wafer. Suppose this wafer is uniformly illuminated and a 1 micron particle is placed on the surface. The surface area of the wafer is 1.77×10^{10} square micrometers in an area while a 1 micron particle has a scattering cross-section of about 0.9 square micron. This means that the particle only scatters 5.09×10^{-11} times the incident illumination.

Now real surfaces will absorb light as will the defects on the surface. Thus the optical cross-section will be larger or smaller than the geometrical cross-section. Nevertheless, it is still possible to determine the size of the particle from the scattering cross-section if we remember that it is *not* the same as the geometrical cross-section.

Light Scattering by Latex Spheres

Light scattering physics is extremely complex in nature. The very simplest problem of a single sphere, suspended in space and illuminated by plane waves was first solved by Gustav Mie in 1908. The solution, known as the Mie theory of scattering, is a series of summations of mathematical functions called spherical harmonics and is well beyond the scope of this discussion. The interpretation of the results, however, is a relatively straightforward process.

Particles such as latex spheres remove (scatter) light from the beam of incident illumination by a combination of the processes of reflection, refraction, absorption and diffraction. Figure 4 is an illustration of the relationship

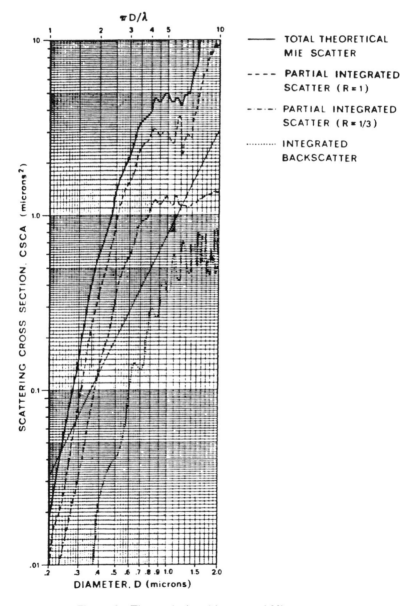

Figure 4: Theoretical and integrated Mie scatter.

between the total scattering cross-section for a spherical particle and the diameter of the particle. The straight line is the geometrical cross-section or simply $\pi D^2/4$ for a sphere. A spherical particle with a diameter much larger than the wavelength of illumination can be modeled using geometric optics. The scattering cross-section is twice the equivalent cross-section area ($\pi D^2/2$). One unit of cross-sectional area is removed by the sum of reflection, refraction and absorption while one unit of cross-section area is removed by diffraction. Geometric optics can be used to calculate the magnitude and direction of the scattered light.

When the diameter of the particle approaches the wavelength of the illumination, ray tracing can no longer be used and Mie Scattering Theory must be applied. In this region, the contribution of diffraction and resonance effects must be included. The net effect is that the scattering cross-section varies from slightly less than twice to considerably more than twice the geometrical value.

As the diameter of the particle is reduced to even smaller size, a final peak occurs in the scattering cross-section at a point slightly more than the wavelength of the illumination (approximately 0.8 micron diameter). Below this point, the cross-section drops rapidly approaching a sixth power of diameter dependence (Rayleigh scattering region) toward zero diameter.

The above discussion applies to particles in free space, but instruments such as the laser scanners are used to detect particles which lie on a wafer surface. The effects of the reflectivity of the substrate upon scattering cross-section must be taken into account. In addition, the geometry of the light collector must also be included. Figure 5 illustrates the scattering pattern from a spherical particle, illuminated by plane waves. The only light that would reach a scattered light detector would be the backscattered component at an angle of ±45 degrees from the incoming beam. Next consider the condition of placing the spherical particle on a perfectly reflecting surface (R = 1) as shown in Figure 6. For a substrate with a reflectivity R, one would expect that the forwardscattered component would be turned back toward the detector with an intensity which is reduced by R. Thus the total detector signal would be the backscatter component plus R times the forwardscatter component. (Note that the particle substrate interactions have been ignored by this simple model.) For particle diameters larger than 0.3 micron, the backscattered component is very small, so the simplified model predicts that for such particles, a laser scanner will indicate a scattering cross-section of R times the free space value. Figure 4 does in fact verify this model.

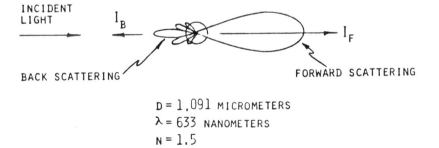

$D = 1.091$ MICROMETERS
$\lambda = 633$ NANOMETERS
$N = 1.5$

Figure 5: Scattering from a spherical particle in space.

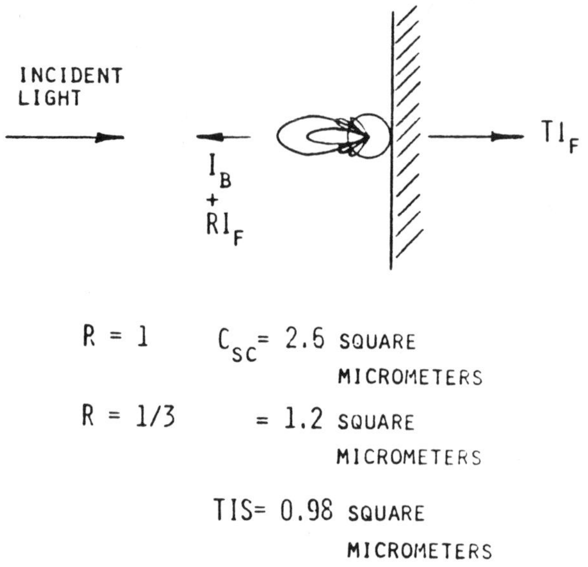

$$R = 1 \quad C_{sc} = 2.6 \text{ SQUARE MICROMETERS}$$

$$R = 1/3 \quad = 1.2 \text{ SQUARE MICROMETERS}$$

$$TIS = 0.98 \text{ SQUARE MICROMETERS}$$

Figure 6: Scattering from a spherical particle on a reflecting surface.

For a 1.091 microns diameter latex sphere on silicon, the predicted scattering cross-section for the collector geometry is about 1.2 square microns while the actual measured scattering cross-section is closer to 0.9 square micron. It is quite clear, then, that the substrate reflectivity must be taken into account. In general, particle sizes would appear to be smaller on "blue" substrates such as certain nitride layers and larger on metallized wafers or chrome masks than on bare silicon.

Table 1 indicates the measured scattering cross-section for various latex spheres deposited on bare silicon and silicon dioxide. Note that for spheres deposited on silicon dioxide, the scattering cross-section is reduced to approximately two-thirds the value for bare silicon. This is expected, as the thickness of the oxide is such that it is an antireflective, 1/4 wavelength coating at the helium-neon laser wavelength (6328 angstroms).

Table 1: Latex Sphere Scattering Cross-Sections

Diameter(μ)	$Si(\mu^2)$	$1085A - SiO_2(\mu^2)$
0.269	0.10	0.07
0.364	0.25	0.16
0.500	0.61	0.37
0.895	0.75	0.48
1.091	0.09	0.64
2.020	2.19	1.18
5.000	7.12	4.64

High Intensity Collimated Light

An inspection system for semiconductor wafers utilizing high intensity colli-mated light and a television camera type detector is illustrated in Figure 7. The television camera contains a high sensitivity photoelectric SIT vidicon tube as the detector. The wafer is illuminated by a high intensity mercury or xenon arc lamp in the range of 200 to 320 foot-candles.

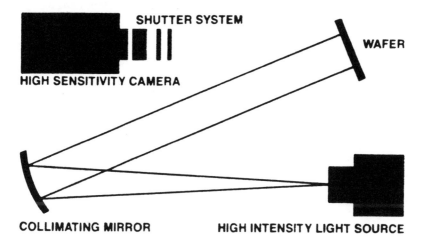

Figure 7: Typical high-intensity collimated light-inspection system.

The light is first reflected by a collimating mirror which directs the beam perpendicular to the plane of the wafer surface. The television camera is placed at an oblique viewing angle of approximately 19 degrees to normal.

The minimum particle size detected depends upon the magnification of the lens, the intensity of the arc lamp and the condition of the wafer surface (back-ground haze). Typical size ranges using latex spheres as standard particles are 0.3 to 0.5 micron. A built-in processor computes the spot size for each particle, counts the number of spots (particles) for each size and outputs the particulate counts in each area of the field of view.

There are several problems associated with this type of system. The greatest difficulty is that the defect or particle must scatter light directly into the camera lens in order to be detected. Some types of defects scatter light over a multitude of angles (isotropic scattering) while other scatter light at a very few preferred angles (nonisotropic scattering). Since the size of the defect must be determined by all the scattered light, incorrect particle sizing results in the former case and complete loss of the particle results in the latter case. These difficulties may be partially overcome by rotating the wafer 360 degrees and simultaneously sum-ming the observed scattered light. The field of view then becomes a 19 degree cone instead of a point. However, unless the camera is simultaneously moved through an arc during the wafer rotation, a large amount of scattered light will still be missed by the television camera.

A second problem occurs when a small particle or defect is located close to a larger one. The large size of the bright spot on the vidicon screen can actually encompass the smaller one providing an incorrect defect count. The image will "bloom" and tend to obscure the smaller, neighboring particles thus limiting spatial resolution.

A third problem is one of "edge effects" on the electrical signals produced by the vidicon tube itself. This is similar to the problem of vidicon based line-width measurement instruments. The electrical transition from dark to light is not sharp but rather is a gradual one with a finite rise time. There will be ambiguity as to where along this rising edge of the pulse to start the measurement and where along the falling edge of the pulse to stop the measurement. For these reasons, new laser scanning techniques have been developed which circumvent these problems.

Scanned Laser Light

Principles of the Method. In scanned laser systems, a small spot of laser light is mechanically moved over the surface of the substrate. Light scattered by a defect is then gathered by a suitable light collector and directed to a low noise detector and amplifier for further analysis. In most systems, the specularly-reflected light is ignored and only light which is scattered out of the specular beam is collected. Thus the method is analogous to dark-field microscopy.

The power of such a seemingly simple technique can be appreciated by a simple calculation.[4] A waterspot of only a few micrometers in diameter with a thickness of 300 angstroms and a refractive index of 1.5 may have a scattering cross-section of only 0.01 square micrometer. (From Figure 4, we see that the latex sphere equivalent diameter for such a scatterer is less than 0.2 micrometer). Such a defect is easily detected with modern laser defect scanners, but one would be hard pressed to see the same defect in a scanning-electron microscope due to the low contrast and small step-height of the spot.

In scanned laser systems, localized defects produce electrical signals which are very short duration pulses with amplitudes which are proportional to the amount of light scattered by the defect into the detector. Typical pulse-widths are shorter than the substrate scan time by factors of 300,000 and more. Large area defects on the other hand give rise to more slowing varying signals. Haze for example may appear as a generalized DC background level with localized defect signals superimposed on the haze signal.

To further complicate the problem, it is necessary to define the defect type in terms of the scanning spot. The reason for this is that the electrical signals produced by the defects are a function of the size of the scanning spot. Defects which are larger than the effective spot size are considered to be large area defects while defects which are smaller than the effective spot size are considered to be localized defects. The effective spot size is determined by the beam-forming optics for scanned laser systems.

Scanned laser instruments utilize Gaussian beam profiles in which the intensity of the illuminating light in the center of the spot is inversely proportional to the square of the spot diameter. The spot diameter is usually taken as that point at which the intensity falls off to $1/e^2$ of the intensity at the center of the spot. If the defect is much smaller than the beam diameter, the maximum pulse height from the detector will thus be proportional to the reciprocal of the spot size squared. This is the usual case for point defects since typical beam diameters are in the range of 25 to 100 micrometers and anything larger than a few micrometers is considered a killer defect in today's processing technology.

Defects which are much larger than the spot diameter produce an entirely different relationship between pulse amplitude and beam diameter. Defects of this type produce a detector signal which depends only on the scattering magnitude measured as a fraction of the illuminating beam. Thus the scattered light signal will be nearly independent of spot size. For this reason, two instruments with different spot sizes will differ in their relationship between localized and large area defect sensitivity. It is not enough to calibrate each machine with either a haze standard or a particle standard since the sensitivity to haze or particles may then differ between the two instruments due to differing spot sizes. The effective instrument spot size is the actual diameter of the scanned beam on the surface in the case of scanned laser systems.

Rotating Polygon Scanners. An example of a flying spot scanner in which a rotating polygon mirror is used to sweep the surface of wafer with a focused spot of light is shown in Figure 8. The wafer to be inspected is passed under the scanning beam at a constant speed. The speed is adjusted to provide overlapping, consecutive, diagonally spaced scans. The scanning beam is produced by a helium-neon (typically 2 to 5 milliwatts) laser. The path is folded in order to reduce overall instrument size and focused by two lenses onto an 18 facet polygon mirror assembly rotating at 3600 rpm. The surface of the wafer that is illuminated by a scan line is viewed by two sets of collection optics. Each collector consists of a fiber optic bundle which is spread into a cylindrical lens and slit at the wafer end and gathered into a circular bundle at the (photomultiplier) detector end. One set of optics collects only light that is specularly reflected from the wafer surface (bright field detection). This collector is designed so that any undulations of the surface cause a loss of light to the photomultiplier tube or bright field detector. A second set of optics is designed so that only light *scattered* from the wafer surface is detected by its photomultiplier tube. This channel is a dark-field detector. The signals from the light channel are analyzed for both frequency and amplitude while the dark channel signal is processed in amplitude for three threshold levels.

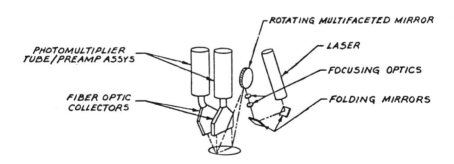

Figure 8: Typical rotating-polygon wafer inspection system.

Oscillating Mirror Scanners. The principle of operation of the scanning mirror technique for substrate illumination is illustrated with a series of schematic diagrams beginning with Figure 9. Light from a two milliwatt linearly-polarized helium-neon laser passes through a double right-angle prism which folds the light path and converts the linearly polarized light to circularly-polarized light. The light then enters the first lens, L1, focusing the light to a spot somewhat smaller than the pinhole in the spatial filter placed at the focal point of this lens. This removes stray diffracted laser light which would otherwise degrade the quality of the final focused spot. The diverging light leaving the spatial filter is turned 90 degrees by the plane mirror, M, and then enters a second lens, L2, which reconverges the light toward its final focal point on the wafer surface.

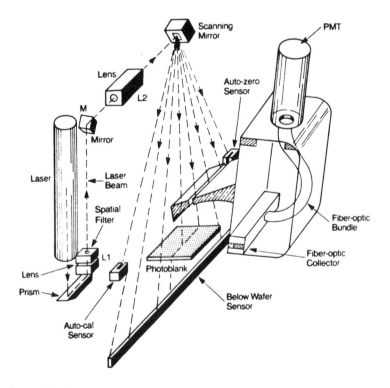

Figure 9: Optical schematic of an oscillating-mirror inspection system.

Before encountering the wafer, however, the beam reflects from the scanning mirror. By oscillating back and forth, the scanning mirror moves the final focused spot rapidly along a straight-line path on the wafer surface. At the same time, the transport system slowly moves the wafer perpendicular to the fast scan so the focused light spot sweeps the wafer surface in a series of adjacent, overlapping scans. During the measurement portion of the scan, the beam sweeps across the wafer in one direction. During the retrace time, the wafer is microstepped 25 microns. Since the wafer is stationary during the measurement scan and microstepped during the retrace, the sweeps are therefore not diagonal across the wafer but are perpendicular to the direction of travel. This ensures that every point on the wafer surface is repeatedly addressed by the light spot and no defects are overlooked. The combination of scanning-mirror oscillation and wafer travel creates an X–Y laser beam raster scan. The signal produced during this raster scan is handled by the signal processing electronics like slow-scan video or facsimile data.

The fundamental limitation of any laser scanner design, such as Figure 8 in which the detector is placed directly above the substrate (wafer or photo-blank), is stray light. Stray light along the scanning beam reflects off the surface and into the light collection/detection system contributing to a large background (optical noise) signal. Stray light effects can be reduced by decreasing the size of the entrance slot of the detector, but this reduces the detector sensitivity and makes the response even more nonuniform.

One technique to minimize stray light is to separate the scattered-light detector from the collector. This usually involves some sort of relay optics such as a reflector, lens or fiberoptic light guide to get the scattered light from a line path up to 8 inches long on the substrate to the entrance window of a detector such as a photomultiplier. This system must also come as close as possible to gathering all the scattered light rays anywhere along that path over a 360 degree azmuthal angle and an elevation angle of from 45 degrees to nearly 90 degrees. Such a system is illustrated in Figure 10. In this system, the laser beam is directed to the surface of the wafer and any stray light follows around the beam.

The optical system itself also has imperfections which will introduce stray light but the geometry of the ellipse tends to reduce the stray light considerably.

Stray light that enters at a given angle of incidence is reflected up into the elliptical mirror. Due to the geometry of this elliptical mirror, the stray light is focused by the mirror at point S. (This light can be ignored since it will be lost in the system.) An elliptical mirror has an interesting property because of the two focal points. With the laser spot focused at one focal point, any light scattered by the defect at that point will always be focused at the other focal point. Conversely, any light which does not emanate from the first focal point (i.e. the stray light) will be focused at other points (point S). The advantage of using this elliptical mirror is to allow only the defect's scattered light to be focused at point F and all other stray light to be focused at point S.

The aperture of the detector is placed at point F, effectively eliminating the beam's stray light by allowing only the defect's scattered light to be measured. This use of the elliptical mirror as the light collector gives instruments, such as Tencor Instruments' Surfscan 4000, about an order of magnitude (10X) increase in the signal to noise ratio, without any changes in the incident laser light.

Figure 11 illustrates the elliptical mirror with a fiberoptic bundle forming a complete light collection system. This system collects light scattered by the laser beam which is focused on the surface of the wafer. This is also the first focal point of the elliptical mirror, as described earlier.

The second focal point is the destination of the light scattered by defects at the first focal point. The aperture of the collection system is placed at this second focal point. The light enters the fiberoptic bundle and is "randomized" before entering the photomultiplier tube (PMT). The end result is a uniform response no matter in what direction the light is scattered.

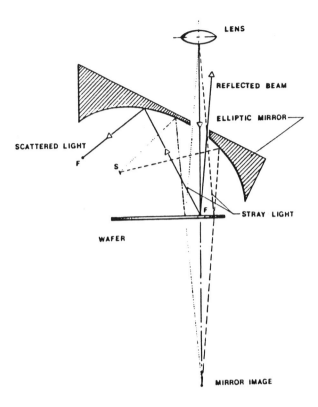

Figure 10: Elliptical-cylindrical mirror light collector.

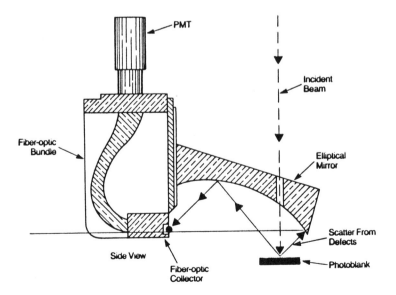

Figure 11: Elliptical mirror with fiberoptic bundle to PMT.

The large collection angle of the elliptical mirror can help avoid problems associated with a small collection angle as shown in Figure 12. If the collector is placed at a certain position and the defect is an epi spike, the particle may be only partially seen. With the light impinging from above, the shape of the epi spike causes the scatter of two relatively large components of light. If the collection angle is narrow (such as with a camera lens for Vidicon tube) the amount of light seen will be approximately 50% (since the other component is missed by the detector). If an elliptical mirror covers the top surface of the wafer, both components will be gathered for measurement. In the next example, the case is described for an irregularly shaped particle. For this case, a small component is seen and the larger component is lost in the other direction. This causes incorrect sizing of the defect. If the wafer is then rotated, the larger component is measured instead. In some cases, the bright component can cause another problem. This problem, associated with running a Vidicon tube at high sensitivity, is "blooming" of the detector's sensitive surface. The particle will appear much larger than it really is and the smaller particles near this "large" particle will not appear (the large component overloads the detector surface).

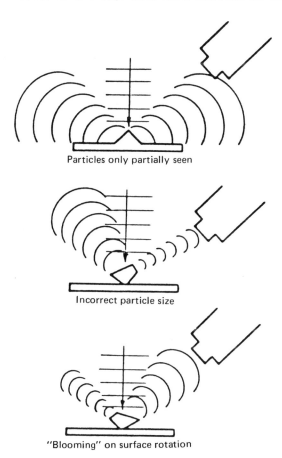

Particles only partially seen

Incorrect particle size

"Blooming" on surface rotation

Figure 12: Limitations of a small collection angle.

Moving Substrate Scanners. This technique utilizes a fixed helium-neon laser and a moving substrate as illustrated in Figure 13. The beam from a two milliwatt helium-neon laser is focused by a converging lens into a spot of approximately 40 microns in diameter on the wafer surface. A typical depth of focus for this configuration is ±0.8 millimeter for a maximum beam intensity variation of ±5%. This depth of focus and the fact that for this configuration the beam is always perpendicular to the wafer surface ensures that the beam intensity remains constant over the surface of the wafer.

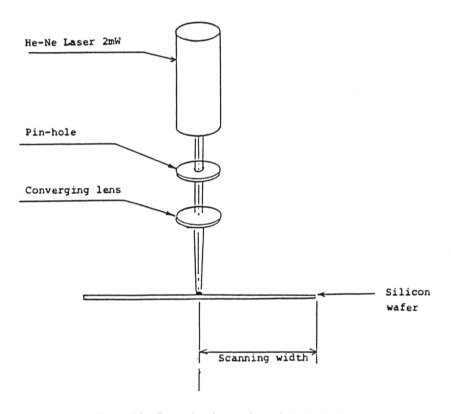

Figure 13: Example of a moving substrate scanner.

The light collection system consists of 12 fiberoptic bundles which are arranged hemispherically around the laser beam. Six of the bundles are equally spaced in a cone at 30 degrees to the plane of the wafer while the remaining six are at 60 degrees to the horizontal as shown in Figure 14. With this circumferential arrangement of the light collectors, most of the scattered light from the particles and defects is collected uniformly around the incident beam. The scattered light, collected by the fiberoptic bundles, is then directed to a combination photomultiplier tube and amplifier for detection and analysis by the signal processor.

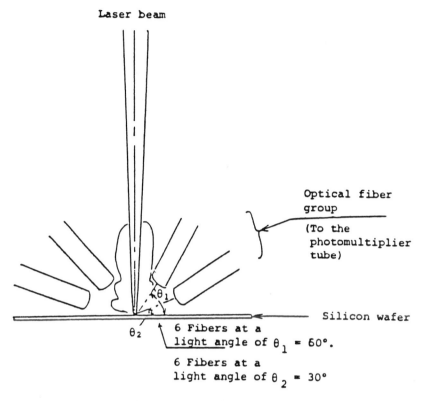

Figure 14: Circumferential fiberoptic light collector.

The actual scanning process consists of rotating the wafer at a constant speed of 1500 rpm and translating the wafer at 0.75 millimeter per second so that the beam spot traces out a spiral scanning pattern from the center to the periphery of the wafer. The beam is stationary during the scan. Since the wafer rotation and radial shift are not synchronized, it is necessary to independently detect the rotation by an angular position detector, and the radial position by a linear position detector. The resolution of the angular position director is 512 divisions per revolution while the radial direction sampling is done at 0.5 millimeter intervals. This is illustrated in Figure 15. It should be noted that the circumferential length of the measurement area increases with the radius, so in order to maintain approximately a 0.5 millimeter square area, the circumferential density is changed in steps from the center toward the periphery. Defects and particulates within each 0.5 millimeter square area are displayed as a single defect (pixel) regardless of the number of defects. There are 17 equally spaced laser beam scans approximately 30 microns apart within each 0.5 millimeter square area. As shown in Figure 16, due to the Gaussian laser beam intensity profile, this results in a maximum peak to valley intensity variation of 32%. A rough size distribution (histogram) across the wafer surface is achieved by dividing the measurement area into four rings and reporting the particulate size within each annular ring.

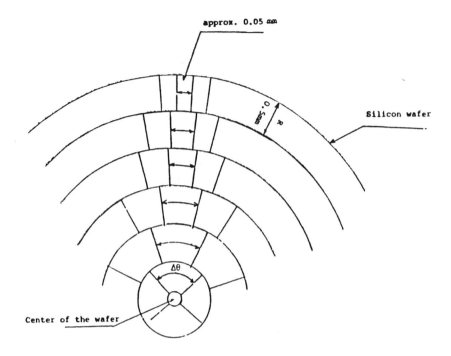

Figure 15: Variation of pixel size with radial distance.

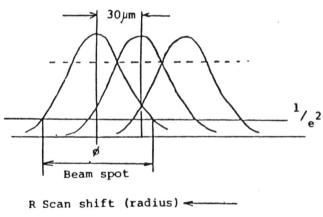

R Scan shift (radius)

$\emptyset$ = 40 μm

do = 30 μm

Figure 16: Overlapping widely spaced Gaussian beam profiles.

Signal Processing Methods. When the scanning laser beam passes over a particle or defect on a substrate, a very small portion of the light within the beam is scattered.

In order to understand the problems associated with designing signal proc-essing circuitry to sort out the myriad of pulses coming from the PMT, one first notes that the scanning laser spot has a Gaussian profile; that is, the intensity of the beam falls off as a bellshaped curve in the radial direction from the center of the spot. The spot "diameter" is arbitrarily chosen to be twice the radial distance at which the intensity falls to $1/e^2$ of its peak value. This corresponds to about 13.5% of its peak value.

When a laser spot of Gaussian cross-section is raster-scanned across a sub-strate with small defects or particles, the output of a perfectly linear, noise-free, infinite bandwidth detector is a Gaussian pulse. The $1/e^2$ width of the pulse equals half of the spot diameter divided by the scanning velocity whenever the spot passes over a defect or particle. Usually the successive scan paths are spaced close enough so that in the worst case, the peak light intensity seen by the par-ticle is at least 90% of the value at the spot center. This means that the spacing should be no more than 1/3 of the beam diameter. For a 100 micron diameter spot, the spacing is typically 25 microns.

Figure 17 illustrates the situation in which a small particle is scanned by a 100 micron diameter laser spot with 25 micron spacing, such as on an oscillating mirror scanner.

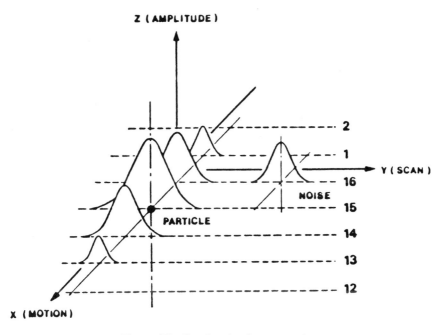

Figure 17: Overlapping laser scanning.

The Y-axis is the scan direction (the motion of the laser beam). The X-axis is the motion of the wafer by the robotic handling system. The numbers shown in the figure represent the various scan lines. For example, starting at the bottom are scan lines 12, 13, 14, 15, and 16. It starts over at 1, 2, etc., because the figure arbitrarily shows the region near the 16th scan line (the boundary of two pixels since there are 16 scans per pixel). If the 16 scans are multiplied by the 25 microns per microstep, the width of a pixel can be seen to be 400 microns on a side.

Imagine that there is a particle located on the wafer where scan line 15 occurs. First, a scan will be taken across scan line 12 but the distance from the bright center of the scan line to the point where the intensity is negligible will be 50 microns (the beam radius). From scan line 12, a distance of 50 microns is on scan line 14. Thus, the edge of the beam is not crossing over the particle, and the Z-axis, representing the PMT output amplitude, will have no putput for scan line 12 (the particle is not seen). After the wafer is microstepped, the beam moves along scan line 13. Now, the 50 micron radius of the beam reaches the particle and is scattered. The PMT thus generates a small output pulse along scan line 13 when the beam passes over the particle.

Another microstep and the beam now moves along scan line 14. With the beam center closer to the particle, a larger pulse is generated by the PMT.

Finally, when on scan line 15, the center of the beam passes directly over the particle and the largest pulse is generated.

Continuing on scan line 16, and the scan lines for the next pixel, smaller pulses are generated by the scanning beam since the intensity of the beam is decreasing as a function of the radius.

If one uses the simple-minded approach of counting pulses with a pulse counter at a fixed threshold, it is obvious that this method will always miscount the number of particles on the substrate. Each particle peak will either be missed altogether if it falls below the threshold, or be counted more than once if it exceeds the threshold by more than a small amount.

The next step might be to divide the scanned area into cells defined by a fixed time interval (clock) in the laser beam scan direction and a specific number of scan lines. As noted earlier, such a cell (pixel) was defined by 16 scans spaced 25 microns apart for a cell size of 400 microns on a side. Having done this, it is possible to attempt to associate successive pulses of a single particle to the cell in which the particle resides. The cell location can be written on a monitor or printer or stored in memory for later use. There are, however, problems associated with such a scheme.

Suppose now, that a pulse detector is attached to the PMT with its threshold set to just below the largest peak pulse from a single particle and that the entire cell is failed and recorded when the pulse is detected. With this arrangement, the particle will be properly located, counted and displayed. However, if the threshold (detection) level is lowered, or a larger particle is encountered, the pulse counter will detect the particle several times. If the particle is physically located near the center of the cell and if the cell is somewhat larger than the scanning spot size, there will still only be one cell count. However, if the PMT pulse from the particle is significantly above the detection threshold, and the particle is located near the edge of the cell, more than one cell count will be associated with the particle. Under worst case conditions in which the particle is located at the intersection of several cells, the cell count could be as great as four times the correct number.

The goal of any laser scanning signal processing system is to sort point defects and particles into several unambiguous size categories with a single wafer scan. This would provide a particle histogram and, at this point, it is obvious that simply connecting a pulse height analyzer to the output of the PMT is not the answer. The multiplicity of pulses per particle would simply smear out the histogram and give an excessively high particle count. A simple minded approach to the problem might be to count clusters of defective cells as a single defect. This method will work if there are sufficiently few defects located more than a few cells from each other on the substrate.

A substantial improvement to the problem of a single threshold can be obtained by digitizing the peak signal levels. An 8-bit A/D converter for example would yield a 256 level histogram. Of course, it is still necessary to perform cluster analysis to extract the essential particle size information. Once the particle size information has been extracted, it is then possible to assign the largest digital value found to the cell in which it was found. This is a step in the right direction, but it is still possible to occasionally fail a neighboring cell as illustrated by the small pulse on scan line 1 of Figure 16.

Little has been said as to how to actually process the data once the peak levels have been digitized. One technique would be to store all of the digitized values in a mass memory and proceed to sift through the magnitudes with various software algorithms. A monumental task at best. A better approach would be to build dedicated hardware which would digitize and analyze the peak signal levels and pulse locations (both X and Y) and in real time "on the fly." Such a scheme has been implemented in the Tencor Instruments' Surfscan 4000 as the Pulse Position Correlator (PPC).

The PPC measures the amplitude of each pulse produced by multiple, overlapping scans of the particle along with both the X (motion and scan) location of each pulse. The PPC then, determines the precise location and correct amplitude to assign to the particle (pulse) being measured. All other pulses are ignored.

INSPECTION METHODS FOR PARTICLES ON PATTERNED SURFACES

Inspection for Particles on Wafers

In the traditional method of laser light scattering to detect foreign particles on wafer surfaces, a focused laser beam is incident at near 90 degrees to the surface. The light scattered by particles and defects is detected by a photosensitive detector such as a photomultiplier tube or photodiode placed at some angle above the sample and out of view of the specularly reflected component. When the sample under test has a surface pattern, these patterns will also scatter light in much the same manner so that it is impossible to distinguish between the light scattered by a defect or particle and the light scattered by the pattern. This method is therefore suitable for patternless samples.

A new technique using polarized laser light has been described by Koizumi and Akiyama.[5] Figure 18 illustrates the basic components of the optical system using polarized laser light. The basic concept is that an analyzer is placed behind the objective lens to cancel the predominantly polarized light from the pattern and not the unpolarized light from the particle. The S-polarized laser illuminates the spot O on the sample at a grazing angle of about one degree. Since the scattering angle from the pattern is small, most of this light is S-polarized. The scattering angle from the particles, however, is large, and some P-polarized light is also included in the scattered light. This scattered light is collected by the objective lens and passes through an analyzer (polarized filter) to eliminate the S-polarized light. Since now only the P-polarized light from the particles passes through the analyzer, this scattered light from the particles can be detected by a photodetector such as a photodiode.

The actual degree of polarization of the scattered light relative to the polarized light from the illuminating beam is a function of the angle of the pattern to the incoming beam. Thus the sample and the analyzer can both be rotated to minimize the light scattered by the pattern which enters the objective lens. The signal-to-noise ratio is maximized when the electric field vector of the incident light is parallel to the wafer surface. The reflected light from the pattern is extinguished when the analyzer is oriented such that the transmission axis is orthogonal to the oscillation of the reflected light. Actual tests have been performed using polystyrene latex spheres (index of refraction of 1.59) with diameters of 0.7 micron and 1.0 micron as reference particles. Test patterns of polysilicon and SiO_2 with edge pattern step of 0.4 micron and refractive indices of 4.2 and 1.45 were employed. Both of these reference particles were easily distinguished from the pattern noise.

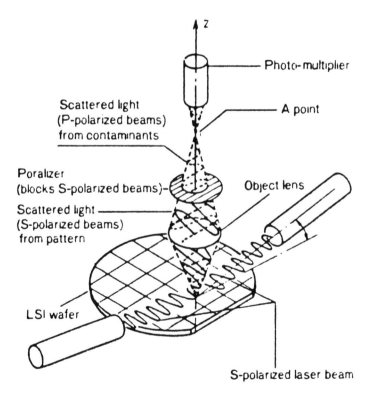

Figure 18: Detection principles for polarized lasers.

Inspection of Particles on Reticles

The traditional method of inspection for soft defects on reticles is to expose a pattern on a dummy wafer and inspect those patterns after developing and etching. This method requires several hours to perform and severely restricts the throughput of the exposure system. Automatic reticle inspection systems have been limited to complex pattern recognition systems requiring long inspection times and large computerized data bases of the individual reticle patterns. A number of systems have recently appeared which augment the exposure system and are designed to provide immediate feedback as to the condition of the reticle to be used for the particular exposure step scattering.

Laser Light, In-Situ Inspection. Light scattering systems have the advantages of being quite simple in principle, having high throughput and requiring no prior knowledge of the reticle pattern. Figure 19 is a block diagram of an in-situ reticle inspection described by Quackenbos[6] et al. The helium-neon laser beam passes through a Pockel cell, a cylindrical horizontal expansion lens, L1, a spherical horizontal focus lens, L2, a scan mirror and a polarizing beam splitter. The beam splitter produces two scanning beams so that both sides of the reticle are scanned simultaneously. Since the particles on the patterned and nonpatterned surfaces have differing effects on the actual printing process, they are inspected by separate optics.

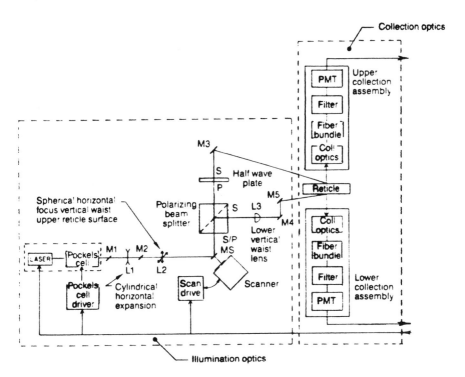

Figure 19: In-situ reticle inspection system.

The Pockel cell passes both the S- and P-polarized helium-neon laser beams simultaneously and synchronizes them with the drive of the scan mirror. The polarized beam splitter reflects only the S-polarized beam through lens, L3, to mirrors M4 and M5 thus scanning the lower surface of the reticle. The P-polarized beam passes through the beam splitter and half-wavelength plate becoming S-polarized to mirror M3, scanning the upper surface of the reticle. The combination of the Pockel cell and the polarized beam splitter, acting as a "shutter" enables the laser beam to scan the upper and lower surfaces of the reticle simultaneously. The one degree grazing angle of the incidence minimizes the scattered light from the reticle pattern to the point where five micron diameter particles can be reliably detected.

Laser Light Scattering Remote Inspection. A system which provides remote inspection of reticles using laser light scattering and intensity monitoring is illustrated in Figure 20. This system was first described by Tanimoto and Imamura[7] and utilizes a number of photodetectors placed at various angles to the reticle under inspection. Detectors PA through PE are arranged both above and below the reticle surface enabling the analysis of the scattered laser light in a number of different directions.

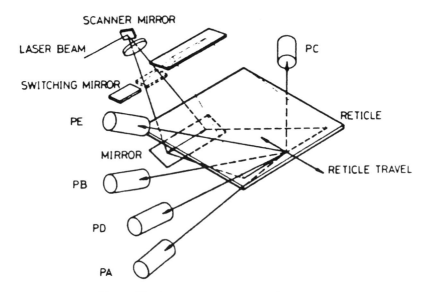

Figure 20: Remote reticle inspection system.

Figure 21 illustrates the difference in the light scattered from the reticle pattern and from particles. The thickness of the chrome or antireflective chrome pattern is between 600 and 1,000 angstroms which is considerably less than the wavelength of the helium-neon laser (6,328 angstroms). The resulting scattering patterns can then be modeled approximately by light diffraction due to the edges of infinitely thin plates. Thus as shown in Figure 21 light scattered by the pattern is directed through air to the lower detector and through glass to the upper detector. Particles are larger than the wavelength of the laser and tend to scatter more of the incident light into the air. Consequently, the light scattered by a particle is stronger in the region of the lower detector than in the upper detector. It should be noted that the number of lower detectors is greater than the number of upper detectors in order to detect smaller particles on the patterned surface than on the upper surface. Particle detection sensitivity is four microns on the patterned (lower) surface and seven microns on the unpatterned (upper) surface.

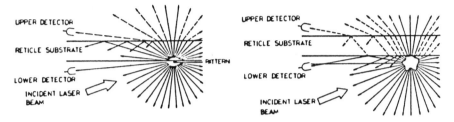

Figure 21: Scattering from reticle patterns and particles.

Laser Light Scattering Inspection of Reticles with Pellicles. A pellicle is a thin transparent membrane made of nitrocellulose and stretched over a metal frame. The surface of the reticle is protected from outside dirt and contamination by the pellicle. In addtion, any dirt which adheres to the surface of the pellicle is away from the plane of the reticle by several millimeters and is thus out of the depth of focus of the projection lens.

A detection assembly for inspection reticles with pellicles in place is shown in Figure 22. This detector is the upper detection assembly (the lower assembly is identical) of a complete inspection system described by Shibaand Koizumi.[8] The principle of the inspection method is illustrated in Figure 23. The reticle surface is illuminated obliquely by an S-polarized helium-neon laser beam. The scattered light is analyzed by varying the rotation angle, B, of the analyzer. Figure 24 illustrates the polarization properties associated with a reticle pattern and a particle. Scattered light produced by patterns is mainly P-polarized while scattered light produced by particles is mainly S-polarized. However, since the particles are irregularly shaped and have rough surfaces, scattered light produced by these particles will contain components of both S- and P-polarizations. However, by utilizing an analyzer designed to eliminate the P-polarization component, it is possible to detect particles down to two microns in diameter.

The system consists of a polarized five milliwatt helium-neon laser focused to a beam spot of 50 micrometers. The laser passes through a beam expander and is transformed into an elliptical beam spot, so as to produce a circular spot on the reticle surface when scanned at an angle of 15 degrees. The 50 micrometer spot is driven by a galvanometer in linear proportion to the scanner angle, O, and when driven by a triangular wave produces a scan in which the velocity and intensity of the laser spot remain constant along a single scan line. The reticle itself is driven along the Y-axis at a constant speed while the beam is scanned in the X-axis in order to inspect the entire surface of the reticle.

Two detectors are placed in such a manner that the system, L, inspects the left half of the reticle and system, R, inspects the right half of the reticle.

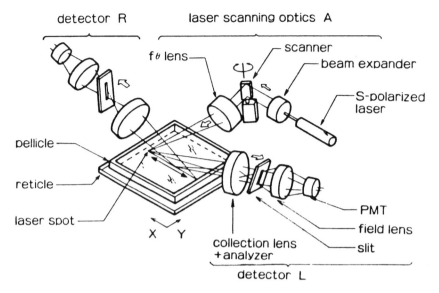

Figure 22: System configuration for a pellicle/reticle detector assembly.

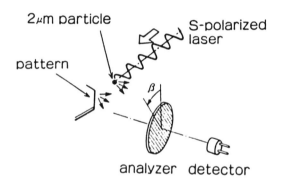

Figure 23: Inspection principle for a pellicle/reticle system.

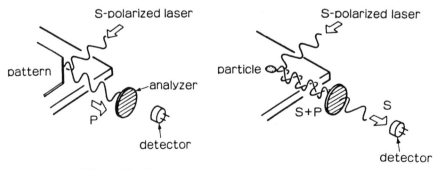

Figure 24: Polarization properties of scattered light.

Similarly, there are two systems (only one shown) for the laser scanning. Each detector consists of a collection lens with analyzer, slit, field lens, and photomultiplier tube. The detector slit is placed in the conjugate plane of the laser scanning line for the collection lens to block off the noisy scattered light.

Classification of particles by size is accomplished by measuring the intensity of the scattered light. Although an infinite number of size classifications are possible, it is unusually convenient to classify the particles into three size ranges: less than 5 microns, 5 to 10 microns and greater than 10 microns.

Standard particles (polystyrene latex spheres) can be used to calibrate such instruments but in general, actual particles produce signals which are stronger than the two standard particles. Particle positions can be mapped by monitoring the angle of the galvanometer mirror and the position of the reticle stage. Using the resulting map, an operator can then perform more detailed analysis and under a conventional optical microscope and precision X-Y stage.

REFERENCES

1. Price, J.E., *Proc. IEEE,* pp. 1290-1291 (August 1970).
2. Oswald, D.R. and Munro, D.F., *Journal of Electronic Materials,* Vol. 3, p. 225 (1974).
3. Bennett, H.E. and Porteus, J.O., *Journal of the Optical Society of America,* Vol. 51, p. 123 (1961).
4. Galbraith, L., Kren, G., Neukermans, A. and Pecen, J., *Solid State Technology* (June 1986).
5. Koizumi, M. and Akiyama, N., *SPIE,* Vol. 538.
6. Quackenbos, G., Broude, S. and Chase, E., *Electronic Imaging* (February 1984).
7. Tanimoto, A. and Imamura, K., *SPIE,* Vol. 470, p. 242.
8. Shiba, M. and Koizumi, M., *SPIE,* Vol. 470, p. 233.

13

Particle Contamination by Process Equipment*

Barclay J. Tullis

INTRODUCTION

In the purchase and use of process equipment, integrated circuit manufacturers are beginning to be concerned by how cleanly this equipment functions.[1-7] They are beginning to break away from the tradition[8] of relying solely on laminar flow environments to guarantee a clean process.[9-12] And they are beginning to specify and control maximum wafer contamination levels caused by the equipment. Vendors would like to compare their equipment's contamination performance with that of competitors, but few have the ability to perform the necessary particle measurements.

This chapter presents a generalized approach to particle measurement that I believe is a good beginning toward achieving comparable results. The method is based on surface particle counts on unpatterned wafers, on the use of statistics to determine confidence limits, and on the recognition that exposure times to particulates in the environments through which the wafers pass can significantly affect results.

Measuring contamination levels produced by process equipment is not an easy task. What is important to controlling die yield is the reduction of contamination reaching wafer surfaces. The first characteristic of a successful measurement of equipment generated contamination or its specification is therefore a focus on measuring surface contamination found on wafers both before and after their passage through the equipment. Subtracting the measure of contamination made before a wafer enters the equipment from that obtained after it leaves the equipment produces an estimate of contamination contributed to the wafer in one pass through the equipment (see Figure 1). If the contamination being measured is limited to counts of particles, the resulting estimate

*Reprinted from *Microcontamination* November 1985, December 1985 and January 1986.

of particles added can be referred to as particles-per-wafer-per-pass, abbreviated as "PWP." This is a cumulative count of all added particles larger than a specified critical size.

Since wafer contamination may be significantly affected by its environment and manner in which the equipment is operated, clear stipulation and tight control must be made of the wafer environment and how the equipment is used, especially the durations wafers spend at equipment input and output elevators and the durations they spend within the individual process steps within the equipment.

Presently, some industry standards exist for measuring and classifying particle contamination,[13-18] but none are directly suitable for characterizing processing equipment. Methods for characterizing the air found within clean rooms and microscope based methods of characterizing particle counts on surfaces exist, but none of these are sufficient for characterizing contamination of wafers as caused by process equipment.

The objective of this chapter is to lay out ground work for a successful practice of measuring cleanliness of process and inspection equipment used for semiconductor manufacturing. The scope is limited to particle contamination. A rating scheme is proposed wherein equipment can be graded against a simple scale of "critical particle counts," abbreviated CPC.

Before proceeding, a comment on aerosol particle counts is in order. Aerosol counts remain extremely efficient and useful in detecting gross particle problems in equipment, and for measuring relative improvements. However, one must eventually resort to surface counts to determine quantitatively the significance of the aerosol counts, and for specification purposes. Since wafer counts require the expenditure of considerably more cost (in consumption of clean wafers and in labor for both processing the wafers and in analyzing the data), it is usually advantageous to use aerosol counts as much as possible in diagnosing the effects of clean-up efforts. Aerosol counts are of little help in detecting particles which reach wafers by surface migration. And they are of little help when the counts measured per sample are less than 10, or in laminar flow situations when contamination evidences itself only in confined wisps that may be difficult to discover by scanning the position of the aerosol probe.

PARTICLES PER WAFER PER PASS (PWP)

As introduced above, the concept of "particles per wafer per pass," or "PWP," is basic. PWP is a direct measure of how many particles a piece of process or inspection equipment adds to a wafer it "handles." As such it is a relatively simple concept to grasp. Figure 1 depicts the particle counts that might be found on a wafer both before and after it is subjected to being passed through a piece of equipment. Subtraction of the surface particle counts measured "before" from the counts measured "after" wafers have been "handled" produces a PWP result.

The word "handled" is placed in quotes in the above paragraph because PWP measurements can't always be obtained by subjecting the test wafers to the full process step of a given piece of equipment. The reason that this is so is that today's surface particle counters are not generally able to count particles on patterned wafers. Also they are not generally calibrated for accurately counting particles on surfaces other than bare silicon, although this might change as particle counter users become more sophisticated in their use. For these reasons, PWP measurements must oftentimes be designed to have the equipment under test carry out a modified process wherein the part of the process that would pattern or alter the wafer surface would be avoided. For example, to avoid depositing a film on the test wafers subjected to "handling" in an oxidation machine, the wafers might be brought to temperature but the active process gas would not be turned on. However, everything else about the process step would be conducted as usual, particularly the amount of time the wafers spend at various stages within the machine would be as close to that of a full process as possible. In a coater, the wafers would be spun and might even be subjected to an edge rinse, but no resist would be dispensed on them so that the difficulties of interpreting the validity of particle counts on a resist film would be avoided.

Let us look at some other details related to the PWP measure. In practice the particles lying within about 8 mm of the edge of the wafer are excluded from analysis. This is because cassettes (and tweezers) can easily contaminate this region of the wafer, and cassettes themselves are usually not considered part of the process equipment. Throughout the rest of this chapter, the term PWP will refer only to the particles added to the central disk area of a wafer, exclusive of this outer edge-zone (see Figure 1). In fact, if additional areas of the wafer are excluded from the analysis, as when the area 8 mm in from a flat on the wafer edge is excluded, then these areas can also be excluded from the PWP measures without loss of generality. Any specification of a PWP value should however be accompanied by a statement of what surface area is included. And, of course, a second detail in the definition of PWP values is the reporting of what particles are included in the count by virtue of their detectability, as determined by instrument settings and limitations.

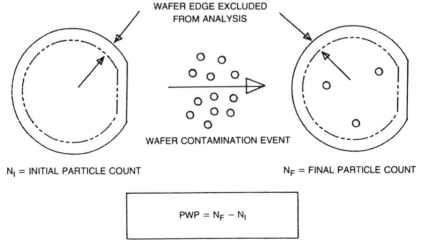

Figure 1: Particles per wafer per pass (PWP).

Sometimes PWP values are best specified separately for wafers positioned in specific slots of a cassette at the input and/or output stations of the equipment being specified. Figures 2 and 3 help to suggest why this is the case. Note with the help of Figure 2 that the particle fallout on a wafer will depend upon the flux of particles in its environment.[19] The sum of particles that land and stick on a wafer depends on particle flux densities near the wafer and on probabilities of impact and retention, each of which can be a function of time. Total particle accumulation will be dependent upon how long the wafer resides in each of what may be a succession of different environments as it proceeds through the equipment. For example, if wafers enter a piece of equipment from a first cassette, with the bottom wafer being handled first, then the top wafer in this cassette is obviously going to be exposed to particle fallout at that location a lot longer than any of the other wafers. Unless the environment of the input cassette is particularly clean compared to the rest of the equipment, this top wafer will most likely end up with a larger PWP value then the others, and for this reason should be analyzed separately.

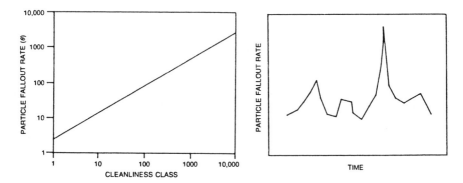

Figure 2: Particle fallout rate on a wafer is principally a function of aerosol particle count density near the wafer, particle flux caused by relative motion between the wafer and its surroundings, dwell time in these surroundings, and gravity. Fallout rate is presented here as particles per square foot per hour for particle sizes above 5 μm. Cleanliness class is particles per cubic foot larger than 0.5 μm. (Fallout data above Class 10 are from Reference 19. Between classes 1 and 10, fallout rates are three times larger for sizes greater than 0.3 μm.)

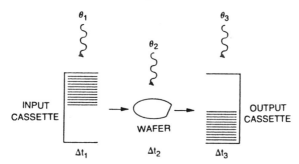

Figure 3: Fallout rates θ_i and dwell times Δt_i vary with wafer location during a process.

In summary, reports of PWP data should be accompanied by (a) the wafer size; (b) the range of particle sizes included by the PWP count; (c) how much of the wafer area near its edge is excluded; (d) the particle count Class[17] and the humidity of the environment in which the equipment is tested, together with air velocities measured at numerous points above the equipment; (e) the duration of individual steps within an equipment pass; (f) how many wafers are handled by the equipment in a pass; (g) the slot positions used in the wafer input/output cassette(s); and (h) the overall durations of an equipment pass and of the entire measurement experiment. If PWP values are obtained by averaging the results of numerous passes, the number of passes in the average should be reported. In reporting PWP values which are averages over wafers in a cassette, whether the end wafers are included in the average or not should be reported. It is also important to mention with the end-wafer data whether the wafers that started out as the top and bottom wafers in the cassette(s) always remained in those positions or alternated positions with each successive pass. Finally, statistical measures of variability and/or confidence should be included. The mathematics of these statistics will be presented and discussed in a section to follow.

In the next section, we will look at some of the variables that relate particle counts on wafers to die yields, and that relate wafer handling contamination levels of process equipment to resulting particle counts on measured wafers. This discussion will amplify the need for careful selection and reporting of how a particle count experiment is conducted and reveal some of the important factors which must be controlled to permit the results to be comparable with those obtained for other equipment or process steps.

PWP Relationship to Yield

Surface particle counts on wafers are believed to be better indicators of potential circuit defects than particle counts in aerosols or in liquids. It is instructive to review a model for how these surface particle counts can be related to yield. Murphy's yield model, and others as well,[20-22] relates die yield Y to defect density D per unit area and to die area A. The variables D and A appear only as a product A x D, the product representing the average number of defects-per-die. Defects-per-die are caused by events which take place in any of a number of individual process steps, so that the net product A x D can be replaced by a sum of products A_i x D_i, where i is a subscript representing a sequence number identifying a particular process step. Each product A_i x D_i can be reexpressed in terms of PWP_i for each process step:

$$(1) \qquad A_i \times D_i = (A_i/A_w) \times PWP_i \times k_i + A_i' \times D_i'$$

where:

A_i = die area susceptible to particle-related defects.

A_w = wafer area minus the particle counter edge-exclusion area.

PWP_i = particles per wafer per pass, i.e., particles added to the wafer area A_w in process step i

that are equal to or greater than the critical particle size.

k_i = kill ratio, or proportion of particles added to a wafer surface by a given process step that will end up causing defects in the finished dice (actual values for k_i are not well known in today's technologies).

A_i' = area of die susceptible to defects that are not particle related.

D_i' = density of defects added that are not particle related.

And note that the ratio PWP_i/A_w is a measure of particle density per unit area of wafer surface and that PWP_i can be expressed in terms of particle accumulation rate θ_i during a process step i which runs from time t1 to time t2:

$$(2) \qquad PWP_i = A_w \int_{t1}^{t2} \theta_i \times dt$$

where:

θ_i = particles per unit area per unit time.

dt = dwell time.

Figure 3 is an illustration emphasizing that accumulation depends upon flux rates θ_i and dwell times for wafers at each segment of a process. Note that some situations may cause a high particle flux but that the time wafers are exposed to this contamination can be so small that accumulated contamination count becomes insignificant.

The importance of the time factor in particle fallout on wafers can not be over emphasized. It is the situations where the wafers must spend significant time in relatively dirty environments for which the equipment designer and user must be most concerned. Considering that individual wafers spend considerably more time in input/output cassettes than they do in active process within equipment stations, one must be conscientious about the quality of the environment at these locations.

Next we will examine the precision and reliability of data taken with surface particle counters. This will help us further appreciate the need for using statistics to remove some of the limitations of the measuring equipment.

Variability of Measurement Tool Data

A precise measurement instrument is one that is repeatable in its measurement results. But with today's particle counters[23-25] rarely do two measurements of the same wafer, even if taken successively on the same counter, produce the same count results. Searching for a 0.3 μm particle on a 100 mm wafer is dimensionally like, on a larger scale, searching for a granule of table salt on an otherwise smooth and clean surface larger than six football fields. But counters of the laser-scanning type and imaging vidicon-tube type all are able to detect particles on wafers with impressive speed and positional accuracy, considering the immensity of the task. These counters are able to detect single particles as small as approximately 0.3 μm in diameter in a matter of seconds anywhere on the polished surface of an unpatterned wafer.

The variability obtained with today's particle counters over several count measurements is primarily a result of limitations in detectivity and particle illumination. These counters detect and "size" particles on the basis of the amount of light they scatter from an incident light source into a detector. Minute differences in the positions of the particles in the illumination beam from one measurement of a wafer to another seem to be the major cause of variances in count results. A second major contributor to count variability is the effect of noise on signals.

Obviously, for meaningful analysis of equipment contamination levels on wafers, design of equipment characterization experiments must include consideration that counts of particles added to a wafer by the equipment should be significantly larger than the uncertainties of the particle counts directly measured. For example, if the uncertainty in a counter's count is ±2 particles for each of the initial and final counts (taken before and after the wafers are handled by the process equipment in question), then it would make little sense to try and detect, with one single-pass experiment, a PWP accumulation of as little as 4 particles, since the uncertainty in count difference would be on the order of 2 + 2 = 4 (or possibly of the square root of the sum of the square of each of the two 2's). If the precision of the particle counter is not good enough to enable determination of an anticipatedly small PWP result, then the only good alternative is to subject the wafers to a succession of equipment passes until a detectable difference or accumulation of particles can be calculated. For example, if equipment adds only 1 particle per wafer per pass, 10 successive passes would accumulate 10 particles on each wafer. This total of particles accumulated per wafer could now be determined with the counter having an uncertainty of ±2 in each measurement, although the final result would still not be very precise. The accumulation measured could then be divided by 10 to calculate accumulation of a single pass. In this simplistic example, the resulting PWP value might be anywhere from 0.6 to 1.4. Also note that the result would represent the average PWP taken over ten passes in the equipment under test, and information would be lost about how much variability existed in particle accumulation from one pass to the next.

A look at the variance characteristics of counters reveals an important fact, that the more particles there are on a wafer the larger will be the absolute differences in particle count from one measurement to the next. Figure 4 shows the type of variability that has been measured on laser scanning particle counters of different types. The standard deviation of particle counts from mean particle counts, obtained from small samples of multiple measurements, is expected to be very small if the number of particles on the wafer is small, and large if the number count is large. As a percentage of count, however, the standard deviation is smaller with increasing count magnitude, but percentage-of-count is rarely the parameter of interest. For these reasons, when it is necessary to measure a small addition of particles caused by a relatively clean process step, it is important to start out with wafers as clean as possible, so that the variability in absolute count due solely to the imprecision of the counter will be minimum.

A good, although only approximate, rule-of-thumb in characterizing the imprecision of today's particle counters is that one standard deviation in count is equal to the square root of one-half the mean count.

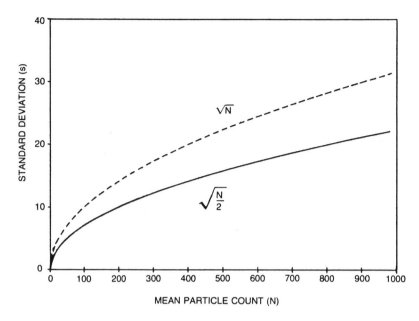

Figure 4: Standard deviations of counter results increase with the number of particles on a wafer. The solid line is in approximate agreement with today's counters for means of small samples. The dashed line is only for reference; it is the standard deviation of a Poisson count distribution.

In the next section, the details of applying statistics to the analysis of particle count data, and to the design of particle experiments is discussed.

USE OF STATISTICS

There are many choices which must be made in carrying out a particle counting experiment to characterize the contamination level of a piece of process equipment. For example, how many times should the same experiment be repeated by the equipment vendor in order to have some confidence in predicting results if the experiment should be done again in the future, as for example during acceptance by a customer? And how many times should a given wafer be measured before the mean of the results is a good representation of what is on the wafer? Also, how can the contamination of the wafers by the measurement tool be accounted for? How can the contamination caused by the handling of the wafers between the process equipment and the measuring equipment be accounted for? And, is factorial design of the experiments practical to determine where the causes of contamination are that might be eliminated and whether the measurements made under one set of conditions are still applicable if these conditions should change? How can one calculate both confidence in the repeatability of measurements as well as confidence in the contamination level deduced for the equipment?

Thoughtful design of experiments with proper control measures, and the use of statistics,[26-31] can bring answers to most of these questions and lead to a successful approach to measuring PWP counts of particles generated by wafer handling in any piece of process equipment or stage in the IC fabrication process.

Particle Counter Repeatability and Contamination by Particle Counters

One of the first things which must be addressed is determining the repeatability of the particle counting tools to be used. The nature of this repeatability was discussed in the previous section. To measure repeatability requires having a set of wafer groups, each group in the set consisting of 5 or more wafers, the wafers in any one group having approximately an equal number of particles, and each group having a significantly different count than the other groups in the set. For example, the set might consist of five groups of 5 wafers each. The particle counts in the first group might be near 10 per wafer; the second group, near 50; the third group, near 100; the fourth group near 400, and the fifth group, near 1000. Each wafer would be measured several times, e.g. 5 times each, and an average and small-sample standard deviation would be calculated for each wafer. These standard deviations would then be plotted along a vertical axis and against a horizontal axis of wafer-average particle counts. A curve can then be fit through these points to obtain an instrument precision graph like the one shown in Figure 4. As discussed in the previous section, this precision data can be used to design particle counting experiments to achieve sufficient numbers of particles on wafers that the imprecision of the counting instrument becomes insignificant to the determination of average particle counts added on the wafer per equipment pass.

After determining the precision of the instrument, it is desirable to determine how many particles on the average accumulate on wafers being measured during the course of the measurement. This determination actually comprises a PWP measurement of the particle counting instrument itself. This is best done by repeatedly remeasuring an initially very clean wafer and determining average and standard error of the PWP estimate during these measurement sequence. (Note that "standard error" is a term used to identify a standard deviation of a calculated result or estimate as opposed to a direct set of measured data.) This approach is most useful when the PWP count is determined as the slope of a regressive fit to a sequence of ten or more measurements, where the abscissa or x-axis includes the integer sequence numbers and the ordinate or y-axis is the particle count. See Figure 5. The equations to use in this situation to calculate average PWP, standard error S of this average PWP, and confidence limits ±CF of this average PWP are as follows:

(3)
$$PWP = \frac{\sum\limits_{i=1}^{N} [(x_i - X) \times (y_i - Y)]}{\sum\limits_{i=1}^{N} [(x_i - X)^2]}$$

$$(4) \quad S = \sqrt{\frac{\sum_{i=1}^{N} [(y_i - Y)^2] - PWP \times \sum_{i=1}^{N} [(x_i - X) \times (y_i - Y)]}{(N-2) \times \sum_{i=1}^{N} [(x_i - X)^2]}}$$

$$(5) \quad \pm CF = S \times t_{1 - \alpha/2, \nu}$$

where

$$X = (1/N) \times \sum_{i=1}^{N} x_i = \text{the average sequence number.}$$

$$Y = (1/N) \times \sum_{i=1}^{N} y_i = \text{the average particle count.}$$

$i = 1,2,3, \ldots N$ = the sequence number for the various measurements taken in sequence.

ν = degrees of freedom, which is two less than N in this case, since the otherwise N independent measurements of particle count have now been related with two regression constants: average count (Y) and slope (PWP).

t = Student's normalized deviate for distribution of small-sample means, where the subscript ν is the degrees of freedom of the measurement made and where the quantity α in the other subscript is the difference between unity and the desired confidence expressed as a fraction of unity. For example, if 95% confidence is desired, then $\alpha = (1 - 0.95) = 0.05$ and $1 - \frac{\alpha}{2} = 0.975$.

For "95% confidence" calculations ($\alpha = 0.05$), the following short table can be used for obtaining values to use for the Student's normalized deviate t.

Table 1: Values for Student's Normalized Deviate t for 95% Confidence Calculations

ν	1	2	3	4	5	25	100
t	12.71	4.30	3.18	2.78	2.57	2.06	1.96

These equations can be applied to the sequential measurements of a single wafer or to the average count of a group of wafers in a cassette. If you are unfamiliar with terms like statistical confidence, standard deviation, standard error, degrees of freedom as applied to a set of numbers, and Student's deviate, then refer to the reference given on statistics.[26-32] Note that "statistical" confidence limits are like tolerances on the accuracy of an estimated parameter. I can use confidence limit calculations to make a statement like this one: "The average PWP count is 1 particle per wafer per pass for particles greater than 0.3 μm in diameter. If I were to redo the set of experiments and measurements 100 times again, I would estimate that 95% of the 100 results would fall within $\pm$CF counts of the value I have already determined for average PWP."

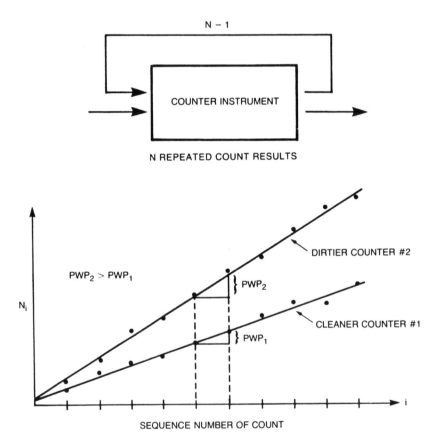

Figure 5: Determining PWP of counters. N_i are accumulated counts of particles per wafer measured sequentially at steps i = 1, 2, 3, . . . N. A counter instrument PWP is the slope of the straight line fit to these measurements. N_i can be the average count over a group of wafers in a cassette or for just a single wafer like the top wafer in a cassette.

EXPERIMENT STRUCTURES FOR MEASURING EQUIPMENT CONTAMINATION

Innumerable choices exist for structuring a particle counting experiment to evaluate a PWP measure for a piece of equipment. The following descriptions are of two structures that have proven most flexible, useful, and practical.

Experiment Structure 1

Structure 1 is most suitable for evaluating equipment that is known to be clean enough that wafers can be passed through it numerous times and still not cause cumulative particle counts to exceed about 100, e.g. 10 particles per wafer per pass for each of ten passes. This structure is presented in Figure 6. Each time wafers are measured they are measured M times and afterwards they are passed through the process equipment under test L number of times. The execution loop depicted in Figure 6 is repeated N - 1 times. And when this loop terminates after N - 1 repetitions, the structure is completed with one additional set of M measurement. The total number of measurements taken is N x M. The total number of equipment passes is (N - 1) x L.

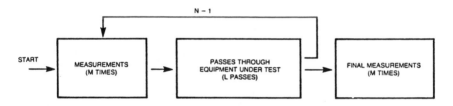

Figure 6: Experiment of Structural Type 1.

With each set of M measurements, the M number of count results can be averaged and replaced with one average value of particle count. This averaging of each set of M measurements has the advantage of reducing the standard error for the calculated result of PWP. These average values can be called y_i, and they can be used with their sequence numbers i = 1,2,3, . . . ,N to calculate results with the equations presented above for PWP, S, and ±CF. These new results we will call PWP', S', and ±CF' since they apply to the average number of particles per wafer added as the result of L passes, not "per pass." The PWP' value calculated can be divided by L to obtain a true PWP result (i.e. "per single equipment pass"). And the standard error S' and confidence limit ±CF' can each be divided by the square root of L to roughly estimate S and ±CF on a per pass basis.

The use of the factor square root of L derives from the assumption that variances with each equipment pass are on the average additive, and that standard error and confidence limits are proportional to the square root of variance. Thus the standard error and confidence limit of the particle accumulation over the total of L number of passes should be approximately equal to the root mean squared of the L number of individual variances of particles added in each pass.

This structure has the advantage that it can be modified according to need by altering the values selected for M, L, and N. In general though, this structure should only be used when N can be made 5 or larger. M should be at least 3, and in fact doesn't really have to be a constant. L can be almost any number except it should be large enough to assure that the uncertainty of the particle counter's PWP is itself not large enough to overwhelm the particle addition added by the equipment. Usually it is best that L have a value of 1 so the true variability of individual passes through the equipment can be detected. In cases where the number of particles added by a single pass through the equipment is not at least an order of magnitude larger than that added by a single pass through the counter, the value of L should be chosen significantly larger than 1.

The PWP value calculated with the above method is in fact the total number of particles (particles larger than a specified threshold size) added to a wafer on the average by not only one pass through the equipment, but also includes the average number added by M passes through the counter. This can be corrected by subtracting, M times over, the PWP of the counter as determined separately (and as described above under particle counter repeatability). In addition, the standard error and confidence limit values for the particle contribution of the equipment alone can be calculated. The appropriate equations follow, where subscripts c and e denote values for the counter alone and for the equipment alone, respectively, and where the value of N_c was the number of measurement repetitions made in evaluating PWP_c:

$$(6) \qquad PWP_e = (PWP' - M \times PWP_c)/L$$

$$(7) \qquad S_e = \sqrt{[(S' \times S') - M \times (S_c \times S_c)]/L}$$

$$(8) \qquad \pm CF_e = S_e \times t_{1-\alpha/2,\nu}$$

where:

$$(9) \qquad \nu = \frac{\left\{ (\dfrac{S' \times S'}{L}) + (\dfrac{M \times S_c \times S_c}{L}) \right\}^2}{\dfrac{(S' \times S'/L)^2}{N+1} + \dfrac{(M \times S_c \times S_c/L)^2}{N_c+1}} - 4$$

and where ν is evaluated as in equation 9 above but rounded upward to the nearest whole integer value. Note that S is calculated as a function of the number N of measurement averages made (N from Figure 6), while S_c was calculated as a function of N_c which was the number of measurements made in evaluating counter repeatability (Figure 5). [Note: Equation 9 has been contrived by the author based on an existing expression in the literature for degrees of freedom when comparing estimates of two small sample means for the case when the standard deviations (or standard errors) about the two are unknown and unequal. The major dissimilarity between the situation here and that for the already established expression is that here we are taking a difference of means which have been determined by respective linear regression, and in this case therefore, the determination of each mean slope has removed two degrees of freedom from each set of data.]

Experiment Structure 2

The second structure for experiment executions is designed for those cases where it is either impractical to execute successive equipment passes because the wafers would be too dirty to use in the successive passes, or because of logistical reasons as when the equipment can not be made available for more than one pass at a time and on only an occasional basis.

Figure 7 depicts one execution of a particle count experiment using this second structure. The structure may be repeated as many times as is practical, but with each execution another sample estimate is calculated for PWP. Each of these individual sampling experiments for PWP begins with repeated measurements of the initial particle count on fresh, clean wafers. Initially, the wafers might be measured P times. If the number of particles found on the wafers is near zero, or at least very much less than the expected count to be added by passing them through the equipment under test, then the value chosen for P may be only 1. Following these initial measurements, the wafers are then passed through the process equipment one or more times (e.g. L times). The value selected for L depends upon how many times it is expected that it will take to accumulate many more particles than represented by the uncertainty of the particle counter's counts. Finally, the wafers are repeatedly remeasured Q times.

Structure 2 is in fact a special case of structure 1, where N = 1 and where M is replaced by P for the initial measurements and by Q for the final measurements. Although it is structurally a special case of structure 1, it deserves to be called a separate structure since it is uniquely suited for making measurements on most pieces of process equipment today. The first of two principle reasons is Murphy's law; it seems in particle measurement work with clean test wafers that something is always going wrong, like a wafer getting hung up in an equipment elevator or on an air track. Since these events easily cause particle accidents to the wafers, the data immediately become suspect and may no longer be representing the PWP of the equipment under test. And it is not always the hardware of the process equipment that may be at fault; many times it is the operator, or the hardware of the particle counter instrument, or the room in which the experiment is conducted. The second principle reason structure 2 is uniquely suited today is that many pieces of equipment can add large enough numbers of particles to wafers in one or only a few passes to make continuing accumulation measurements on these wafers imprecise and impractical.

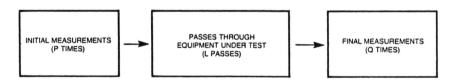

Figure 7: Experiment of Structural Type 2.

If this structure can be executed many times, say R times, then the equations are relatively simple that are needed to calculate the average PWP, the standard error of this average, and the confidence limits around this average. First calculate the average of the initial P measurements. Then calculate the average of the final Q measurements. Subtract the first result from the second to get a PWP' measurement that still includes some of the particles added by the counter. If the impacts $P \times PWP_c$ and $Q \times PWP_c$ are small compared to this total PWP', then for simplicity of computation, PWP_e can be approximated by the result: $PWP_e = PWP'/L$.

If the contamination PWP_c caused by the counter is significant, it should be removed from the final result. This can be done either of two ways. One way, of course, is to assume the impact of the counter's contamination is a constant and is known, this can be determined separately as described previously. If this is the approach taken, then the corrected PWP_e can be obtained as

$$(10) \qquad PWP_e = \left(PWP' - \left\{[(P+Q)/2] \times PWP_c\right\}\right)/L$$

The second approach to mathematically removing the contamination added by the counter is a little more involved. In Figure 8, the variable along the x-axis or abscissa is measurement sequence number. Note that the distance along this axis between the last of the initial measurements and the first of the final measurements is equivalent to one jump in sequence number. Note too that a linear regressive fit has been made to the three (P = 3) initial measurements, and one also has been made to the three (Q = 3) final measurements. Finally note that an estimate for $L \times PWP_e$ is taken as the difference along the y-axis (particle counts) between a forward-extrapolated point from the initial counts and a backward-extrapolated point from the final counts. An advantage of this method is that it makes a correction based on counter contamination occurring during the actual measurements of this one experiment. However, if counter contamination is very low, the slopes of the two linear fits, which are to represent the contamination rate of the counter per measurement, may be more a property of the counters variability in count (precision) than on counter contamination. Which of the above two correction methods should be used in any one case will have to be left to the judgment of the experimenter.

Once a number of values for PWP_e have been obtained using repetitions of the experiment structure number 2, i.e., obtained from multiple experiments, and if necessary once they have each been corrected to remove the effect of contamination by the counting instrument, then they can be averaged to obtain a summary result $PWP_{e, avg}$. And calculations can also be made for estimates of the standard error in this average S_{avg} (deviation between averages) and for confidence limits $\pm CF_{avg}$ within which repeated collections of experiments like the first would be expected to fall 95% of the time. (For a Gaussian distribution, 95% confidence limits work out to be approximately two standard deviations to either side of the average, whereas 99% confidence equates to approximately three standard deviations to either side.) The following equations can be used:

$$(11) \qquad PWP_{e,avg} = \frac{1}{R} \times \sum_{i=1}^{R} PWP_{e,i}$$

$$(12) \qquad S_{avg} = \sqrt{\frac{1}{R} \times \sum_{j=1}^{R} \frac{(PWP_{e,j} - PWP_{e,avg})^2}{R-1}}$$

$$(13) \qquad \pm CF_{avg} = S_{avg} \times t_{1 - \alpha/2, \nu}$$

$$(14) \qquad \nu = R - 1$$

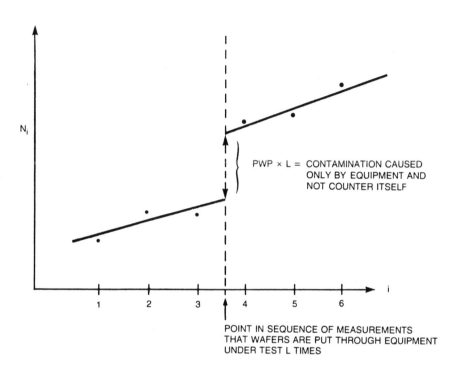

Figure 8: One method of removing the contamination contribution of particle counter instruments from the measurement of PWP (for process equipment under test using Structural Type 2 testing) is by extrapolating linear regression fits made independently to initial and final measurements. N_i is the individual particle count per wafer (for a single or group of wafers). The abscissa i is the sequence number of the measurement. Counts 1, 2, and 3 are initial measurements. Counts 4, 5, and 6 are final measurements.

One last bit of useful statistics can be applied to single executions of experiments of the form given by structure 2. Since data taken in a single execution of this form cannot tell us anything about how much variability there is in the contamination contributed to wafers by the equipment, calculating confidence in the calculated PWP result may at first seem inappropriate. But we can question how much uncertainty there is in the measurements themselves (assuming for the time being that the counter is accurately calibrated). Recall that we would have measured the wafers P times before additional particles had accumulated on them and then Q times afterward. The scatter in values among the first P values and the scatter among the final Q values cause uncertainty whether or not the difference in their averages is a correct measure of what was added to the wafers. It is possible, knowing the number of times the wafers were measured and the scatter in their data, to estimate confidence limits in the repeatability of the counting instrument to determine whatever particles may have been added. It should be emphasized that this is not the confidence in the repeatability in the number of particles caused by the equipment under test; and for this reason we will call it CF_c instead of CF_e. The required equations, assuming for simplicity that $PWP_c = 0$ and $L = 1$, are given as follows:

$$(15) \qquad \bar{N}_p = \frac{1}{P} \times \sum_{p=1}^{P} N_p, \text{ average initial count}$$

$$(16) \qquad \bar{N}_q = \frac{1}{Q} \times \sum_{q=1}^{Q} N_q, \text{ average final count}$$

$$(17) \qquad PWP_e = \bar{N}_q - \bar{N}_p, \text{ estimate for particles added}$$

$$(18) \qquad s_p = \sqrt{\sum_{p=1}^{P} \frac{(N_p - \bar{N}_p)^2}{P - 1}}, \text{ standard deviation of initial counts}$$

$$(19) \qquad s_q = \sqrt{\sum_{q=1}^{Q} \frac{(N_q - \bar{N}_q)^2}{Q - 1}}, \text{ standard deviation of final counts}$$

$$(20) \qquad \pm CF_c = \sqrt{\frac{s_p^2}{P} + \frac{s_q^2}{Q}} \times t_{1 - \alpha/2, \nu}$$

where the degrees of freedom ν is calculated as the closest integer value equal to or larger than that given by:

$$(21) \qquad \nu = \frac{[(\frac{s_p^2}{P}) + (\frac{s_q^2}{Q})]^2}{\frac{(s_p \times s_p/P)^2}{P + 1} + \frac{(s_q \times s_q/Q)^2}{Q + 1}} - 2$$

DESIGN OF EXPERIMENTS

Figure 9 shows a typical example of what can be expected of the magnitude of confidence limits which bracket the average of a group of measurements. As the number of experiments in the group increases, confidence improves, rapidly at first and then more slowly as the number of measurements gets larger. The equations above for confidence limits [Equations (5), (8), (13), and (20)] all reflect this same trend. Notice that as the number of degrees of freedom in the data increases (i.e., ν increases) the value of the t factor in these equations drops rapidly from $\nu = 1$ to say $\nu = 4$ or 5, and thereafter decreases more slowly. In addition, in the expression for confidence limits around sample means, the squareroot of the number of samples in the mean appears in the denominator. That is, as the number of measurements in the group increases, one factor decreases as the inverse squareroot of the number of measurements.

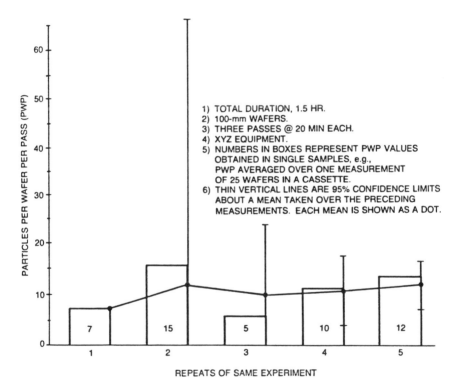

Figure 9: Repeated PWP measurements permit calculation of a mean value. As the number of PWP measurements is increased, the confidence ranges placed about the means diminish. With only a single sample, there can be no confidence that the PWP measure obtained represents a repeatable result.

These relations allow one to calculate in advance how many measurements need to be made to determine an average to within any arbitrarily narrow limits—provided that the standard deviation of the group of measurements is known a-priori. In such cases, equations for confidence limits can be solved iteratively for the correct number of measurements. Usually though, when measuring contamination to wafers caused by process equipment, the standard deviation of the equipment's contamination characteristics is not known beforehand. In other words, a numerical calculation of how many experiments must be made cannot be performed until at least one group of measurements is taken to determine the standard error of $PWP_{e, avg}$.

It is very important to perform control measurements to determine if any of the required wafer handling steps during execution of a particle measurement experiment is making a significant contribution to the total number of particles accumulated on the wafers. For example, the counter repeatability experiment is one such experiment. Another example is running an experiment to determine the effect of carrying wafers back and forth between the equipment and the counter where particles from the environment along the way could fall out on the wafers. This experiment might proceed as shown in Figures 6 or 7 but the passing of the wafers through the equipment would be omitted and the wafers would be returned to the counter from the equipment just as soon as they had reached the equipment.

Another, similar, control experiment would consist of carrying wafers to the equipment, placing them in the proximity of the equipment under test, letting them sit there for the length of time that would normally be required for the wafers to be put through the equipment, and then proceeding back to the counter for counting again. This would reveal how many particles accumulate on the wafers because of the time they are exposed to the environment surrounding the outside of the equipment.

The above examples cover some basic controls. However, much more can be found about the causes of equipment contamination by designing experiments in which only one factor is changed at a time. In this way, comparing PWP results with and without each factor can reveal the causes of contamination.

Incidentally, statisticians might suggest what are called screening experiments[32] in which more than one factor at a time are either included or excluded in each experiment; this reduces the number of experiments needed to test all factors for their influence. But screening experiments are of little use until all of a set of preplanned experiments have been performed. For this reason, I have found them of limited application to particle experiments. The reason is that what is learned with each experiment execution is most often cause to significantly change any plans for the experiments which are to follow. With this dynamic state of affairs, it is not always practical to predesign a large array of experiments.

CALIBRATION

No manner of statistics can make up for lack of inaccurate instrument calibration. Although detailed description and analyses of calibration techniques and performances have been published for aerosol particle counters,[33-36] nothing comparable has appeared on surface particle counters.

Today's surface particle counters detect and count particles on the basis of the particles' abilities to scatter incident light to a photosensor. If a particle is too small and cannot scatter sufficient light intensity onto the detector, then the particle cannot be detected and counted. If two or more such small particles lie close enough together on the surface of the wafer being analyzed, then together they may scatter sufficient light to be detected as one particle. All of these counters, in one way or another, classify the signals they detect on the basis of their amplitudes. Often this classification is interpreted to be a classification of particle size, but as will be discussed in paragraphs to follow, this size-interpretation is not strictly correct and can lead to interpretation errors if not appreciated.

Two calibration requirements exist with particle counters. One is to calibrate the ability of the counter to count a group of particles accurately. The second requirement is closely related to the first; it is to accurately discriminate what is to be counted from what is not to be counted. These two requirements, though different, are really so closely interrelated that they must be handled together. The reason for this derives from the nature of the relationship between particle sizes and their light scattering amplitudes; this relationship, expressed with scattering amplitude as the independent variable, is not linear or single valued.

Calibration of Gain and Offsets

Figure 10 shows what we would be dealing with if the relationship between scattered signal amplitudes and some particle parameter like its projected area were linear, and therefore, single valued. (Actually, in this figure, the independent variable shown along the x-axis is called "particle scattering cross-section" and is only roughly related to a cross-sectional area of the particle. For the time being, let's assume the two are linearly related.) In this figure, it can be seen that a change of instrument gain can cause the count results to change.

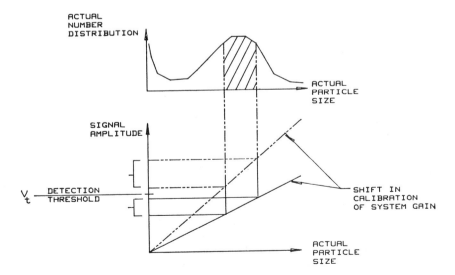

Figure 10: Effects of gain change on particle counts.

For example, if the voltage limit of detectability is set at a voltage of V_t, and the gain in the detector increases (from the solid sloped line to the dashed sloped line in the figure), then particles that once fell below the detection threshold and couldn't be counted could now be counted.

Obviously then, holding gain settings constant is critical to meaningful particle counts. And the best way to do this is to adjust gain using reliably repetitive signals generated from some kind of reference scattering target like calibrated latex spheres on a wafer surface or accurately grown or etched features on a wafer surface (or on any surface whose light interaction properties remain constant over time). Alternatively this calibration signal could be achieved with any device which redirects a calibrated proportion of the incident light into the detector.

Calibration of Count Accuracy Versus Size

Once this calibration of gain relative to threshold settings is fixed, what remains is to ensure that all legitimate signals occurring above the desired thresholds are counted, and only these, and that some physical meaning related to particles is ascribed to the threshold settings.

One can detect whether or not detectivity is constant by observing the positions of detected particles on a wafer-map display (most counters have these) as the wafer is rotated (and shifted if possible) and remeasured.

To ensure that only what should be counted is counted, reference wafers with known artifacts are indispensible. The author has had good luck with fluorescent latex spheres dispersed with automated particle dispensing equipment, using appropriate heaters to dry the spheres before they reach the wafers, and using very dilute suspensions of particles in the generator's feed supply and by inspecting the deposited spheres under a dark-field optical microscope. These fluorescent spheres make the difficult task of finding focus on the surface of polished wafers relatively easy. They also greatly simplify searches over the wafer surface to find the particles. Finding particles is unavoidably a time consuming labor of love when particle counts on the wafers are low enough not to cause a high incidence of multiple particles within pixels of the particle counter. In short, these fluorescent spheres have proven very successful in being able to compare the automated particle counting counts with visually counted numbers of very monodisperse distributions of particles.

Ascribing some intuitively meaningful particle attribute to particle detection thresholds is still another matter. Figure 11 depicts the situation at a wafer surface where incident light is made to interact with a particle (and the surface it is sitting on) to redirect a portion of this incident light into directions other than the direction the light would normally reflect from the bare wafer surface had the particle not been there. This redirection of the light is referred to as "scattering of the incident light."

There is no existing mathematical solution to this problem of relating the portion of incident light, which is scattered by a particle into a particle counter's detector, to the particle size (not to mention particle shape, index of refraction, wavelength, surface texture, etc.). Elegant solutions do exist for scattering from very small particles (Rayleigh scattering) and from perfect spheres suspended away from other physical surfaces.[37] They also exist for spheroidal shapes in free space,[38] and for rough convex particles in free space.[39] They even exist for scattering from two neighboring spheres in free space,[40] from cylindrical and spherical bosses on perfectly conducting surfaces, and from cylinders on nonabsorbing planes.[41,42] But they don't exist for single particles of arbitrary shapes, or even for spheres, on wafers; so the signals which are obtained by automated particle counters cannot be related closely to theoretical models.

Thus, counts obtained by surface particle counters can rarely be directly interpreted as the number of particles that lie on a wafer and have some physical attribute, such as size, that is larger or equal to some identifiable threshold value. The attribute that can best be related to the counter thresholds is an optical rather than a physical one and is called the particle's scattering amplitude. This represents the fraction of incident light a particle will scatter and that the particle counter can detect, normalized by the amount of incident light that intercepts the cross-sectional area of the particle.

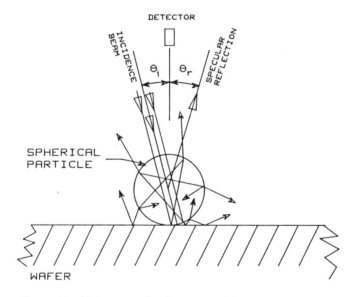

Figure 11: Light scattering from a particle on a substrate.

Light Scattering by Particles on Surfaces: It is reasonable to try to get at least a qualitative appreciation of what scattering cross-sections might look like, as a function of particle size, for scattering from a particle on a surface.[43] This can be done by inspecting the results of the theoretically calculated relationships for simpler cases. Figure 12 shows the relationship between sphere diameters in free space and "scattering amplitudes" from what is called Mie theory. Gustaf Mie in 1908 exactly solved the electromagnetic boundary value problem for scattering from a perfectly shaped homogeneous sphere in free space. Figure 12 shows some results of this theory for particle sizes up to 20 μm in diameter, where the material is latex and the light is of the popular He-Ne wavelength of 633 nm. Although difficult to see all four plots, two of the curves represent scattered light that is incident on the particle with plane-polarized light, and the other two, with parallel-polarized light. The vertical axis is a quantity that is related to scattering cross-section by a constant of proportionality divided by the square of the distance from the particle to the detection point. Two of the curves represent how much light would be scattered in a direction 15 degrees from the direction of the original incident light, and the other two curves represent how much of the light would be scattered 165 degrees from the direction of the incident beam (which is 15 degrees from the backward direction).

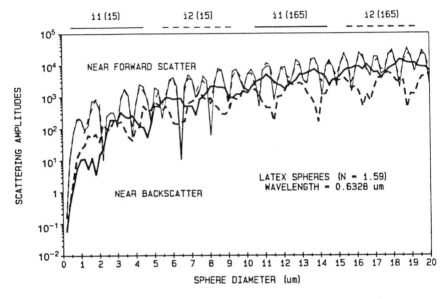

Figure 12: Mie scattering (particle diameters, 0.2 to 20.0 μm).

With reference to Figure 11, it is easy to imagine how much more complicated these relations might become when another physical boundary like the surface of a wafer is brought in proximity to the sphere doing the scattering. The multivalued nature of these relationships might change in their detailed structure and even in average magnitude, but they should retain much of the same nonlinear character. It is true that today's counters integrate the light scattered in many directions other than just the two shown in this example, and that counters which use white light rather than monochromatic light will tend to average out a lot of the variations shown, but not all.

Thus, what a particle counter "sees" is not easy to describe. If particles of sizes that will scatter light both above and below the instrument's threshold are present, and furthermore, if they are of a variety of shape and material composition, then the instrument will most likely detect some smaller particles of one type and only larger ones of another type. This predicament means that it is conceivable that counts could go up if particles of one type are replaced with an equal number of particles of the same size but of different shape or material composition. Or the count might go down in a similar case but where the materials and shapes are different and the sizes and numbers remain the same. Fortunately, experience seems to indicate that when a calibrated particle counter's count increases or decreases that this is a true representation of the correct change in the particle population.

In summary, although it may not be possible to have a particle counter's detection thresholds closely related to particle sizes of what will be counted, it is possible using latex spheres to calibrate the instrument response so that it remains unchanging with time, and be able to say that particle counts obtained are of particles whose optical cross-sections equal or exceed that of the calibrating latex sphere size.

Calibrations on Films. On top of all the ambiguities which arise in giving physical meaning to the characteristics of particles that can be detected with a given set of calibrated thresholds, another complication can arise if it is necessary to ascertain counts of particles lying on top of a dielectric film, like nitride or oxide.

Whenever counting is to be done on a surface, calibration is necessary using calibrating artifacts lying on the type of surface in question. Figure 13 shows the optical reflection characteristics of an oxide film on silicon. This plot has been generated from an exact mathematical model for reflection from a homogeneous dielectric film of uniform thickness lying on a flat dielectric substrate.[44-46] The general solution can handle any number of films in stratified layers, and any of them including the substrate may be partially absorbing. What should be observed from this and other plots like it for other films is that the reflectivity of the wafer changes very significantly as the film thickness changes. Reflectivity oscillates between maxima and minima as the thickness of the top film grows; the period of oscillation is defined as thickness changes equal to one-half of the wavelength of the light used, evaluated within the film media.

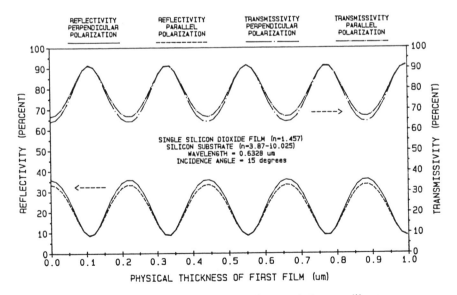

Figure 13: Thin-film reflection and transmission on silicon.

Within the film, the wavelength will be a little shorter than in free space, and the cosine of the angle of incidence is also a factor. Away from normal incidence, the amplitude of the oscillation also depends on the polarization of the incident light. In short, to get dependable counts on films, since light scattering by the particle is going to depend also on film thickness and its dielectric constant, threshold calibrations should be made using latex spheres on each combination of material and thickness of interest.

If particles that lie wholly within a film or underneath a film are to be counted, then calibrations to handle these cases will also be a requirement. However, if particles lie in all of these different situations (on, in, and below films), then it is not possible to achieve a calibration that is consistent with all three at once. Thus, it would appear that such situations with films are going to be very ambiguous when it comes to interpreting particle count data. Perhaps particle counts can accurately indicate trends toward cleaner or dirtier wafers in these cases, but cannot accurately be ascribed to a size threshold.

Pixel Saturation. A final consideration regarding calibration of count should be noted. If there are sufficient numbers of particles present on a wafer, then there is increasingly higher probability that some of these particles will lie close enough together that the particle counter will detect them as being a single particle of larger size. It then follows that the total reported count may be undervalued. For counters with many thresholds for the purpose of determining "size" distribution, this effect will also shift counts toward the larger "sizes."

If one can define a fixed pixel size within which a counter cannot discriminate two or more particles, then a simple equation expresses the probability of obtaining more than one particle within a pixel, given the pixel size and the total number of particles landing on a wafer. This will be presented in the following paragraphs, complete with a couple of examples of undercounting that can occur.

Suppose a counter analyzes a wafer into a number P of pixels. And suppose there are a number N of particles on this wafer, distributed randomly. Then the probability of finding a number T of particles within any one pixel can be expressed as Q_T, where:

$$(22) \qquad Q_T = {}_NC_T \times (1/P)^T \times [1 - (1/P)]^{N-T}$$

In this expression, the factor ${}_NC_T$ represents the number of combinations that one could select T of the particles from the set of N particles; this can be evaluated from the following equation:

$$(23) \qquad {}_NC_T = \frac{N!}{(N-T)! \times T!}$$

The number of pixels with T particles in them is then most probably $P \times Q_T$. If we sum up all of the particles that are likely to be found together with other particles in common pixels (where T represent how many are in a pixel), the following equation results for calculating the number N' of pixels which have at least one or more particles:

$$(24) \qquad N' = N - \left\{ P \times \sum_{T=2}^{N} [(T-1) \times Q_T] \right\}$$

where the summation term can be terminated at T = N or when the term diminishes to less than unity, whichever occurs first.

Consider for example a 3 inch wafer with a 10/32 inch edge exclusion which is covered with 4536 pixels in the counting area, each pixel measuring 1/32 inch on a side. If this wafer has 2000 particles on it, randomly distributed, then there will likely be 1287 pixels containing a single particle each, 284 pixels containing 2 particles each, 42 pixels containing 3 particles each, and 5 pixels containing 4 particles each. This means that the counter would report that it has detected a signal in exactly 1287 + 284 + 42 + 5 = 1618 pixels, that is it would report a particle count of 1618 instead of 2000. This is an error of just over 19 percent. If accurate particle measurements are necessary, as they are when taking a difference in count between wafers before and after a process step, then it is best to begin with clean wafers so that the final count is kept as low as possible—in this case much below 2000 counts. Figure 14 is a plot of this example case over a very wide range of particle counts.

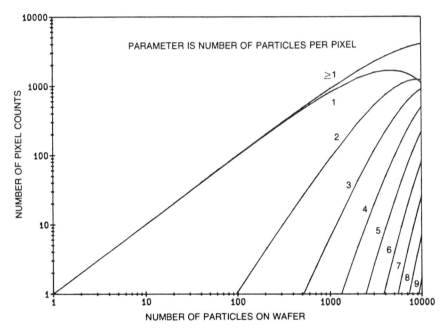

Figure 14: Statistics of multiple particles per pixel.

An equation of the same form as Equation 22 can be derived to represent the probability of two or more particles lying within an area the size of a scanning illumination beam. This occasion would account for loss of counts caused by insufficient spatial resolution of the beam itself. An equation of this same form can be derived for the probability of two particles straddling adjacent pixels to correct overcounting caused by counting such particles twice. All of these results can then be combined to find the net counting error caused by the above three effects.

In general, the counting errors just described relate to an instrument's pixel size and laser beam spot size, or both, depending on the type of counter used. These errors can be kept to an insignificant percentage of the total count if the number of particles on the wafer is small relative to the total number of pixels or resolution elements spanning the wafer.

MEASUREMENT PLANNING

Before beginning a careful experiment, consider every detail of how the experiment will be conducted. If this is done, a lot of false starts and wasted experimentation can be avoided. Similarly, be very clear about just what questions the experiments are to answer.

The following is a checklist of things to consider:

1. Consider how much of a process performed by the process equipment under test can be conducted and still permit wafers that exit the equipment to be read by your particle counter. And if some of the process steps have to be left out, consider just how this be accomplished.

2. Decide how many process passes to execute between wafer measurements.

3. Decide how many wafers to process at a time, where they will be placed within cassettes, what changes will occur in slot locations during the experiment and if this could cause certain wafers to accumulate more particles than others.

4. Carefully predict the durations wafers will undergo different environments so that these can be kept constant or purposefully changed for comparison purposes.

5. Carefully think through every aspect of wafer handling by both the operator and the equipment as each of these is a potential contributor of particles.

6. Anticipate reversals in the sequential order of wafers within cassettes to keep track of the top and bottom wafers (or end wafers in general) since some of these might be expected to capture a greater number of environmental particles during the time cassettes are sitting in input and output elevators, both in the equipment and in the particle counter itself.

7. Analyze the need to and, if necessary, calculate separate results for certain wafers within the cassette. Top and bottom wafers are one example.

8. Define control experiments so that PWP results can be compared to other performances like the contamination caused by the counter or by contamination from the environment when wafers sit out in the same kind of environment as the equipment for the same length of time as a process/measurement cycle.

9. Plan on placing wafers as stationary monitor wafers on top and around the equipment to collect environmental particles, and particles injected into the environment by the equipment, during the course of the equipment processing steps. (This can be thought of as another control experiment.)

10. Decide which computational facilities will be used to aid in the analysis of the collected data. Select the format in which raw data will be recorded and how it will be input into automated computational tools.

Measurement Records

As an aid in record keeping during particle counting experimentation, the following is a checklist of things to record:

1. Classes of the environment at the particle counter, at the site of the equipment under test, and in the path between the two.

2. Air flow directions and velocities over, around, and perhaps even through the equipment under test.

3. Temperature and humidity of the environment.

4. Presence of operators and notes about their potential impact on the cleanliness of the wafers.

5. Instruments used, their states of calibration, and their settings. Also locations of aerosol particle counter probes and locations of monitor wafers. (Be cautious of the possible influence of makeup air which enters equipment as the result of volumes of air being removed by aerosol particle counters sampling relatively closed spaces.)

6. Timing of both counting and processing activities, and snafus in wafer handling.

7. The location and identity of potential particle sources (e.g. pneumatic switches, cylinders, motors, frictional parts, and triboelectrically active material combinations).

8. Locations of cooling fans, scrubber exhausts, and thermally induced convection. (These can move dirty air from the floor and the bowels of equipment upward toward wafer handling regions of the equipment.)

Measurement Execution

Great care should be exercised in performing and documenting minute details of every execution of an experiment. In particular, note all unexpected and unusual occurrences—these notes can prove very valuable, after the fact, to explain anomolies discovered at some later time in the results when it may be too late to go back and repeat the work. An example of detail necessary might be recording whether plastic or metal cassettes were used, what scrubber exhaust flow rate was in effect, and that the operator had to assist wafer number 3 in the second pass. Prepare to deal calmly with the results of Murphy's law, to readily redesign an experiment on the basis of preliminary results, to repeat the experiment sufficient times to determine a useful confidence limit in the statistical validity of the results. Above all, plan in advance to patiently persevere. The road is not easy, but you can pass over it more easily if you know this in advance.

A computer interface to today's particle counters is almost mandatory if extensive particle counting investigations are to be performed. Manual entry of particle count data for numerous wafers in a cassette that is to be measured many times can become quite tiresome and error prone. Special purpose software or even analysis tools like computer spreadsheets having built in statistical functions are a must.

Once reduced PWP data has been obtained, then a standard of comparison is needed. This subject is addressed next.

SPECIFICATION STANDARD

Particle contamination classifications in the past have tacitly assumed larger particles are more harmful than smaller ones.[13,17,47] This may be appropriate for situations which are concerned with cleanliness of surfaces and gas environments where optics and delicate mechanical assemblies are of concern. Larger particles can more easily foul a precision mechanical clearance as in disc drives, and larger particles can cause a greater blemish and degradation of image in lenses or even blockage of optical pinholes and slits. Federal Standard 209B for air classifications and 1246A for surface classifications were both founded on these assumptions.

It is not at all obvious, however, that such classifications are appropriate for yield-related concerns in the manufacture of integrated circuits. On the contrary, weighting factors cannot yet meaningfully be applied to particle sizes above the minimum size of particle which might cause a defect in a particular IC technology. Not enough is known about kill ratios as a function of size, not to mention weighting their varying effect as a function of the material they are composed of and as a function of individual process steps (e.g. the relative potency of particles introduced in lithography compared to those introduced in diffusion). Kill ratio is used here to define the ratio of particles that land on a critical area and cause a defect to those that land on critical areas and may or may not cause a defect. In fact, uncertainty even remains about how small a particle can be and still cause a defect. Some think that the critical particle diameter (major diameter) is 1/10 to 1/3 of the smaller of linewidth and line spacing. Some think it is the thickness of a gate. Others could argue that the critical particle size is much smaller—the size of particle whose material can cause a defect by diffusing into a gate, channel, or other critical device geometry.

Federal Standard 1246A is itself not very useful in any situation since it is based on comparing measured distributions against a set of standardized distributions all of which somewhat arbitrarily have been chosen to have a median size or geometric mean of 1 μm and a geometric standard deviation of between 3 and 5 μm (depending upon how closely you expect log-normal distributions to fit the straight lines on the 1246A standards). Measurements of realistic particle distributions on actual wafers reveal significantly different distributions than those chosen for defining 1246A categories.

I propose that the surface contamination rating scheme used to rate PWP measurements of IC equipment be based on the total particle count detected above a specified critical size, and these rating values be called *critical particle count* (CPC) values. It should not in any way be concerned with the size distribution which may or may not be known above the selected critical size. Note the relative simplicity of such a rating scheme (see Figure 15). Log-log, log-log-squared, log-normal, and other such graphs are not required. All that is required is a single number, and that number is derived by taking the measured PWP results and rounding the PWP number upwards to the nearest decade count and second-decimal-place fraction thereof. For example, a PWP result 1.76 particles-per-pass larger or equal to 0.3 μm, per 100 mm diameter wafer would be classified as meeting a CPC of "1.8 for 0.3 μm size limit on 100 mm wafers." A PWP result of 153 particles-per-pass larger or equal to 0.5 μm, per 125 mm diameter wafer would be classified as meeting a CPC of "160 for 0.5 μm size limit on 125 mm wafers." An edge exclusion of roughly 8 mm would be implied.

Consider an equipment vendor who measures a particle distribution over a wide range of sizes and finds that this equipment adds only one 10 μm particle to each wafer and adds nothing else. The cumulative particle count for sizes equal to and above an anticipated critical size of 0.3 μm is therefore equal to 1. But the cumulative count of sizes equal to or above 10 μm is also 1. The above proposed CPC rating scheme would rate this equipment as 1, i.e. 1 for 0.3 μm size limit on the size wafers the vendor used. Classification schemes analogous to those adopted by others for particles in air and for particles on equipment surfaces[13,47] would classify the performance as causing particle counts-per-wafer orders of magnitude higher than 1, and this would be unfair in this circumstance.

This proposed rating system is much different than classifying cumulative distributions on the basis of standardized distributions. With standardized distributions, the counts of the measured distribution must everywhere lie below the counts of one of the members of a family of standard distributions to be rated as equal to or better than the classification value assigned to that member (see Figure 15).

Note too that this proposed rating system is based on counts-per-wafer and not per-square-centimeter or other unit area. This is for convenience in handling numbers. Today's clean equipment can produce PWP results below 1 count/cm^2 and this leads to expressions of decimal fractions of a particle (a little awkward). In terms of particles added per wafer, however, most equipment will produce PWP and CPC numbers ranging from 1 to 1000. In particularly clean equipment like a SMIF-equipped inspection station, numbers below 1 (e.g. 0.01) may result, but this is not as awkward as an approximately equivalent number of 0.0001 per square centimeter on a 100 mm wafer.

Ratings should be based on an average PWP count, where this average has been taken over a minimum of three individual PWP counts. Each individual PWP measurement may itself be an average taken over a number of successive passes of the test wafers through the equipment, with measurements taken only before and after the particular sequence of equipment passes; this is often necessary to obtain a large enough count of particles accumulated on the wafers to permit a statistically accurate count of particles added. When a single PWP measurement involves multiple successive passes through the equipment, PWP is calculated by dividing the total of particles accumulated on the wafer by the number of equipment passes executed between measurements; in such cases, as mentioned earlier, some information about the variability of particle accumulation between single passes is lost, but this may be unavoidable.

In addition to the average PWP data, 95% confidence limits should be calculated based upon the standard errors of the multiple PWP values obtained. If these confidence limits aren't given, then at least the standard deviation and number of PWP measurements included in the average PWP should be given so others can calculate the confidence limits at 95% or whatever percentile is desired.

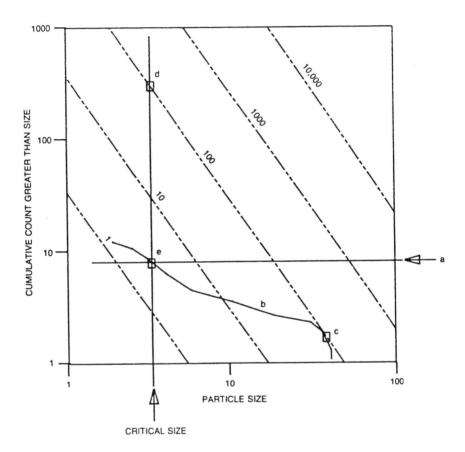

Figure 15: A cumulative count distribution (b) is better compared (e) to a critical particle count (a) than to a classification distribution (dashed lines). The critical particle count shown (a) is slightly less than 10. The measured distribution (b) would traditionally be classified as 100 particles per wafer greater than the critical size because of the intersection (c), but this implies a classification for which cumulative count at the critical size threshold is considerably worse (d) than actual fact.

THE FUTURE

The path toward lowering the particle contamination levels of process, handling, storing, and inspection equipment will be shortened if equipment vendors and users alike adopt standards in the measurement and rating of particle performance based on wafer surface particle counts. I hope the material presented above will accelerate the move toward such standard practices. One important step in this direction would be accomplished if particle counter vendors would automate the particle size calibration of their products and provide data-analysis capabilities to calculate PWP values and confidence limits easily from a collected data base of initial and final particle counts taken before and after wafers are passed one or more times through equipment under test.

Additionally, in the near term, enhancing particle counters to automatically determine particle size distributions, complete with meaningful statistical parameters of the distributions (like geometric mean sizes and geometric standard deviations in cases where the counter's resolution limit reaches below the mode of the distribution) may lead to rapid associations of these parameters with the natures of the particle sources and, therefore, the equipment components or subsystems that cause them.

In the future there will be a need for better and better particle counters, counters that will enable more accurately size-calibrated particle counts on a variety of substrate materials, on films of various complexities, within films, and under films, and on patterned wafers. This ability will greatly improve the methods available to measure particle contamination contributed to wafers by the complete process step performed within the equipment and not by just the wafer handling and chemically inactive aspects. A real need exists to be able to rapidly and objectively scan a wafer having fewer than 10 or 100 particles, to not only detect their presence but also determine the material of which they are comprised.[48-55] Perhaps with such improvements, the particle hurdle will be surpassed and a truly clean wafer fab process may be achieved effectively. Then, production quality IC technology may reach its ultimate physical and electrical limits.

Acknowledgements

The author wishes to acknowledge Hewlett Packard Co. for funding and supporting fastidious research efforts to evaluate, reduce, and control particle generation by process equipment, IC fab facilities, and wafer handling in general. These efforts have not only generated the above practices in particle measurement, but also the SMIF approach to clean wafer handling, and valuable insights into the particle generation mechanisms that exist within process equipment. Special recognition is given to the author's associates and co-contributors in these works; they are, in alphabetical order, Shanti Gunawardena, Ulrich Kaempf, David Thrasher, and John Vietor. For forward-looking guidance and managerial stimulation, recognition is also due Patricia Castro, Ulrich Kaempf, and Mihir Parikh.

REFERENCES

1. Burnett, J., *Microcontamination,* Vol. 3 (9), pp. 16-18, 98 (Sept. 1985).
2. Baker, E., *Microelectronic Manufacturing and Testing,* Vol. 8 (7), pp. 20–22 (July 1985).

3. Hardegan, B. and Lane, A., *Solid State Technology,* Vol. 28 (3), pp. 189-195 (March 1985).

4. Logar, R. and Borland, J., *Solid State Technology,* Vol. 28 (6), pp. 133-136 (June 1985).

5. Laneuville, J., Marcoux, J. and Orchard-Webb, J., *Semiconductor International,* Vol. 8 (5), pp. 250-254 (May 1965).

6. Tolliver, D., *Proceedings of the Electrochemical Society, Conference on VLSI Technology,* Toronto (1985).

7. Beeson, R., *Hardware Cleaning and Sampling for Cleanliness Verification,* Designated ISBN:0-915414-72-4, Mt. Prospect, IL, The Institute of Environmental Sciences, 34 pages (1983).

8. Morrison, P. and Yevak, R., *Semiconductor International,* Vol 8 (5), pp. 208-214 (May 1985).

9. Gunawardena, S., Haven, R., Kaempf, U., Parikh, M., Tullis, B. and Vietor, J., *The Journal of Environmental Sciences,* Vol. 27 (3), pp. 23-30 (May-June 1984).

10. Parikh, M. and Kaempf, U., *Solid State Technology,* Vol. 27 (7), pp. 111-115 (July 1984).

11. Parikh, M. and Bonora, A.C., *Semiconductor International,* pp. 222-226 (May 1985).

12. Gunawardena, S., Kaempf, U., Tullis, B. and Vietor, J., *Microcontamination,* Vol. 3 (9), pp. 54-62, 108 (Sept. 1985).

13. *Product Cleanliness Levels and Contamination Control Program,* Designated MIL-STD-1246A, Philadelphia, The Naval Publications and Forms Center (1967).

14. *Standard Method for Measuring and Counting Particulate Contamination on Surfaces,* Designated F24-65(76), Philadelphia, American Society for Testing and Materials (1976).

15. *Sizing and Counting Airborne Particulate Contamination in Clean Rooms and Other Dust-Controlled Areas Designed for Electronic and Similar Applications,* Designated F25-68(79), Philadelphia, American Society for Testing and Materials (1979).

16. *Procedures for the Determination of Particulate Contamination of Air in Dust-Controlled Spaces by the Particle Count Method,* Designated ARP-743A, Warrendale, PA, Society of Automotive Engineers (1966).

17. *Clean Room and Work Station Requirements, Controlled Environments,* Designated Federal Standard 209B, Washington, DC, General Services Administration Service Center (1973) (Amended 1976).

18. *Compendium of Standards, Practices, Methods and Similar Documents Relating to Contamination Control,* Designated IES-CC-009-84, Mt. Prospect, IL, Institute of Environmental Sciences (1984).

19. Hamberg, O., *Journal of the Environmental Sciences,* Vol. 25 (3), pp. 15-20 (May/June 1982).

20. Gulett, M., *Semiconductor International,* Vol. 4 (3), pp. 87-94 (March 1981).

21. Monkowski, J.R. and Zahour, R., "Failure Mechanism in MOS Gates Resulting from Particulate Contamination," San Diego, Reliability Physics Symposium (March 1982).

22. Monkowski, J.R., *Technical Proceedings of the Semiconductor Processing and Equipment Symposium,* Semiconductor Equipment and Materials Institute, Mountain View, CA, pp. 304-310 (1983).

23. Iscoff, R., *Semiconductor International,* Vol. 5 (11), pp. 39-40, 42, 44, 46, 48, 50, 52, 53 (1982).
24. Gise, P., *Microcontamination,* Vol. 1 (3), pp. 41-44, 62 (1983).
25. Galbraith, L., "Advances in Automated Detection of Wafer Surface Defects," *Proceedings of the American Society for Testing and Materials,* Philadelphia (1982).
26. Kimble, G., *How to Use (and misuse) Statistics,* New Jersey, Prentice-Hall (1978).
27. Gibra, I., *Probability and Statistical Inference for Scientists and Engineers,* New Jersey, Prentice-Hall (1973).
28. Box, G., Hunter, W. and Hunter, J., *Statistics for Experimenters, an Introduction to Design, Data Analysis, and Model Building,* NY, John Wiley & Sons (1978).
29. Waclawiw, M. and Jurkowski, J., *Basic Statistical Techniques for Environmental Engineers,* Mt. Prospect, IL, Institute of Environmental Sciences (1984).
30. Alvarez, A., Weltera, D. and Johnson, M., *Solid State Technology,* pp. 127-133 (July 1983).
31. Weisbrod, S., *Semiconductor International,* pp. 261-265 (May 1985).
32. Bryce, G. and Collette, D., *Semiconductor International,* pp. 71-77 (March 1984).
33. Lui, B., Szymanski, W. and Ahn, K., *The Journal of Environmental Sciences,* pp. 19-24 (May/June 1985).
34. "Recommended Practice for Equipment Calibration or Validation Procedures," IES Designation IES-RP-CC-013, Mt. Prospect, IL, Institute of Environmental Sciences, [pending approval, as cited in *Microcontamination,* Vol. 3 (9), pp. 47-52, 104-106 (Jan 1985)].
35. *Standard Practice for Secondary Calibration of Airborne Particle Counter Using Comparison Procedures,* ASTM Designation F649-80, Philadelphia, Naval Publications and Forms Center (1980).
36. *Standard Practice for Determining Counting and Sizing Accuracy of an Airborne Particle Counter Using Near-Monodisperse Spherical Particulate Materials,* ASTM Designation F328-80, Philadelphia Naval Publications and Forms Center (1980).
37. Van de Hulst, H.C., *Light Scattering by Small Particles,* NY, John Wiley & Sons (1957).
38. Asano, S. and Yamamoto, G., *Applied Optics,* Vol. 14 (1), pp. 29-50 (Jan 1975).
39. Schiffer, R. and Thielheim, K.O., *Journal of Applied Physics,* Vol. 54 (7), pp. 2437-2454 (April 1985).
40. Trinks, W., *Annalen der Physik,* Vol. 22, pp. 561-590 (1935).
41. Twersky, V., *Journal of Applied Physics,* Vol. 23 (4), pp. 407-414 (April 1952).
42. Twersky, V., *Journal of Applied Physics,* Vol. 22 (6), pp. 825-834 (June 1951).
43. Twersky, V., *Journal of Applied Physics,* Vol. 25 (7), pp. 859-862 (July 1954).
44. Dobrowolski, J.A., Chapter 8 in *Handbook of Optics,* pp. 42-43, W.G. Driscoll (ed.), NY, McGraw-Hill Book Co. (1978).

45. Hecht, E. and Zajac, A., *Optics,* pp. 312-314, Menlo Park, CA, Addison-Wesley Publishing Co. (1974).

46. Born, M. and Wolf, E., *Principles of Optics,* NY, Pergamon Press (1965).

47. Bardina, J., *Microcontamination,* Vol. 2 (1), pp. 29-34, 78 (Feb/Mar 1984).

48. Barth, H.G. (ed.), *Modern Methods of Particle Size Analysis,* NY, John Wiley & Sons (1984).

49. Allen, T., *Particle Size Measurement,* NY, Chapman and Hall (1981).

50. McCrone, W. and Delly, J., *The Particle Atlas,* Vols. I-VI, MI, Ann Arbor Science Publishers Inc., 1973-1980.

51. Bakale, D.. and Bryson, C., *Microcontamination,* Vol. 1 (3), pp. 32-35, 63 (1983).

52. Linder, R., Bryson, C. and Bakale, D., *Microelectronic Manufacturing and Testing,* Vol. 8 (2), pp. 1, 9-13 (Feb 1985).

53. Muggli, R.Z. and Andersen, M.E., *Solid State Technology,* Vol. 28 (4), pp. 287-291 (April 1985).

54. Gavrilovic, J., *Solid State Technology,* Vol. 28 (4), pp. 299-302 (April 1985).

55. Humechi, H., *Solid State Technology,* Vol. 28 (4), pp. 309-313 (April 1985).

14

Wafer Automation and Transfer Systems

Mihir Parikh

INTRODUCTION

This chapter surveys automation systems for wafer movement and wafer-cassette transport, with an emphasis on contamination control. Since automation by itself does not "process" wafers but rather improves the yield and efficiency of the manufacturing process, the contamination considerations relating to automation are different than those of processing equipment. For instance, purity of chemicals and gases are generally of secondary importance with respect to automation systems where particle control is concerned.

The first section of this chapter discusses the driving forces for automation, in particular those affecting yields, processing errors and operating cost reduction. In the second section, automation technologies for individual wafer handling are surveyed; these will provide a basis for automated cassette-to-cassette operation of processing equipment. Interequipment transport of wafer cassettes is outlined in the third section, with considerations for an automated transportation system. Automated facilities incorporating transportation systems are described.

THE NEED FOR AUTOMATION

The automation of handling of individual wafers or batches of wafers in a cassette is driven by several forces or needs within the semiconductor industry. As expected, all of these needs have an economic incentive based on profitability through improvement of yields, reduction of operating costs, or both.

Among these forces are the need to minimize wafer breakage; the need to minimize scratching or marring of the sensitive surface of the wafer; the need to reduce particles or other unwanted contaminants on the wafer surface; the

445

need to organize wafers and to minimize misprocessing; the need to more fully utilize processing equipment; and the need to reduce operating costs relating to facilities, labor or other economic inputs.

In order to develop effective automation systems to address these needs, it is necessary to determine the magnitude of the problems caused by manual handling of wafers and wafer cassettes. This has been attempted via some simple estimates.

Wafer Damage

Consider the loss due to wafer breakage during manual handling of wafers. The manual handling of wafers is performed typically by a gloved human holding an individual wafer via a tweezer or a vacuum wand.

If we assume that each wafer is manually loaded or unloaded from a piece of equipment approximately 400 times in the course of a typical 200-step "process," then there are as many as 10,000 manual pickups and placements during the processing history of a 25-wafer lot. Assuming a probability of 0.1 percent for wafer breakage, we obtain the devastating result of 10 wafers being broken during the manufacturing process. Thus only 15 wafers would remain intact out of the 25-wafer lot!

Consider wafer damage due to scratches caused by tweezers. Assuming a probability of 1 percent for wafer scratching or marring (even if only along the periphery) due to the human handling of individual wafers, we obtain an estimate of 100 scratches on 25 wafers, or four scratches per wafer!

It appears that automation would be ideally suited to minimize these problems. In fact, the invention and subsequent development of wafer cassettes and automated wafer-indexers took place primarily to address these problems.

Wafer Contamination

Unwanted contamination may be defined as an external "agent" that is detrimental to the wafer during the manufacturing process. Such agents can come from many sources. Some are: organic and inorganic species (films, particles, etc.); electromagnetic radiation (visible, IR, UV, X-Ray, charged particles, etc.); or mechanical vibration, to name a few.

For the purpose of this chapter, we will consider only particles as the primary form of contamination. There are many sources for the generation and subsequent transmission of particles during the IC manufacturing process. Humans, even in specially designed clean room garments ("bunny" suits), are a major source of particles. The equipment used in the IC manufacturing process is in some instances even a greater source of particle contamination. Facilities with vertical laminar flow (VLF) of HEPA filtered clean air, for example, provide a medium for transmission of particles as well as their removal.

We will further discuss particle contamination as related to wafer and cassette handling by humans or robots as well as other automated mechanisms. Later, we will estimate and correlate the fallout of particles onto wafers from the handling operation and the resultant defect density contribution.

Table 1 shows the fallout of particles from a human performing various activities.[1] In spite of the large particle fallout, clean rooms with Class 10 VLF air moving at 90 feet per minute can be designed to minimize the probability of particles landing on wafers. As many as 6,000 particles per cubic foot per minute are known to emanate from a human in full clean room clothing.[2] For the purposes of estimating defect density caused by particles on wafers, we will assume that only two particles will actually land on a 150 mm wafer during human handling of wafers or wafer-cassettes.

Table 1: Fallout of Particles from Human Performing Various Activities

	Times Increase Over Ambient Levels (Particles, 0.2 to 50 μm)
Personnel Movement	
Gathering together 4–5 people at one location	1.5 to 3
Normal walking	1.2 to 2
Sitting quietly	1 to 1.2
Laminar flow work station with hands inside	1.01
Laminar flow work station—no activity	None
Personnel Protective Clothing (Synthetic Fibers)	
Brushing sleeve of uniform	1.5 to 3
Stamping on floor without shoe covering	10 to 50
Stamping on floor with shoe covering	1.5 to 3
Removing handkerchief from pocket	3 to 10
Personnel Per Se	
Normal breath	None
Breath of smoker up to 20 min. after smoking	2 to 5
Sneezing	5 to 20
Rubbing skin on hands and face	1 to 2

What is the effect of such an apparently small particle fallout on the defect density or on the yield of integrated circuits?

The effect of two particles landing on the surface of a 150 mm wafer is not always deleterious; there is a large probability, perhaps as high as 70 percent, that the particle would land on a "safe" region of the integrated circuit and thus would not affect circuit performance. Assuming then that there is a 30 percent probability of a particle causing an electrical defect, one can arithmetically add the number of defects that occur during each handling operation. However, not all process steps are defect sensitive; post-exposure, resist bake and develop, for example, are very insensitive to particle induced defects; while prediffusion clean is very particle sensitive. Assuming that only 20 percent of all handling operations are defect sensitive, then there are 80 defect-sensitive operations out of a possible 400 handling operations in a typical 200-step VLSI process.

Defect Density Contribution

Based on these assumptions, we can calculate that the defect density caused by the two-particle fallout on a 150 mm wafer is a product of:

(1) the probability of causing a defect per particle (0.3),

(2) the number of defect-sensitive operations (20% of 400) during a typical 200-step VLSI process,

(3) the surface area of a 150 mm wafer (176.7 square cm)

This leads to a defect density contribution due to particle fallout, of 0.27 defect per square cm. This number is an estimate based on the assumptions noted above. Those assumptions are probably conservative, however, and thus the calculated defect density contribution is probably an underestimate of the true defect density contribution. Of course the probability of catastrophic defects increases as device geometries decrease.

The impact of such a defect density contribution on yield is evident through Murphy's Yield Model.[3] A set of yield curves as a function of chip area are shown in Figure 1. The possible improvement in yield through reduction or elimination of the defect density contribution due to human handling can be estimated from such curves. For example, if the total defect density is 0.1 per square cm and the chip area is 0.2 square cm, yield improvement of approximately 5 percent is possible. As expected, a greater contribution to yield improvement is possible at lower absolute defect densities, which, of course, is imperative for large area circuits.

The use of appropriate wafer and wafer-cassette automation systems undoubtedly reduces the defect density contribution to the VLSI manufacturing process. In later sections we will survey robotic handlers, mechanisms, vehicles and other systems that attempt to address this need.

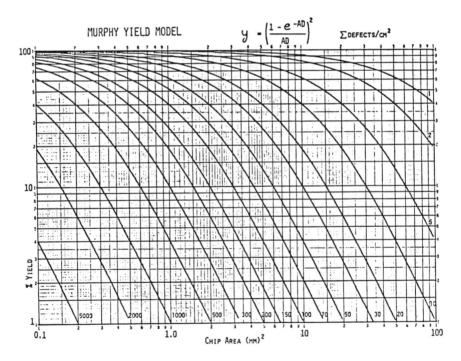

MURPHY YIELD MODEL $y = \left(\dfrac{1 - e^{-AD}}{AD}\right)^2$ Σ DEFECTS/CM^2

Figure 1: Murphy's yield model; dependence of yield on chip area and defect density.

Wafer Misprocessing and Equipment Utilization

Automation of wafer handling can also address the problem of misprocessing of wafers, which necessitates the reworking or scrapping of the affected wafers. If one assumes a 0.1 percent probability of error in the manual loading or unloading of equipment, the magnitude of the resulting misprocessing may be approximately the same as that for wafer breakage. In other words, potentially 10 wafers out of a 25-wafer lot could be misprocessed due to errors in manual loading.

The use of wafer cassettes and wafer indexers has minimized misprocessing by automating the sequencing of wafers into and out of equipment. However, the possibility of human error in bringing the wrong cassette to the equipment, and thus misprocessing an entire 25-wafer cassette, still remains.

Utilization of process equipment can be improved through automated wafer handling. Cassettes and indexers make it possible to load individual wafers into equipment on demand and at the correct rate, thereby minimizing equipment idle time. However, if a cassette is not brought to the equipment when it is needed, the equipment becomes idle and its utilization is below optimum.

Interequipment wafer-cassette control and "transport," which guarantees that the correct wafer-cassette is brought to the appropriate equipment at the correct time, is necessary for optimal performance in a fully automated facility.

Facility Operating Cost

Facility operating costs may be reduced by any of three means, or by a combination of them: labor cost reduction; improvement in the efficiency of facility utilization; or reduction in the amount of clean space required.

Although automation makes it possible to reduce the amount of direct, unskilled labor required in a wafer fab, it is unlikely that overall labor costs can be reduced in this way. While the amount of unskilled labor time required in the IC manufacturing process can be reduced through automation of wafer handling and transport, the very process of automation itself requires the hiring and utilization of more highly skilled and more highly paid workers such as engineers, technical support personnel, equipment maintenance staff and other highly trained employees.

Ordinary line workers will need more skills (such as, knowledge of set-up and maintenance) in an automated facility, and will, therefore, require more training and probably command higher wages than workers in a nonautomated fab.

Thus, although less direct labor might be required in an automated fab, it seems unlikely that total labor costs will be reduced by automating.

Automation should improve the equipment utilization rate and result in better control and tracking of work-in-process. Use of such techniques as just-in-time manufacturing, which calls for cassettes and materials to arrive at work stations just as they are needed and not before, should enhance not only operating efficiency but bottom-line financial performance. Such techniques can also potentially improve process and contamination control by reducing the wafer aging phenomenon.

Devices such as bar code label readers linked to central computers, as well as other intelligent lot-tracking techniques, can further augment the improvement in equipment utilization.

Finally, construction and operating costs of clean rooms can be affected by the design of the facility. By localizing the environment immediately around the processing equipment, either through clean air "tunnels," modules or enclosures (with dedicated clean air supplies), the quantity of clean air required can be reduced significantly—as much as 10 to 20 times. The interface into this localized clean equipment environment, could be via a standardized interface.

A concept called the *Standard Mechanical InterFace* (SMIF), to be discussed in the SMIF System portion of this chapter, can provide such an interface and affect localization of clean space. Estimates for construction cost reduction range from $100 to $200 per squre foot, resulting in $1 to $2 million savings for a 10,000 square foot facility; operating cost savings of $500K per year are also estimated, since a significant reduction in Class 10 clean space is attained.[4]

INDIVIDUAL WAFER HANDLING

Most processes in the fab ultimately require the handling of individual wafers either by human beings or machines. In the early days of the semiconductor industry, all wafers were handled individually by human operators. Today, fab engineers can choose whether to automate the handling of individual wafers at some or all steps of the IC manufacturing process.

Manual Handling

Manual handling of individual wafers typically is accomplished by use of specially designed tweezers to grasp the periphery of the wafer or by use of vacuum wands, which generate a suction to lift and hold the wafer, usually from the backside.

Whatever method is used, be it tweezers or vacuum wands, the manual placement and alignment of wafers onto the loading area of the processing chamber of a piece of equipment will inevitably involve a high risk of wafer scratching or marring, as discussed in the wafer damage portion of this chapter. Further direct particle and/or nonparticle contamination from the tweezer or vacuum wand is possible.

Up to about the early or mid-1970s, wafers were manually loaded onto and off of flat trays, on which they were carried from one process station to the next. At that time, the cassette was invented as a means for storage and sequential ordering of wafers. Wafer cassettes became the natural choice for a carrier for the movement of wafers from one piece of equipment to another.

Automated Handling of Individual Wafers

Automation of the handling of individual wafers—the automated movement of individual wafers in a preprogrammed way—is the simplest form of automation in the fab process, in contradistinction to the much more highly sophisticated concept of interequipment wafer-cassette transport.

A simple device, the 'elevator' or the 'indexer,' was invented (see Figure 2), to automate the extraction (or replacement) wafers from (or into) cassettes without human intervention, and thereby, kept safe and in sequence during processing.

2660

Figure 2: Photograph of a Siltec 'elevator' or an 'indexer' for automatically loading/unloading wafer cassettes.

Today, devices such as Siltec Corporation's Send/Receive Elevator (Figure 2),[5] offer a variety of features in conjunction with automated sequential loading and unloading of wafer cassettes. The elevator accurately lowers (or raises) the cassette by the distance between individual wafer slots in a cassette, thereby making it possible to extract (or replace) a wafer from (or into) each cassette slot. The elevator can be equipped with a flipper arm for loading wafers onto or off a belt transport. The elevator has a wafer sensing lens that automatically detects the presence of a wafer made from a wide variety of semiconductor materials, including GaAs. Via a minibus communications interface, the elevator can be linked to an external controller.

Nonsequential, or so-called random loading and unloading of wafers to and from cassettes can be achieved by use of a random access indexer. Brooks Automation, for example, manufactures a versatile elevator that can operate in either a sequential or random access mode.[6] However, it necessitates a wafer extracting robot, to operate in random mode.

Brooks Automation's Orbitran Wafer Handling Robot (Figure 3), is an example of one type of wafer extracting robot which features a versatile transport arm that moves in a radial pattern to service multiple work stations. Such a robot does not use belts or air tracks, which could cause contamination.

The robot lifts and places the wafer in position without sliding or pushing. Additionally, it can be used in conjunction with many types of machines to place and remove wafers from inspection microscope stages, equipment process chambers, or "lazy susan" style platens.

Figure 3: Photograph of Brooks Automation's orbitran wafer handling robot.

Another special purpose robot (Figure 4), Proconics International's Automatic Wafer Transport System, is designed to transfer wafers between plastic wafer carriers and quartz wafer carriers, with minimum operator involvement.[7] The system has the ability to sense the locations of the slots in the quartz rails, thus permitting loading without sliding contact between the quartz boats and the wafers.

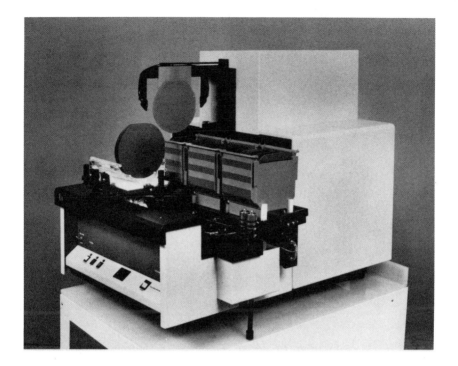

Figure 4: Photograph of Proconics International's automated wafer transfer system between quartz and plastic wafer carriers.

General purpose robots, which can be adapted to clean room applications, for individual wafer handling are available from several suppliers: Precision Robotics, Inc., Microbot, Inc., for example.[8] Such robots have many axes of articulated movement and via specialized gripper mechanisms, perform various types of wafer loading/unloading functions within and between processing equipment.

A different concept for movement of individual wafers is a track-based wafer transport system (Figure 5). IBM at its QTAT *(Quick Turn-Around Time)* facility at East Fishkill, New York used such a concept.[9] The system involved an enclosed, self-centering air track to automatically transport individual wafers. Each wafer was surrounded with Class 50 air as it moved along the track. The wafers were propelled by pressurized jets of air emanating upward to angles from holes in the horizontal surface along which the wafers moved. The system maintained the wafers in a localized Class 50 environment, even when the external environment was Class 10,000 or worse. However, the system was found to be inflexible and was sensitive to failure within individual pieces of processing equipment.

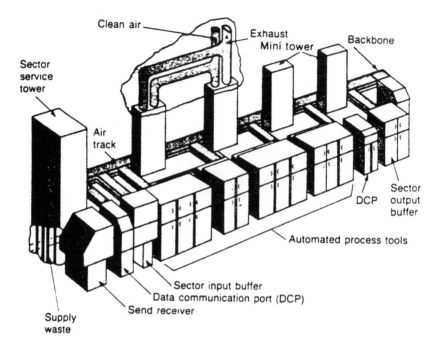

Figure 5: Artist rendition of an automated section of photolithography section of IBM Sindelfingen QTAT line.

Particle Control Effectiveness

The effectiveness of the automated handling of individual wafers can be measured in terms of the reduction of breakage or damage to the wafers as well as in terms of external contamination that is added to the wafer surface. The former is clearly a tacit requirement for a viable automation system; the latter is an equally important requirement, whose magnitude needs to be ascertained.

In the case of a Brooks Automation's wafer handling robot, published data[10] shows in Figure 6, better than Class 10 effectiveness. These data, measured with an airborne particle measuring probe located below the moving gears of the robot's pick and place mechanism, indicate an exponential increase in particle generation with decreasing particle sizes below 0.1 μm. However, this is still better than that specified in the proposed Federal Standard 209C for "Class 10." The "fallout" of particles on a wafer was measured qualitatively and thus could not provide explicit particle contribution to wafers.

The general purpose robot, such as the Microbot, has been measured for its effectiveness in a clean room. Data (shown in Figure 7) shows that Class 10 effectiveness was obtained with a modified robot (Alpha II) and a further improvement was obtained when the robot was "dressed" in a specially designed static-free clean room fabric constructed garment.[11] This resulted in a better than "Class 1" effectiveness, as measured by the airborne particle measurement system.

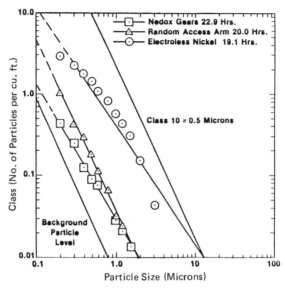

Figure 6: Airborne particle distribution for various kinds of gears at the end of a Brooks Automation's wafer handling robot (see Reference 10 for details).

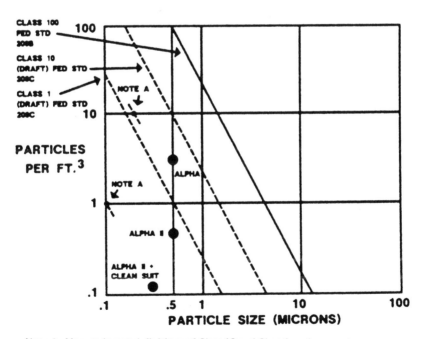

Note A—More stringent definitions of Class 10 and Class 1 environments are preferred by critical semiconductor organizations, such as the Institute of Micro-contamination Control (MC).

Figure 7: Airborne particle distribution for Microbot robots, showing a steady progression to increasing cleanliness (see Reference 11 for details).

While each of the above described measurements have shown Class 10 level of airborne particle effectiveness, the more critical measurement is the particle addition or accumulation measurements on wafer surfaces. This evaluation appears to have been limited by the availability of on-surface particle measurement technology.

INTEREQUIPMENT WAFER-CASSETTE TRANSPORT

Thus far we have discussed automation of the handling of individual wafers during the IC manufacturing process. This, as noted, can be accomplished by a variety of mechanical devices that removes wafers from cassettes and place them into a processing chamber or environment.

Another important requirement for an automated facility is interequipment wafer-cassette transport—the movement of cassettes from one piece of processing equipment to the next.

The control of interequipment wafer-cassette transport can help meet some of the same needs achieved through automation of individual wafer handling: reduction of particulate contamination and concomitant improvement of manufacturing yields; reduction of misprocessing; and possible labor cost reduction.

The main reason for upgrading wafer-cassette transport in any fab is, yield improvement. As noted in the discussion of individual wafer handling, it seems unlikely that total labor costs will be reduced by automating, although less direct labor might be required in an automated fab. The author believes this is also true of wafer-cassette transport.

Three Kinds of Wafer-Cassette Transport Systems

Three types of systems have been devised for interequipment wafer-cassette transport: robotic carts; tunnel-track networks; and SMIF systems.

Robotic carts or Automated Guided Vehicles (AGV) are battery-powered floor moving vehicles that navigate through a fab along a route usually marked by a special stripe. The AGV can carry several cassettes at a time, and each cassette may be either enclosed in a protective container or open to the clean room environment. At appropriate stops along the programmed path, cassettes are transferred to and from processing equipment by special robotic arms mounted on the AGV.

Tunnel-track networks transport wafer cassettes through the fab on miniature "flatcars" that move along tracks within a HEPA filtered tunnel. Each mini-flatcar carries a single cassette from one piece of processing equipment to the next. The system can be interfaced to processing machines by means of equipment—specific robotic mechanisms.

SMIF systems use compact, sealed, ultraclean containers for storing as well as transporting wafer cassettes through the fab. Each SMIF container is equipped with a special door designed to interface with a compatible door on an ultraclean enclosure at the entry and at the exit ports of each piece of process equipment. This maintains the integrity of the environment within the container and canopy. The containers are transported within the fab by human operators or by an appropriate automated transport system.

Automated Guided Vehicles. Two major developers of AGV (Figure 8) for use in a wafer fab environment are Flexible Manufacturing Systems, Inc. (FMS) and Veeco Integrated Automation, Inc.[12] Both FMS and Veeco market their AGV in conjunction with intelligent work-in-process cassette storage stations and other automation tools that are designed to function in unison as both a wafer-cassette handling and wafer lot tracking system.

Each AGV communicates with a central computer via an infrared link. The tracking, scheduling and routing of AGV and wafer cassettes through the fab is handled by a central computer.

Cassettes are transferred to and from processing equipment by special robotic arms mounted on each AGV. Cassette pick-and-placement accuracy is, therefore, determined by the alignment accuracy of the AGV with respect to the equipment. Special techniques employing equipment based docking modules (sensors) are used to accurately align and "teach" the AGV its precise location.

Although AGV are dockable or alignable vehicles, their use necessitates the creation of elaborate and foolproof collision avoidance systems to ensure that an AGV does not collide with anything that might interfere with its path. Additionally, if an AGV is immobilized due to an obstruction or an equipment undergoing repairs, it may be necessary to reroute other AGV around it to avoid "traffic jams" in the fab.

Underlying the robotic cart approach to wafer-cassette transport, is the impetus to isolate wafers from human beings by removing operators from the fab. The use of AGV systems therefore depends on remote supervisory control of process equipment operation and material movement.

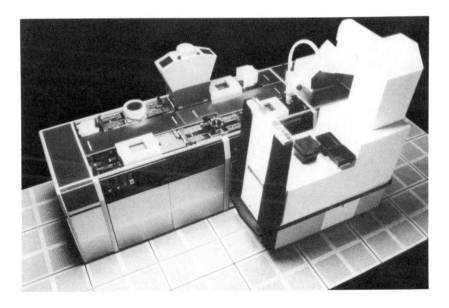

Figure 8: Photograph of a Veeco AGV for automatic loading of equipment.

Estimates have been calculated for the benefits and the Return-on-Investment (ROI) of a typical IC facility with AGV based material handling systems.[13] The model used there is dependent on relationships between yield, productivity and efficiency (cycle-time) improvement and automation. ROI of less than 1 year have been claimed.

Tunnel-Track Networks. Tunnel-track networks (Figure 9) transport cassettes on miniature motorized flatcars that move along tracks in HEPA filtered tunnels and elevators. Each mini-flatcar which carries a wafer cassette through the clean-air network can communicate with a central computer that handles tracking, scheduling and routing. The carts are powered by low voltage direct current supplied through the rails on which the carts move. Turntable "track switching" mechanisms are built into the network at locations where tunnels intersect.

Wafer cassettes are moved between the mini-flatcars and processing equipment by equipment-specific robotic mechanisms. Between process steps, cassettes can be moved on flatcars to work-in-process storage areas outfitted with sources of laminar flow air, or they can be shunted to a resting place in a "switchyard" within the tunnel network.

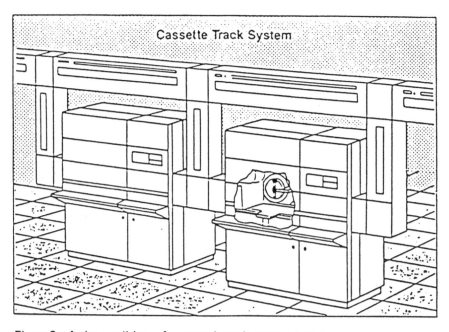

Figure 9: Artist rendition of a tunnel track system interfaced to a processing equipment.

Since the tunnel-track system is a sequential system, it is highly inflexible. It is logistically difficult to route the flatcars to inventory storage areas and then reintroduce them into the process line. In addition, the tunnel-track system is dependent for its successful implementation on the design and effectiveness of the various equipment-specific robotic interfaces at each piece of process equipment. Such interface mechanisms are often difficult and costly to design, and their performance may be less than optimal.[14]

The essential idea that gave rise to the tunnel-track concept—creating an enclosed ultraclean space within which wafer cassettes could be transported, while reducing clean air requirements throughout the remainder of the fab—was nevertheless a meritorious concept. This has been embodied in the SMIF concept.

The SMIF System

The SMIF system involves small, sealed, containers for storing and transporting wafer cassettes. These containers serve, through a specially designed door to transfer cassettes between the container and the equipment. The environment around the equipment is isolated via an enclosure, and thus becomes a small particle-free environment specific to the equipment.

Each container is fitted with a specially designed "door;" each enclosure surrounding a piece of equipment with a mating door and "port." This mating arrangement is the *Standard Mechanical InterFace* or SMIF. The two doors are opened simultaneously so that particles that may have been on the outside surface of either door are trapped in the space between the doors. In addition, the equipment must be fitted with a special mechanism which operates the doors and transfers the cassette between the containers and the equipment's indexer. This transfer has to be such that the external environment does not contaminate the wafers.

The SMIF system provides a high degree of flexibility. The containers can be transported by human operators or by an automated transportation system. In fact, the systems described earlier, the cassette-track and robot-cart systems, are even more effective with SMIF containers, as SMIF provides an additional level of protection during transport above that provided by an automated transport system. Thus, the SMIF system provides an *evolutionary path* for automation.

Finally, the SMIF system provides an especially unique opportunity for controlling and maintaining "minimum-volume" clean environments around the wafers and equipment. This is possible via a localized Class 10 or better environment provided by equipment enclosures or isolation of the equipment behind a wall. The control of "cleanliness" in the exterior aisle space is less stringent.

The ramifications of a SMIF implementation throughout an IC facility can be significant, in terms of clean space volume reduction as well as energy savings. Estimates have been made to assess the magnitude and long term potential.[4]

An integrated SMIF system has three major components: SMIF-Pod, SMIF-Arm, and Equipment Enclosures with dedicated clean air supplies.[15]

The SMIF-Pod (Figure 10), is a specially designed sealed container with two unique features: a gasket that seals the port door and a disposable liner which totally encompasses the wafer cassette. Using materials with specific properties, the liner can provide a specific local environment to the wafers as well as minimize the need to clean the SMIF-Pod. The security of wafers in the cassette is ensured via an active retainer, which retracts automatically to allow wafer cassettes to enter/leave the pod. The locking of the wafers has to be performed in a manner that minimizes particle generation.

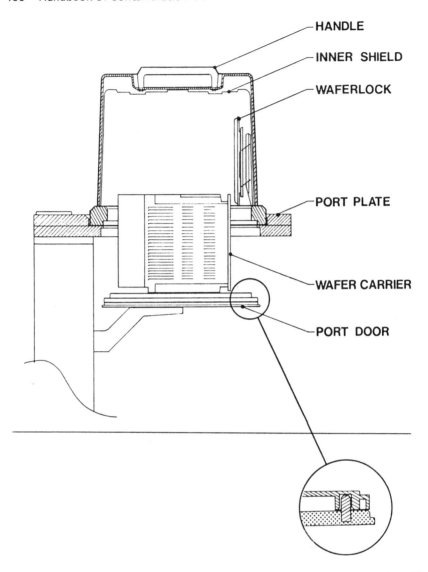

Figure 10: Cross-sectional view of the Standard Mechanical InterFace (SMIF), showing a wafer cassette being loaded into the SMIF-Pod. The detail shows the SMIF "sealing" between the port door and the SMIF-Pod door.

The SMIF-Arm (Figure 11) consisting of a specially designed robotic mechanism, provides fully automated extraction and placement of cassettes on equipment indexers. This minimizes particle fallout on wafers during the cassette transfer process. For adaptability to a large variety of processing equipment, the system has versatile cassette placement capabilities—reach height and cassettes—orientation. A small footprint is also required for easy retrofit into existing facilities.

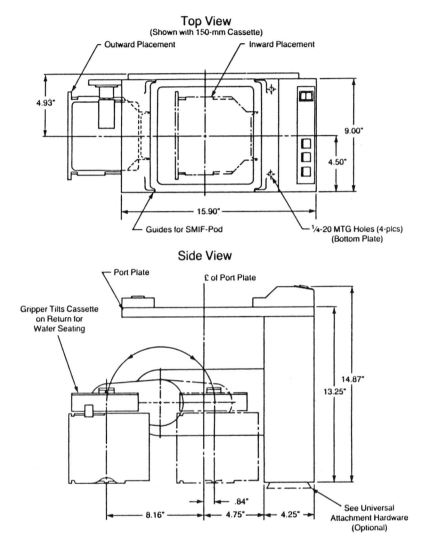

Figure 11: Cross-sectional view of a SMIF-Arm, a robotic mechanism, capable of extracting a cassette from a SMIF-Pod and placing it on an equipment indexer.

Isolation of equipment from external environment is possible through dedicated enclosures (Figure 12) with local air supplies or specially designed chambers and walls isolating equipment (Figure 13) in ultraclean areas. Environments external to such areas can be kept at Class 100 or worst, while the isolated area of the equipment can attain better than Class 10. The SMIF-Arm, through the SMIF port, provides a means for wafer cassettes to enter into the isolated equipment environment without any contamination from the external environment affecting either the wafers or the isolated clean environment of the equipment.

Figure 12: Photograph of a SMIF interface to a wafer-track processing equipment with dedicated enclosures (courtesy, Eaton Corp.).

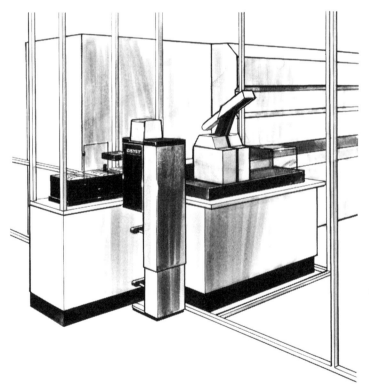

Figure 13: An artist rendition of a "through-the-wall" isolation of equipment with SMIF system interface.

Effectiveness of SMIF System Components. The effectiveness of the SMIF System components or the integrated system can be judged via careful measurements of particle addition, as measured by an airborne monitor or a scanning laser wafer surface inspection machine. In this section we will describe results of measurements performed on various components of the SMIF System.

The effectiveness of the SMIF-Pod in sealing and maintaining cleanliness under different external environments has to be measured.[16] Thus, SMIF-Pods with clean wafers in cassettes were left under different types of environment for several weeks. Their sealing effectiveness was determined by counting the number of particles adhering to the wafer surface with a wafer particle measurement machine: Hitachi DECO, Model 200C. Shortly before the experiments were performed, calibration was carried out by the manufacturer, Hitachi Electronics Engineering Co., Ltd., using uniform latex spheres at six predetermined sizes: 0.220 μm, 0.312 μm, 0.481 μm, 1.09 μm, 2.02 μm and 2.95 μm.

The two types of environments used in this experiment were at extreme ends of the spectrum: an office room with tobacco smoke (Class 100,000 or more at 0.5 μm or more particle size), and a clean room (Class 10 or less). In each case, ten 150 mm wafers were stored in a wafer cassette which can accommodate 25 wafers and kept in each environment for about two weeks. Measurements were performed (ten times, to obtain reliable statistics) at the beginning, at certain times during, and at the end of the storage periods. The wafer cassettes were stored for 12 days in the "office" environment and subsequently the 15 days in a Class 10 environment.

The results are shown in Figure 14. The abscissa depicts the number of days and the ordinate the number of particles accumulated on the surface of a 150 mm wafer. As shown in Figure 14, there is a small change in the number of particles accumulating on wafers in the SMIF-Pod stored in the clean room. However, in the office area the number of small particles of size between 0.22 μm and 0.48 μm increased by approximately 50 particles on a wafer after 12 days of storage.

The cleanliness of SMIF-Arms has a significant impact on the effectiveness of the integrated SMIF system. Particle generation by the SMIF-Arm can be measured by airborne particle sensors. Results have shown essentially *zero* particle emission for particle sizes of 0.5 μm or greater.[16] However, measurements of particle accumulation on wafers during the cycling of the SMIF-Arm can provide an even more relevant measure of SMIF system effectiveness.

For two specific types of motions, denoted in Table 2, the SMIF-Arm was cycled continuously over 300 times with a cassette of wafers. The conditions of the experiment were: (1) Ambient particle level in the room was better than Class 10 as measured by a MET-One Airborne particle monitor with a sensitivity of 0.3 μm size particles. (2) Each motion (noted in Table 2) involved the movement of a wafer cassette containing 23 wafers; each 100 mm wafer had initial particle counts between 5 and 20, of 0.5 μm size or larger. (3) All particle measurements were performed on the Aeronca WIS-150 with thresholds calibrated (via VLSI Standards No. ASC-1420) to detect 0.5 μm or larger sized particles.

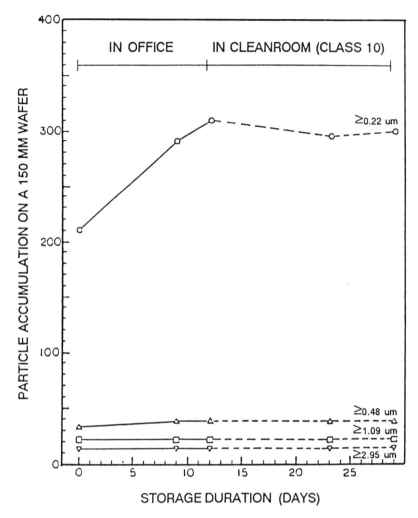

Figure 14: Measurement of particle accumulation inside a SMIF-Pod over a 27 day period, in two environments: An office area (12 days) and a Class 10 clean room (15 days).

Table 2: SMIF-Arm Experiment

Motion	Time Per Pass (seconds)	PWP*
From SMIF-Pod	17	0.19
From SMIF-Pod Base onto equipment extender	11	0.01

Note: *PWP: particles per wafer per pass; pass is a single motion of cassette as described above.

The results in Table 2 show approximately 0.2 PWP addition in the combined motions of the SMIF-Arm. This indicates that the motion of the SMIF-Arm contributes negligibly to particle addition on wafers. Other measurements indicate that the SMIF-Arm is at least as clean as other clean room robots, which are typically one hundred times cleaner than a fully clean room garbed human.[17]

Effectiveness of an Integrated SMIF System. In order to assess the true effectiveness of the integrated SMIF system, a series of experiments were conceived and performed to compare the effect of the external environment on the SMIF system and the effectiveness of conventional handling in a Class 10 room. Figure 15 shows schematically the five experimental conditions.[16]

1. SMIF system used in a Class 10 clean room (0.5 μm) environment.

2. SMIF system used in a Class 1000 clean room environment.

3. SMIF system used in a Class 20,000 "clean room" environment.

4. Conventional wafer cassette handling by human operators, using "blue-box" type containers for ambient protection in a Class 10 clean room environment.

5. Conventional wafer cassette handling by human operators, without any additional cassette protection in a Class 10 clean room environment.

In each case, the initial wafer surface contamination was measured with a Hitachi DECO HLD-200C in a Class 10 environment. These wafers were then transported 30 meters to a simulated process equipment station and in cases 1, 2 and 3 put through a load/unload cycle using the SMIF-Arms or in cases 4 and 5, manual loading/unloading and one minute residence time. The wafers were then transported back to the surface inspection machine and measured again. This constituted one cycle and for each of the five experimental conditions, *one hundred* repetitions of the cycle were performed to ascertain the rate of particle gain and the level of data repeatability.

In all tests, operators were fully gowned, including hat, mask and gloves and careful attention was given to minimize particle generation caused by the human operator. Wafers used were 150 mm diameter, with ten wafers per cassette used for measurements. As mentioned earlier, the Hitachi DECO HLD-200C was calibrated before the experiments were initiated. Typical particle accumulation data (for case 4) are shown in Figure 16 which depicts the number of handling cycles on the abscissa and the number of particles on a particular 150 mm wafer on the ordinate. Such data was collected and used for analysis for each of the ten wafers in the cassette. The fluctuation of measured values is typical of particle measurement testing and indicates the need for numerous data points in order to extrapolate statistically valid conclusions.

To determine representative values for the *rate* of particle gain, the nominal slope of the graphical data was determined for each test condition, thus providing a numerical value termed *particles per wafer per pass* or PWP.

The PWP values for each of the four particle sizes (0.22 μm, 0.48 μm, 1.09 μm and 2.95 μm) were obtained for each of the wafers. The arithmetic average of the PWP for the 10 wafers is plotted against particle sizes in Figure 17.

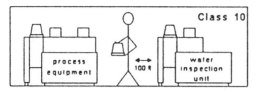

SMIF system in a Class 10 environment.

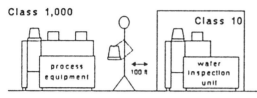

SMIF system in a Class 1,000 environment.

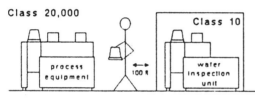

SMIF system in a Class 20,000 environment.

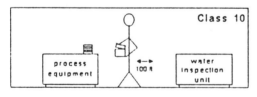

Conventional handling in a Class 10 environment, cassette transported in a standard box.

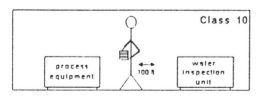

Conventional handling in a Class 10 environment, open cassette transport.

Figure 15: Schematic showing five types of experiments that were performed to evaluate the SMIF system handling (types 1, 2, and 3) in different clean room conditions and conventional handling (types 4 and 5).

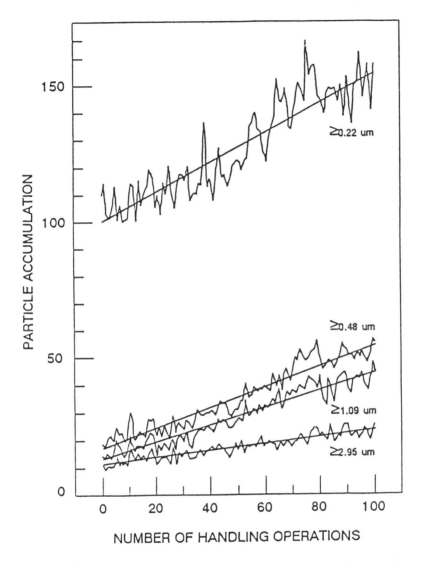

Figure 16: Particle accumulation on a particular 150 mm wafer (number 5 out of 10 in a cassette) during 100 handling operations in the case of experiment type (4), shown schematically in Figure 15.

The PWP results indicate a distinct advantage for the SMIF system, particularly at decreasing particle sizes. At 0.5 μm particle size, the SMIF performance in the Class 20,000 ambient was superior even to conventional cassette handling with conventional cassette boxes in a Class 10 ambient under the test conditions described previously. The results at 0.22 μm particle size are even more favorable for the SMIF system, indicating a 2 to 1 margin of superiority in a Class 10 ambient.

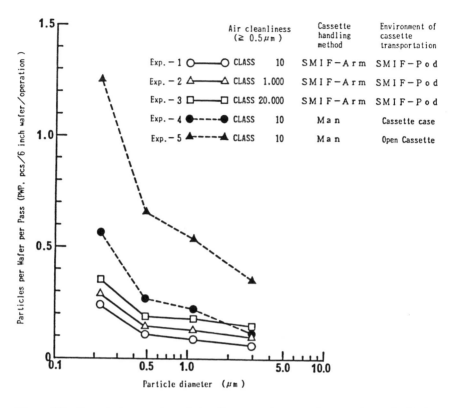

Figure 17: Cumulative results of the five experiments, shown schematically in Figure 15, shown in terms of Particles per Wafer per Pass (PWP) as a function of four different particle sizes: 0.22 μm, 0.48 μm, 1.09 μm and 2.95 μm.

The particle control and effectiveness of the SMIF system is dependent on the level of "cleanliness" existing within the processing equipment itself. The benefits of the SMIF system can be overwhelmed by any major contamination sources within the process or the equipment. Thus carefully designed and optimized equipment/processes, integrated with the SMIF system can yield better than Class 10 effectiveness—even in an ambient environment that is much worse than Class 10.

AUTOMATED FACILITIES

Two types of automated facilities can be identified; ones that involve interequipment transportation automation and others that use a combination of manual and automated interequipment wafer/cassette transportation system. The former typically involve AGV or tunnel-track systems; the latter involve SMIF systems.

An early example of a "fully" automated facility is IBM's QTAT for final metalization level(s). As described earlier and shown in Figure 5, the QTAT facility involved tunnel air-tracks that automatically transported individual wafers between and into/out-of processing equipment. This unique automated facility combines physical movement automation with isolation of product (wafers) from the external environment.

The effectiveness in terms of improvement in product yield and quality via the QTAT facility compared to another "Business-As-Usual" BAU facility is shown in Figure 18.

Productivity Ratio

$$\frac{YQ}{YB} = \frac{\text{No. good chips to stock per wafer start in QTAT}}{\text{No. good chips to stock per wafer start in a conventional line}}$$

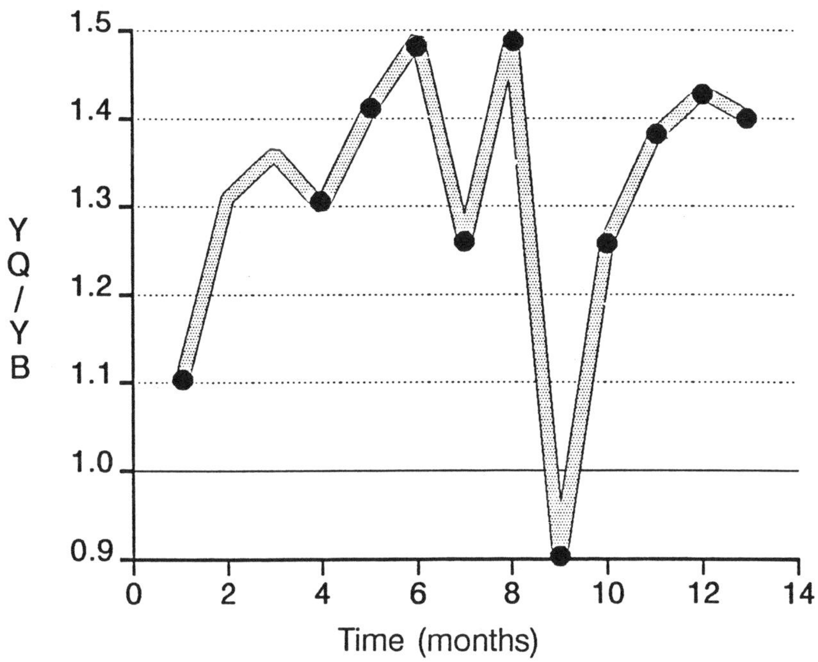

Figure 18: Effectiveness of IBM's QTAT as measured by the Productivity Ratio over 14 months.

Automated wafer-cassette transportation system facilities have been described by Mitsubishi Electric Co. and by NMB Semiconductor, Inc. in Japan, among others. Both facilities use an AGV system to transport and load wafer-cassettes onto equipment.

The Mitsubishi Saijo factory[18] has been described as a major DRAM manufacturing facility using AGV. The AGV move along the "main street" and load cassettes onto branch stations. A moving robot carries cassettes between the branch stations and processing equipment which are located in "side streets." Figure 19 shows a schematic of the layout at the Mitsubishi factory.

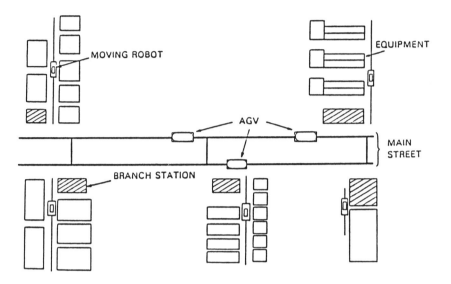

Figure 19: Schematic of the layout at Mitsubishi Saijo factory with AGV and moving robots.

To achieve this level of automation required a comprehensive network of controllers, and an elaborate scheduling system. All of the AGV's, host and cell computers/controllers and the software were developed at Mitsubishi.

While no explicit yield improvement data has been published, it has been estimated by DataQuest[19] that Mitsubishi obtains 85 to 90% yields with defect density of 0.1 defects/mask-level/cm^2 for the 64K DRAM product.

A different approach was taken by NMB Semiconductor.[20] Their facility design was based on zones of "cleanliness" class 1—active area where cassettes of wafers were transported by robots on tracks and loaded into processing equipment; class 30 area for equipment maintenance and class 300 area for equipment operators. Each zone is separated from others by floor-to-ceiling glass walls. Figure 20 shows a schematic plan view of the facility. The class 1 "subtransfer line" houses the moving robot while the class 1 main transfer line has an overhead tunnel track system for moving wafer cassettes between stockers in each module and to the main stocker.

Figure 21 shows a photograph of the facility with glass walls separating class 300 and class 1 areas. Figure 22 shows a photograph of the overhead tunnel-track cassette transport system.

Significant automation, including software, was developed by NMB in conjunction with robotics and automation suppliers. All operator activity in terms of monitoring and controlling the process equipment has been remoted and controlled by remote controllers and closed circuit TV monitors.

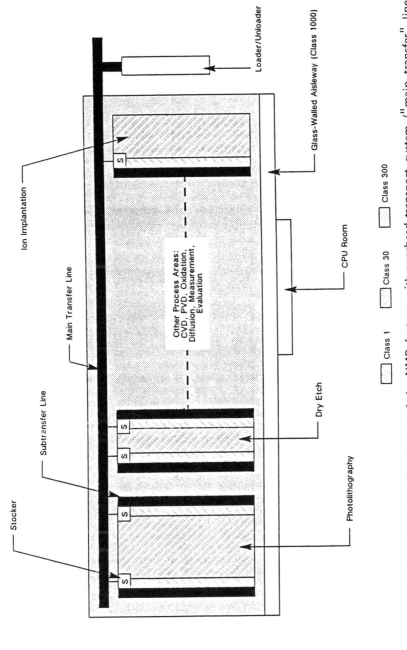

Figure 20: Schematic of the layout of the NMB factory with overhead transport system ("main transfer" line) and moving robot system ("sub-transfer")

Figure 21: NMB's facility interior with glass walls separating class 1 areas from class 300 areas.

Figure 22: NMB facilities' overhead transport system.

While no yield results have been published, it is estimated by NMB that this automation system has had a major impact on yield and has reduced cycle time.

A fully automated facility utilizing the SMIF concept has been announced[21] by VTC Inc., a subsidiary of Control Data Corporation. The design concept is similar to NMB in terms of isolation and zoning of different areas of the facility. Figure 23 shows a cross-section of the facility with separation into 3 zones: class 1 for wafer-cassette loading/unloading; class 10 for operator activity and class 100 for maintenance.

Unlike the NMB facility, the class 1 area is interfaced with SMIF system. Thus wafer-cassettes can be transferred into the class 1 region, via automated extraction from SMIF-Pods by the SMIF-Arms. The operators (in class 10 areas) move SMIF Pods between SMIF-Arms to different processing equipment. This approach provides the cleanliness via separation of zones and without the complexity of introducing automated transportation. In addition, it requires the minimum possible class 1 volume.

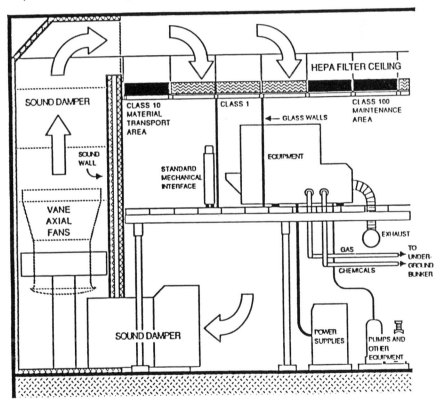

Figure 23: Cross-sectional view of VTC's two level facility with three zones of cleanliness: class 1, class 10 and class 100.

Figure 24 shows SMIF-Arm interfaces to process equipment load/unload indexers in class 1 clean space separated by glass walls from the class 10 space of operators.

Figure 24: Photograph of SMIF interfaces to process equipment at VTC's facility.

Facilities with side-wall air return are typically classified as class 100 facilities. Implementation of a well designed SMIF system can provide class 1 effectiveness at the equipment. Figure 25 shows plexiglass panels directing clean air from a HEPA filter ceiling; in addition, the panels isolate the internal space of the equipment from the external class 100 space using SMIF interfaces, wafer cassettes can be introduced inside the clean space and loaded onto the equipment indexers.

The effectiveness of the SMIF system in a manufacturing facility has been described in detail elsewhere.[22] Based on a comprehensive analysis of defect density data over a period of 6 months, it has been reported that approximately 13% reduction in defect density was achieved using the SMIF system. This was obtained for a variety of 2.5 micron CMOS products in an existing class 10 facility. This translates to between 5 to 10% improvement in yield.

Figure 25: Photograph of SMIF system interfaced to a coat/spin/bake track with plexiglass panels directing clean air from the HEPA filters in the ceiling.

CONCLUSION

This chapter has reviewed wafer and wafer-cassette automation systems. The need for automation is driven by requirements for minimizing wafer damage, contamination and misprocessing. Automation of single wafer handling using wafer-cassettes and indexers dramatically reduces the problems due to misprocessing and damage. Single-wafer robotic handlers reduce contamination problems.

Interequipment transport of wafer-cassettes is driven by similar requirements—minimization of wafer-cassette contamination and misprocessing. Using AGV and/or SMIF systems such problems can be addressed. The SMIF system, in addition, separates all external environments from the environment of the wafers and the equipment. Thus, it provides an opportunity for reduced (minimum) class 1 clean volume compared to conventional clean rooms.

Automated facilities using either AGV and/or overhead track systems or using the SMIF systems have been built. A facility that utilizes both technologies has been envisioned as shown in Figure 26. The future will be determined by the effectiveness of such systems in terms of their cost, payback, implementation and operational efficiency.

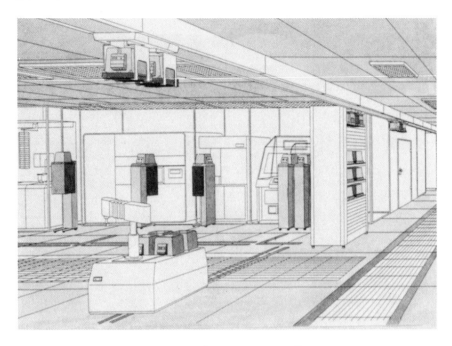

Figure 26: Artist rendition of a "fab-of-the-future" with a combination of automated movement and SMIF technologies.

REFERENCES

1.   ICE Status 1984 Report; Integrated Circuit, Engineering Corp., Scotts-dale, AZ.
2.   Dalhstrom, M.S., *Semiconductor International,* p. 110 (1983).
3.   Murphy, B.T., *Proc. IEEE,* p. 1537 (1964).
4.   Parikh, M. and Bonora, A.C., *Semiconductor International* (1985).
5.   Siltec Corporation, Cybeq Division, Menlo Park, CA.
6.   Brooks Automation, Inc., N. Billerica, MA.
7.   Proconics International, Inc., Woburn, MA.
8.   Precision Robots, Inc., Woburn, MA.; Microbot, Inc., Mt. View, CA.
9a. Deininger, C.R. and Mozer, D.T., *Proc. Semicon.* (May 1980).
9b. Brunner, R.H. et al., *Electronics* (Jan. 1981).
10. Hardegen, B. and Lane, A.P., *Solid State Technology* (March 1985).
11. Rhodes, G.W. and Cheng, A., "Towards Zero-Defect Processing: Class 1 Clean Room Robots," Conference on "Robots in Clean Room Applica-tions," Robots International of SME (Society of Manufacturing En-gineerings) (Sept. 1985).
12a. Flexible Manufacturing Systems, Los Gatos, CA.
12b. Veeco Integrated Automation, Dallas, TX.
13. Harper, J.G. and Bailey, L.G., *Solid State Technology* (July 1984).
14. Nakayama, S., quoted in "SEMI Officials Tour Automated Factory in Japan," *SEMI News* (May 1986).

15. Asyst Technologies, Inc., Milpitas, CA 95035.

16. Harada, H. and Suzuki, Y., "SMIF System Performance at 0.22 μm Particle Size," *Solid State Technology* (December 1986).

17a. Suzuki, Y. et al., "Generated Particles From Worker Clothed in a Clean Room Garment, No. 2 Generated Particles From Various Parts of The Worker," *Proc. of No. 5 Annual Technical Meeting on Air Cleaning and Contamination Control* (April 9, 1986).

17b. Harada, H., Kase, T. and Suzuki, Y., "Effects on Clean Room Caused by Industrial Automatic Equipment No. 2 (Particle Generation Rate From Scaler Type Robot Hand)," *Proc. of No. 5 Annual Technical Meeting on Air Cleaning and Contamination Control* (April 9, 1986).

18. Komiya, H., "Automation in LSI Manufacturing," presented at the Dataquest SEMIIS Conference October 14, 1985; H. Komiya, "Totally Automated IC Manufacturing," Automated IC Manufacturing III Symposium, Honolulu, Hawaii (1987).

19. Bogert, H. and Greiner, J., Dataquest Research Newsletter (August 1985).

20. NMB Semiconductor Co. Ltd., *Technical Perspectives* (August 1986).

21. Workman, W., "VLSI/ULSI Automated Factory, Design, Construction and Implementation," Automated IC Manufacturing III Symposium, Honolulu, Hawaii (1987).

22. Titus, Steve and Kelly, P., "Defect Density Reduction in a Class 100 Fab Utilizing the Standard Mechanical Interface (SMIF)," Automated IC Manufacturing III Symposium, Honolulu, Hawaii (1987).

Glossary

Reproduced from *A Glossary of Terms and Definitions Related to Contamination Control* with permission of the Institute of Environmental Sciences.

ABRASION, WIPER
The tendency of a wiper to lose integrity as a result of wiping another surface.

ABRASION RESISTANCE, GLOVES
The ability of the glove or finger cot to resist abrasion while in use, judged by one of the following criteria:
1. Retention of Barrier Integrity
2. Particle Generation

ABRASIVITY, WIPERS
The tendency of a wiper to damage the surface of the object or device being wiped.

ABSORBENT
A material which, due to an affinity for certain substances, extracts one or more such substances from a liquid or gaseous medium which it contacts and changes physically and/or chemically, and/or both, during the process. (Calcium chloride is an example of a solid absorbent, while solutions of lithium chloride, lithium bromide, and ethylene glycols are liquid absorbents.)

ABSORPTION
A process whereby a material extracts one or more substances present in an atmosphere or mixture of gases or liquids accompanied by the material's physical and/or chemical changes.

ACCEPTANCE TEST
A test made upon completion of fabrication, receipt, installation, or modification of a component unit or system to verify that it meets the requirements specified.

ACCURACY
The extent to which the value of a quantity indicated by an instrument under test agrees with an accepted value of the quantity.

ACTIVATED CARBON (CHARCOAL)
An adsorbent of porous structure manufactured by carbonization of organic material and treated by controlled oxidation to increase its adsorptive properties.

ADSORBENT
A material which has the ability to cause molecules of gases, liquids, or solids to adhere to its internal surfaces without changing the adsorbent physically or chemically. (Certain solid materials, such as silica gel and activated alumina, have this property.)

ADSORBER
A device or vessel containing an adsorbent.

ADSORBER CELL
A modular replaceable adsorber element.

ADSORPTION
The action, associated with the surface adherence, of a material in extracting one or more substances present in an atmosphere or mixture of gases and liquids, unaccompanied by physical or chemical change in the material. Commercial adsorbent materials have enormous internal surfaces.

AEROSOL
A gaseous suspension of solid or liquid particles about 100 μm or smaller in size.

AEROSOL GENERATOR
A device for generating an aerosol.

AEROSOL PHOTOMETER
A light scattering mass concentration indicator. (Instruments of this type have a threshold sensitivity of at least 10^{-3} microgram per liter for 0.3 micrometer diameter DOP particles and are capable of measuring concentrations over a range of 10^5 times the threshold sensitivity).

478

AGGLOMERATION
A process of contact and adhesion whereby particles form clusters of increasing size.

AIRBORNE PARTICLES
Particles suspended in air. (See AEROSOL).

AIRBORNE PARTICULATE CLEANLINESS CLASSES
A set of particle concentration levels based upon the number of particles greater than or equal to a specified size which are present in a unit volume of air.

AIRBORNE PARTICULATE CLEANLINESS LEVELS
The number of particles equal to or larger than a given size per unit volume of air.

AIRBORNE PARTICULATES
Airborne particulates are discrete particles having measurable physical boundaries in all directions and of such size and mass as to remain suspended in air long enough to be sampled and measured (usually 100 micrometers or less except for lint fibers). Particulates are distinguished from particles which may have the connotation of atomic or sub-atomic matter.

ANEMOMETER
An instrument for measuring air velocities.

ANEMOMETER—HOT WIRE TYPE
See Thermoanemometer.

ANEMOMETER—VANE TYPE
An instrument for measuring air velocities which mechanically converts the momentum of the air by rotation or deflection.

AS FOUND DATA
Data comparing the response of an instrument to known standards as determined without adjustment after the instrument is made operational.

ASPIRATOR
Any apparatus such as a squeeze bulb, fan, pump, or venturi, that produces a movement of a fluid by suction.

BAFFLE
A nonperforated member oriented substantially perpendicular to the direction of air flow, connected to a wall or divider of the cell, and having the purpose of preventing wall effect and/or channeling.

BLANK-BLANK AREA
A nonperforated area within the perforated portions of a perforated sheet or screen.

BYPASS
(a) The diversion (or facilities for diversion) of air around an air cleaning unit or component.
(b) The inadvertent leakage or diversion of air around an internal component of the air cleaning unit or component, resulting in the release of uncleaned air.

CALIBRATION
Comparison of a measurement standard or instrument of unknown accuracy with another standard or instrument of known accuracy to detect, correlate, report, or eliminate by adjustment, any variation in the accuracy of the unknown standard or instrument.

CALIBRATION, FIELD
Calibration test performed in the field in accordance with the manufacturer's recommendation and/or accepted industry practices.

CALIBRATION ON LINE
Calibration performed using the reference system built into the instrument, in accordance with manufacturer's recommendations and/or accepted industry standards.

CASE, CASING
The frame or cell sides of a modular filter element.

CERTIFICATE OF COMPLIANCE (CONFORMANCE)
A written statement, signed by a qualified party, attesting that the items or services are in accordance with specified requirements, and accompanied by additional information to substantiate the statement.

CHALLENGE CONCENTRATION
The concentration of an aerosol or gas of known characteristics used to expose a filter, adsorber, or other air cleaning device, under specified conditions, for the purpose of testing. The test aerosol or gas is the challenge aerosol or challenge gas, respectively.

CHANNELING
The greater flow of a gas or vapor through passages or areas of lower resistance which may occur within a bed due to nonuniform packing, segregation or irregular sizes or shapes of granules, displacement of granules by direct impingement or high-velocity air, or causes other than mechanical leaks. Also, the displacement of sorbent granules within the sorber bed which results in channeling.

CHEMICAL COMPATIBILITY
The interaction of a material with a chemical substance with which it has come into contact. A minimum interaction is desirable.

CHEMICAL DEGRADATION
Changes in material when in contact with chemicals. Undesirable forms of degradation are swelling, loss of tensile strength, deformation, and loss of abrasion resistance. See also: Permeability.

CHEMISORPTION
Adsorption, especially when irreversible, by means of chemical forces in contrast with physical forces.

CHIMNEY EFFECT
A phenomenon consisting of a vertical movement of a localized mass of air or other gases due to temperature differences.

CLEAN-AIR DEVICE
A clean bench, clean work station, downflow module, or other equipment designed to control particulate air cleanliness in a localized working area and incorporating, as a minimum, a HEPA filter and a blower.

CLEAN-AIR DEVICE, LAMINAR FLOW
A clean bench, clean work station, wall or ceiling hung module, or other device (except a clean room) which incorporates a HEPA filter(s) and motor-blower(s) for the purpose of supplying laminar flow clean air to a controlled work space.

CLEAN-AIR SYSTEM
An air cleaning system designed to maintain a defined level of air cleanliness usually in terms of a permissible number of particles in a given size range per unit volume, within an enclosed working area.

CLEAN AREA
A defined space within which the airborne particulate level is controlled to specified limits.

CLEAN ROOM (CLEAN FACILITY)
A room (facility) in which the air supply, air distribution, filtration of air supply, materials of construction, and operating procedures are regulated to control airborne particle concentrations to meet appropriate cleanliness levels as defined by Federal Standard 209. (See also Laminar Airflow Clean Room, Turbulent Flow Clean Room and Mixed Flow Clean Room.)

CLEAN ROOM (FACILITY)—AS BUILT
A clean room (facility) which is complete and ready for operation, with all services connected and functional, but without production equipment or personnel within the facility.

CLEAN ROOM (FACILITY)—AT REST
A clean room (facility) which is complete and has the production equipment installed, but without personnel within the facility.

CLEAN ROOM (FACILITY)—OPERATING
A clean room (facility) in normal operation with all services functioning and with production equipment and personnel present in the facility.

CLEAN ROOM—LAMINAR AIRFLOW
A clean room in which the filtered air makes a single pass through the work area in a parallel flow pattern with a minimum of turbulent flow areas. Laminar airflow rooms have a minimum of 80% of the ceiling (vertical flow) or one wall (horizontal flow) producing a uniform and parallel HEPA filtered airflow.

CLEAN ROOM—MIXED FLOW
A hybrid clean room consisting of a combination of laminar airflow clean room and turbulent flow clean room.

CLEAN ROOM—OPERATIONAL MODE
(See Clean Room As Built, At Rest, and Operating.)

CLEAN ROOM—TURBULENT FLOW
A clean room in which the air enters the room in a non-uniform velocity or turbulent flow. Such rooms exhibit a non-uniform or random airflow pattern throughout the enclosure.

CLEANABILITY
The ultimate limit of cleanliness obtainable for a material by a cleaning process.

COATING
Paint or other protective surface treatment applied by brushing, spraying, or dipping (does not include metallic plates).

COLLECTOR
A device for removing and retaining contaminants from air or other gases. Usually this term is applied to cleaning devices in exhaust systems.

CONDENSATION NUCLEI
Small particles, normally within the size range from 0.001 to 0.1 μm radius, upon which water vapor condenses in the atmosphere.

CONTAINED SPACE
A building, building space, room, cell, glove box, or other enclosed volume in which the air supply and exhaust are controlled.

CONTAMINANT
Any unwanted substance present in or on a material.

COT, FINGER COT
A covering or sheath for a finger.

DAMPER
An operable device used to control pressure or flow by varying the air path area.

DECONTAMINATION
The removal of contamination from air, other gases, surfaces, or liquids.

DECONTAMINATION FACTOR
The ratio of the concentration of a contaminant in the uncleaned (untreated) air to its concentration in the clean (treated) air.

DESORPTION
The process of freeing from a sorbed state.

DIFFUSION, MOLECULAR
A process of spontaneous intermixing of different substances, attributable to molecular motion and tending to produce uniformity of concentration.

DISPERSE SYSTEM
A two-phase system consisting of a dispersion medium and a disperse phase.

DISPERSION
The most general term for a system consisting of particulate matter suspended in air or other gases.

DISPERSOID
Matter in a form produced by a disperse system.

DOP
Dioctyl phthalate.

DOP AEROSOL
A dispersion of dioctyl phthalate (DOP) droplets in air. (See DOP Generator; Air Operated Gas-Thermal and Thermally Generated DOP.)

DOP, AIR GENERATED
An aerosol generated by blowing air through liquid dioctyl phthalate at room temperature. When generated with a Laskin type nozzle, the approximate light-scattering mean droplet-size distribution is:
　99 + percent less than 3.0μm
　50 + percent less than 0.7μm
　10 + percent less than 0.4μm

DOP GENERATOR, AIR OPERATED
A device for producing a DOP aerosol, operated by compressed air at room temperature, equipped with Laskin nozzles to produce a heterogeneous DOP test aerosol.

DOP GENERATOR, GAS-THERMAL
A device for producing DOP aerosol operated by using an inert compressed gas to discharge a regulated quantity of liquid DOP into a heated area where the DOP is vaporized and reconstituted into a heterogenous DOP aerosol.

DOP GENERATOR (Thermally Generated DOP)

An aerosol generated by quenching (condensing) vapor that has been evaporated from liquid dioctyl phthalate by heat. The aerosol has a light-scattering mean diameter of about 0.3 micrometer with a geometric standard deviation of about 1.4.

DUST

Small, solid particles which may be present on a surface or in a gas.

FIBER

A particle having a length 100 micrometers or greater, and an aspect ratio of at least 1:10.

FILTER, CONTROLLED PORE

A filter of various plastics or metals having a structure of controlled uniform pore size. Sometimes referred to as a membrane or molecular filter.

FILTER, EXTENDED MEDIUM

A filter having a pleated medium or a medium in the form of bags, socks, or other shape to increase the surface area relative to the frontal area of the filter.

FILTER, FLOW RESISTANCE

Resistance offered by a filter medium to fluid flow; the pressure difference required to give unit flow of a fluid of unit viscosity through a unit cube of filter medium. See also Resistance.

FILTER, MECHANICAL

A device which mechanically removes particulate matter from a fluid by the mechanism of impaction, settling, screening, inertia, diffusion, or any combination of these.

FLAMMABILITY, GARMENTS

Materials which comply with Federal Standard CS 191-53 and NFPA 702 requirements for nonflame-resistant wearing apparel. (Compliance with the referenced standards does not preclude the materials burning when exposed to an ignition source.)

FLOWMETER

An instrument for measuring the rate of flow of a fluid moving through a pipe or duct system. The instrument is calibrated to give volume or mass rate of flow.

GARMENTS, CLEAN ROOM

Special items of clothing designed to be worn to protect clean room atmosphere from contaminants released by workers. Special clothing apparel includes footwear or covers, and head covers.

GARMENTS, FLAME RESISTANT

Wearing apparel made of flame resistant materials which comply with the requirements of Federal Standard CS 191-53 and NFPA 702. (Compliance with the referenced standards does not mean that the materials will not burn when directly exposed to an open flame.)

GAS

One of the three states of aggregation of matter, having neither independent shape or volume and tending to expand indefinitely.

GAS CHROMATOGRAPH

An analytical instrument used for quantitative analysis of extremely small quantities or organic compounds whose operation is based upon the adsorption and partitioning of a gaseous phase within a column of granular material.

GAS RESIDENCE TIME

The calculated time that a contaminant or test agent theoretically remains in contact with an adsorbent, based on active volume of adsorbent and air or gas velocity through the adsorber bed.

GLOVE

A covering for the hand having separate sections for each of the fingers and the thumb, often extending part way up the arm.

HEPA (HIGH EFFICIENCY PARTICULATE AIR) FILTER

A replaceable extended-media dry-type filter in a rigid frame having minimum particle-collection efficiency of 99.97% for 0.3 micrometer thermally-generated dioctyl phthalate (DOP) (or specified alternative aerosol) particles, and a maximum clean-filter pressure drop of 2.54 cm (1.0 in) water gage when tested at rated airflow capacity.

HETEROGENEOUS DIOCTYL PHTHALATE (DOP)

An aerosol having the approximate light-scattering mean droplet size distribution as follows:

99+% less than 3.0 μm
50+% less than 0.7 μm
10+% less than 0.4 μm

IMPACTION

A forcible contact of particles of matter; a term often used synonymously with impingement.

IMPACTOR

A sampling device that employs the principle of impaction (impingement). The "cascade impactor" refers to a specific instrument which employs several impactions in series to collect successively smaller sizes of particles.

IMPINGEMENT

A gas sampling procedure for the collection of particles in which the gas being sampled is forcibly directed against a surface.

IMPINGEMENT, DRY

The process of impingement carried out so that particulate matter carried in the gas stream is retained upon the surface against which the stream is directed. The collecting surface may be treated with a film of adhesive.

IMPINGEMENT, WET

The process of impingement carried out within a body of liquid, the latter serving to retain the particulate matter.

IMPREGNATED CARBON

Activated carbon to which one or more chemicals have been added to improve retention of radioactive iodine compounds from air or gas.

IN-PLACE LEAK TEST
A test of an installed component or bank of components, as opposed to a pre-delivery or pre-installation test of individual components. In-place leak tests of filters and adsorbers are of two types: (1) a gross test which is designed to challenge all parts of the installation, including possible bypasses of the bank; and (2) a shrouded test in which portions of the bank are shrouded or blanked off so that only a limited portion of the bank is subjected to the challenge aerosol or gas at one time, thereby limiting the total quantity of challenge aerosol or gas released to the total bank.

INCHES OF WATER GAGE (w.g.)
A unit of pressure or pressure differential (1 in w.g. = 0.036 psi).

INDUCTION LEAK TEST
A procedure to evaluate potential particle intrusion into a clean space by induction through unsealed construction joints, piping/utility penetrations or by back-streaming from work space openings.

ISOKINETIC
A term describing a condition of sampling, in which the flow of gas into the sampling device (at the opening or face of the inlet) has the same flow rate and direction as the ambient atmosphere being sampled.

ISOKINETIC SMOKE GENERATOR
A small air generated source such as a chemical smoke tube connected to a length of tubing and with a precision metering valve for adjusting the discharge speed to within ± 1.52 mps (5 fpm) of the surrounding airflow velocity. Tubing outlet should be less than 6.4 mm (.25 inch) in diameter.

LAMINAR AIR FLOW
Air flow in which essentially the entire body of air within a confined area moves with uniform velocity along parallel flow lines.

LAMINATE
A material formed by bonding together two or more layers in order to achieve an effect not otherwise attainable from each of the component layers separately.

LASKIN NOZZLE
A nozzle used for the generation of a hetrogeneous DOP aerosol by compressed gas. (As defined in Air Generated DOP.)

LAUNDERABILITY
The ability of an article such as a glove or a garment to be laundered to a specified level of cleanliness.

LEAK, HEPA FILTER
A gap or void in filter media or gaskets which permits unfiltered air to penetrate into the clean room or clean work station.

LEAK-TIGHTNESS
The condition of a component, unit, or system where leakage through the pressure boundary is less than a specified value at a specified differential pressure.

MARGIN
An unperforated area at the side, end of, or around the perforated area of a perforated sheet or screen.

MASS CONCENTRATION
Concentration expressed in terms of mass per unit volume of gas or liquid.

MECHANICAL LEAK
The measure of the direct leakage through metal parts of the cell or its gasket due to manufacturing defects, by-passing of the sorbent due to settling of the sorbent in the bed, or inadequate cell design. The results are reported as percent refrigerant penetration.

MEDIUM (PLURAL, MEDIA)
The filtering material in a filter.

MICROMETER
A unit of measurement equal to one-millionth of a meter or approximately 0.00003937 inch (e.g. 25 micrometers are approximately 0.001 inch).

MOUNTING FRAME
The structure to which a filter unit is clamped and sealed.

NUMERICAL APERTURE
A meaure of the light gathering and resolving power of a microscope objective.

ODOR
That property of a substance which affects the sense of smell; any smell, scent, or perfume.

OUT-GASSING
The liberation of a gas from a solid material or a liquid.

PARALLEL AIRFLOW
Unidirectional airflow, as demonstrated by introduction of an isokinetic smoke stream which exhibits a measured dispersion of not more than 14° from straight line flow.

PARTICLE (CONTAMINATION CONTROL)
A minute quantity of solid or liquid matter.

PARTICLE COUNTER, AIRBORNE
An instrument for continuous counting of airborne particles larger than a given threshold size. The sensing means may be optical, electrical, aerodynamic, etc.

PARTICLE COUNTER, OPTICAL
A light scattering instrument with display and/or recording means to count and size discrete particles in air, as defined by ASTM F-50-83.

PARTICLE DIAMETER (COUNT MEAN)
The arithmetic, or count mean, diameter is defined by equation:

$$d_a = \frac{1}{N} N_i d_i \quad \text{where } N_i \text{ is}$$

the number of particles with diameter d_i and N is the total number of particles.

PARTICLE DIAMETER (MASS MEDIAN SIZE)
A measure of particle diameter based on the particle mass. For the mass median size, one-half of the particle mass is contributed by particles with a size less than the mass median size, and one-half of the particle mass by those particles larger than that size.

PARTICLE DIAMETER (MATHEMATICAL RELATIONSHIP)

Mathematical definitions of the average diameters of particulate substance are related by the following equations:

$$1n\ d_g = 1n\ d_m - 3.0\ 1n^2\ S$$

$$1n\ d_g = 1n\ d_s - 3.1\ 1n^2\ S$$

Where d_g is the count mean diameter, d_s is the light scattering mean diameter, d_m is the mass mean diameter and S the geometric standard deviation of the particle-size distribution.

PARTICLE SIZE

The maximum linear dimension of a particle as observed with an optical microscope or the equivalent diameter of a particle detected by an instrument. The equivalent diameter is the diameter of a reference sphere having known properties and producing the same response in the sensing instrument as the particle being measured.

PARTICLE SIZE DISTRIBUTION

The relative percentage by weight or number of different particle size fractions.

PARTICULATE

A substance which consists of particles. (ex. flour, rouge, clay).

PARTICULATE CLEANLINESS LEVEL, FLUIDS

The number of particles equal to or larger than a specified size which are present in a unit volume of fluid.

PARTICULATE CONCENTRATION, FLUIDS

The number of particles per unit volume of fluid.

PARTICULATE CONTAMINANT

A contaminant consisting of particles.

PARTICULATES, AIRBORNE

See Airborne Particles.

PENETRATION

The exit concentration of a given gas from an air cleaning device, expressed as percentage of inlet concentration.

PENETROMETER

A self-contained instrument for the determination of penetration characteristics of very high efficiency filter media and filter units with thermally generated DOP in accordance with MIL-STD-282.

PERMEABILITY

The process whereby a fluid or gas passes through a barrier at the molecular level. Passage of these materials through defects such as holes or tears does not constitute permeability.

PLENUM CHAMBER

A compartment designed to ensure an even distribution of flow in a gas system. Examples are, air distribution in air-conditioning and ventilation; the inlet to a filter system to create an even flow of gas to each filter unit.

POLYURETHANES

Synthetic plastics formed by action of diisocyanates on dihydric alcohols, polyesters or polyethers. As open cell foams are often used in air-conditioning filters.

PRECIPITATION, ELECTROSTATIC

The separation of particulate matter from air or other gases under the influence of an electrostatic field.

PRECISION

The degree of agreement of repeated measurements of the same property, expressed in terms of dispersion of test results about the mean result obtained by repetitive testing of a homogenous sample under specified conditions. The precision of a method is expressed quantitatively as the standard deviation computed from the results of a series of controlled determinations.

PREFILTER

A filter unit installed ahead of another filter unit to protect the second unit from high dust concentration or other environmental conditions. The prefilter usually has a lower efficiency than the filter it protects.

PRESSURE DIFFERENTIAL

The difference between two pressure gage readings.

PRESSURE, GAGE

The difference in pressure existing within a system and that of the atmosphere. Zero gage pressure is equal to atmospheric pressure.

PRESSURE, STATIC

The pressure of a fluid at rest, or in motion, exerted perpendicularly to the direction of flow.

PRESSURE, TOTAL

The pressure representing the sum of static pressure and velocity pressure at the point of measurement.

PRESSURE, VELOCITY

That pressure caused by and related to the velocity of the flow of fluid; a measure of the kinetic energy of the fluid.

PROBING OR SCANNING

A method for disclosing leaks in filter units in which the probe nozzle of an aerosol photometer is held approximately one inch (25.4 mm) from the area to be tested and moved at a rate of not more than 10 ft/min (3.05 m/min) across the test area.

PRODUCTION AND SUPPORT EQUIPMENT AND HARDWARE

Equipment, tools and devices whose primary purpose is to perform an operation on a product or material. Production and support equipment shall include filters, conveyors, jigs, work tables and desks that are located in clean room or laminar flow clean air device. Equipment whose primary function is to provide an appropriate manufacturing environment for the product (air supply and exhaust equipment) is not considered production and support equipment.

RECOMMENDED PRACTICE (IES)

Recommended Practices are issued either in the form of guidelines or a handbook and describe preferred technical methodologies and procedures. Recommended Practices are the only authorized vehicle by which technical guidance or philosophy may be published or presented in the name of the IES.

REFRIGERANT-11

Trichloro (mono) fluoromethane in accordance with ANSI/ASHRAE 34, "Number Designation of Refrigerants."

REFRIGERANT-12
Dichlorodifluoromethane in accordance with ANSI/ASHRAE 34, "Number Designation of Refrigerants."

REFRIGERANT-112
1, 1, 2, 2—Tetrachlorodifluoroethane in accordance with ANSI/ASHRAE 34, "Number Designation of Refrigerant."

RESIDENCE TIME
The calculated time that a contaminant gas or vapor remains in contact with the sorbent at a specified volume flow rate, based on the net unbaffled screen area and thickness of the bed.

RESISTANCE (FILTER)
The pressure drop across a filter at a stated flow and under given conditions; generally expressed in millimeters water gauge or, in SI units, as N/m^2 or Pascals. See also Filter, Flow Resistance.

REUSABILITY
The ability to use a product more than once. (Includes the ability to clean or launder a product to a definable level of cleanliness).

ROUGHING FILTER
A prefilter with high efficiency for large particles and fibers but low efficiency for small particles, usually of the panel type.

SAMPLING
A process consisting of the withdrawal or isolation of a fractional part of a whole. In air or gas analysis, the separation of a portion of an ambient atmosphere with or without the simultaneous isolation of selected components.

SAMPLING, CONTINUOUS
Sampling without interruptions throughout an operation or for a predetermined time.

SAMPLING, INSTANTANEOUS
Obtaining a sample of an atmosphere in a very short period of time such that this sampling time is insignificant in comparison with the duration of the operation or the period being studied.

SAMPLING, INTERMITTENT
Sampling successively for a limited period of time throughout an operation or for a predetermined period of time. The duration of sampling periods and of the intervals between are not necessarily regular and are not specified.

SCALE DIVISION
On a photometer with numbered major scale divisions (0, 1, 2, 3, 4, 5), "one scale division" means the first intermediate division following the 0.

SENSOR
A device designed to respond to a physical stimulus (as temperature, illumination, and pressure etc.) and transmit a resulting signal for interpretation, or measurement, or for operating a control.

SEPARATORS
Corrugated paper or foil (usually aluminum alloy or plastic) used to separate the folds of a pleated filter medium and to provide air channels between them.

SETTLING VELOCITY
The terminal rate of fall of a particle through a fluid as induced by gravity or other external forces.

SHEDDING
The generation of particles as a result of mechanical action on a material.

SOAP TEST
A procedure to evaluate potential particle intrusion into a clean space through leakage from a high pressure source by applying a soap solution to potential leakage locations.

SOLUTION
A homogeneous substance, usually a single phase mixture of two or more materials.

SOLVENT
A substance which dissolves another to form a solution.

SORBENCY
A collective term describing the tendency of a wiper to hold fluids, whether by absorption (within the capillaries or pores of the wiper) or by adsorption (as a surface phenomenon).

SORBENT
A liquid or solid medium in or upon which materials are retained by absorption or adsorption.

SORBER, BED
A layer or sorbent contained between two perforated sheets spaced at a specified distance. Also, the assembly of perforated and nonperforated members that comprises the volume into which the "sorbent" is packed.

SORBER-CELL/CELLS
A modular container for a sorbent, with provision for sealing to a mounting frame, which can be used singularly or in multiples to build a system of any airflow capacity.

SORPTION
A process consisting of either absorption or adsorption.

SPECIFICATION – DESIGN
A concise document defining technical requirements in sufficient detail to form the basis for a product or process. It indicates when appropriate, the procedure that determines whether or not the given requirements are satisfied.

SPECIFICATION – PERFORMANCE
A concise document which details the performance requirement for a product. The performance specification includes procedures and/or references for testing and certification of the product.

STANDARD AIR
Air at 50 percent relative humidity 70°F and 29.92 in. Hg (21°C and 760 mm Hg). These conditions are chosen in recognition of the data which has been accumulated on air-handling equipment. They are sufficiently near the 25°C and 760 mm Hg commonly used for indoor air contamination work that no conversion or correction ordinarily need be applied.

STANDARD AIR DENSITY
Air having approximate density of .075 lb/ft^3 (1.201 kg/m^3). This corresponds to air at a pressure of 29.92 in Hg (760 mm Hg) at temperature of 69.8°F (21° C) with a specific volume of 13.33 ft^3/lb (0.832 m^3/kg).

STRENGTH

The properties of tensile strength, elongation, and modulus of materials. Accelerated aging may be included in the specification for strength.

SURFACE FINISH, GLOVE

The surface properties of a glove or finger cot which may be chemically or mechanically applied, affecting the adhesion, the abrasivity, the hand or feel, or the hardness of the covering.

THERMOANEMOMETER

An instrument for measuring air velocities that contains an element heated by a regulated electric source and cooled by the air flow, and converts the temperature of the element to a velocity reading.

THROUGH BOLT

A bolt or other fastener which passes through the sorbent bed.

ULPA (Ultra Low Penetration Air) FILTER

A throw-away extended-media dry type filter in a rigid frame having minimum particle-collection efficiency of 99.999% for particulate diameters $\geqslant 0.12\ \mu$m in size.

UNIFORM AIRFLOW

Uni-directional airflow pattern in which the point-to-point readings are within plus or minus 20% of the average airflow velocity for the total area of the laminar flow work zone.

VALIDATION

Establishing documented evidence that a system does what it purports to do.

WALL EFFECT

Partial gas stream by-pass of the sorbent which occurs along an unbaffled metal-to-sorbent interface.

WATER GAGE

The measure of pressure expressed as the height of a column of water in inches or micrometers.

WORK ZONE

That volume within the clean room which is designated for clean work and for which testing is required. The volume shall be identified by an entrance and exit plane normal to the air flow (where there is laminar airflow).

Index

486